Chiral Nanoprobes for Biological Applications

# Chiral Nanoprobes for Biological Applications

*Edited by Chuanlai Xu*

WILEY-VCH

**Editor**

***Prof. Chuanlai Xu***
Jiangnan University
Food Science and Technology
1800 Lihu Da Road
Binhu District
214122 Wuxi
China

**Cover Image:** © Chuanlai Xu



**Library of Congress Card No.:** applied for

**British Library Cataloguing-in-Publication Data**
A catalogue record for this book is available from the British Library.

**Bibliographic information published by the Deutsche Nationalbibliothek**
The Deutsche Nationalbibliothek lists this publication in the Deutsche Nationalbibliografie; detailed bibliographic data are available on the Internet at <http://dnb.d-nb.de>.



**Print ISBN:** 978-3-527-34942-5
**ePDF ISBN:** 978-3-527-83358-0
**ePub ISBN:** 978-3-527-83359-7
**oBook ISBN:** 978-3-527-83360-3

**Typesetting** Straive, Chennai, India
**Printing and Binding** CPI Group (UK) Ltd, Croydon, CR0 4YY

Printed on acid-free paper

C115082_280422

# Contents

**Preface** *ix*

1 **Introduction** *1*
*Chuanlai Xu, Hua Kuang, and Liguang Xu*
1.1 Chirality and Asymmetry *1*
1.2 Chiral Origin of Nanomaterials *2*
1.2.1 Chiral Geometries of Nanoparticles *2*
1.2.2 Structural Origins of Chirality *6*
1.3 Relationship with Known Chiral Materials *10*
1.3.1 Similarity to Chiral Inorganic and Organic Molecules *10*
1.3.2 Similarity to Biological Nano- and Mesoscale Compounds *11*
1.3.3 Hierarchical Chirality *11*
1.4 Characterization Methods *12*
1.4.1 Optical Spectra of Chiral Nanostructures *12*
1.4.2 Methodological Aspects for Computational Studies *14*
1.4.3 Selected Computer Modeling Study *17*
References *20*

2 **Chiral Nanocrystals** *27*
*Chuanlai Xu, Liguang Xu, and Si Li*
2.1 Synthetic Methods *28*
2.2 Chiral Semiconductors *29*
2.2.1 Semiconductor Nanoparticles *29*
2.2.2 Semiconductor Helicoids *38*
2.3 Chiral Metal Nanostructures *40*
2.3.1 Metal Nanoparticles/Nanorods *41*
2.3.2 Metal Nanohelices *50*
2.3.3 Metal Nanorings and Superstructures *54*
2.4 Chiral Metallic Compound Nanoparticles *59*
2.4.1 Chiral Oxide Nanoparticles *59*

2.4.2 Chiral Sulfide/Selenide Nanoparticles *64*
2.4.3 Ceramic Nanohelices *69*
References *71*

**3 Chiral Nanoassemblies** *79*
*Chuanlai Xu, Xiaoling Wu, and Changlong Hao*
3.1 Twisted Nanorod Pairs *80*
3.2 Chiral Nanoparticle Dimers *85*
3.3 Tetramer Assemblies *90*
3.4 Nanoparticle Helices *100*
3.5 Chiral Nanochains *111*
3.6 Core–Satellite Nanoassembly *113*
3.7 Nanoplatelets and Chiral Films *129*
3.8 Chiral Supraparticles *134*
References *139*

**4 Chiral Nanostructures for Biorecognition and Bioanalysis** *149*
*Chuanlai Xu, Yuan Zhao, and Xiaoling Wu*
4.1 Chiral Analysis Strategy *149*
4.2 DNA Detection *150*
4.3 Biomarker Bioanalysis *156*
4.4 Enzyme Bioanalysis *164*
4.5 Metal Ion Bioanalysis *170*
4.6 Small Molecule Detection *175*
4.7 Chiral Recognition *184*
4.8 Future Prospects and Outlook *193*
References *193*

**5 Chiral Nanomaterials for Emerging Biological Effects** *199*
*Chuanlai Xu, Maozhong Sun, and Yuan Zhao*
5.1 Photothermal Therapy *199*
5.2 Photodynamic Therapy *209*
5.3 Multimodal Therapy Methods *220*
5.4 Photochemical Effects *233*
5.5 Future Prospects and Outlook *237*
References *237*

**6 Chiral Nanomaterials for Biocatalysis** *241*
*Chuanlai Xu, Wei Ma, and Hua Kuang*
6.1 Chiral Nanomaterials Based Enantioselectivity in Biocatalysis *241*
6.2 Biocatalysis of Chiral Drugs and Related Chemical Intermediates Based on Chiral Nanomaterials *249*
6.3 Biocatalysis of Peptides Based on Chiral Nanomaterials *257*
6.4 Biocatalysis of Biological Macromolecules Based on Chiral Nanomaterials *261*

6.5 Biocatalysis Based on Enzyme-Encoded Chiral Complex Nanomaterial System *264*
6.6 Chiral Nanomaterials Based Biocatalysis In Situ for Emerging Biological Applications *269*
References *282*

**Index** *287*

# Preface

Chirality has always been one of the core research topics in multidisciplinary fields including materials science, chemistry, biology, physics, and so on. Of particular interest, recent advances in chiral nanomaterials represent one of the most dynamic areas and have emerged as a new "growth point" of nanoscience. Recent breakthroughs in chiral fabrication have led to the study of many interesting properties of chiral nanocrystals, and some have demonstrated realistic application potentials that may address key challenges at the heart of nanoscience. Nevertheless, only few chirality-related books have been published in recent years, most of which essentially focus on the introduction of different types of chiral materials, chiral nanophotonics and spintronics, and chiral separation techniques. However, the available books usually ignore or barely focus on introduction of chiral nanomaterials in biological applications, which is attracting increasing theoretical and applicative interests.

These reasons made us to realize that there is a need for a book that is capable of:

(i) presenting the techniques and strategies in a clear and concise manner to understand chiral fabrications, chiroptical mechanisms, structure manipulation, and chiroptical enhancement;
(ii) illustrating biological applications based on intense chiroptical signals, including chiral-encoded recognition, bioanalysis, photochemical effects, and biocatalysts; and
(iii) encouraging the researchers to the forefront of developments in chiral nanoscience, such as advanced chiral-related multidisciplinary techniques for innovative fabrication methods, chiral-related optical and physicochemical properties, futuristic visions of the light–matter interaction.

It is timely and necessary to write a comprehensive book to introduce the state-of-the-art advancements in strong chiroptical nanomaterials, their biological applications and to discuss their future avenues of research. To address such emerging advances, we present the book *Chiral Nanoprobes for Biological Applications*. In this book, we have attempted to provide developments in chiral nanoscience, with special emphasis on the milestones achieved in research on key chiral nanoprobes in the recent past five years, focused on biological applications, and provided some specific future perspectives in the field.

Being actively involved in chiral nanoscience and having published more than 100 research articles and reviews, we have included six distinct chapters in the book. Chapter 1 gives an introduction of chiral nanomaterials from their biological counterparts, chiral mechanism and theory. The origin of chiral materials brings new insights into understanding homochirality on Earth. The fundamental concepts about the nature of chirality is divided into four types – the inorganic core, chiral ligands, chiral patterns, and electronic structures. The general principles, modeling, and methodology characteristics for computational studies of chirality have been presented for better understanding of nanoscale chirality.

Chapters 2 and 3 introduce the fabrication technologies and strategies for fabricating individual chiral nanomaterials and chiral assembled nanostructures, with a focus on tailoring and enhancing chiroptical effects. Major efforts have been made to realize high-purity chiral nanomaterials consisting of ligand molecule moieties, semiconductors, metals, and other materials. Individual nanoparticles and their superstructures can display intensive chiroptical and chiral-coupled physicochemical or electrochemical secondary properties, as well as unique and useful functions. Strategies for chiral amplification including enantiopure nanocrystal synthesis, and highly spontaneous symmetry breaking or well-controlled chiral geometrical assemblies (pyramidal, helical, etc.), as well as high-order enantiomeric superstructures.

Chapter 4 is focused on analysis of chiral nanomaterial-based ultrasensitive biomolecules (DNA, biomarkers, enzymes, etc.). Chiral nanoprobes offer a new potent platform for bioanalysis featuring low sensitivity associated with biosensing, biolabeling, and bioimaging using individual chiral nanostructures or chiral assemblies. Chapter 5 extends chiral nanoscience in emerging biological effects including photothermal and photodynamic therapy as well as photochemical effects. Chiral nanomaterials selectively absorb photons and can convert light into heat or reactive oxygen species to enhance phototherapeutic or photochemical-related biological effects. Chapter 6 illustrates the fundamental concepts and the characteristic approaches related to chiral nanomaterials for biocatalysis. Chiral nanomaterials demonstrated as promising biomimetic catalysts afford facile preparation, tunability of catalytic activity, and high stability. Chiral molecules or chiral polymers are considered effective chiral selectors of nanocatalysts for asymmetric biocatalysis.

Finally, we would like to summarize future trends and perspectives in chiral nanoscience. Robust artificial chiral fabrication techniques and multilevel hierarchical chirality control for precise tuning of polarization rotation in inorganic nanostructures will enable development of multidisciplinary chiral nanoscience in future. (i) Synthesizing or assembling complex nanostructures with ultrastrong chiroptical effects is an intrinsic challenge for all nanoarchitectures without exception, where. extreme control on enantiomeric purity to achieve high anisotropy factors is in great demand. (ii) It is crucial to integrate multiple techniques from various fields such as supramolecular chemistry, inorganic synthesis, self-assemblies, and related large-scale construction (lithography, selective deposition, glancing angle evaporation, etc.) for homochiral fabrication in large quantity with a higher degree of structural chirality. (iii) The theoretical simulation can be used to uncover the chiral growth dynamics for predicting potential routes for chiral induction and amplification, which may include how the Ångström- and nano scale chirality being induced and transferred to the scales of meso-, micro-, and macro inorganic structures? Through theoretical simulation of growth dynamics, the facile design and fine-tuning of chiral nanostructures with predesigned biological functions will be available. (iv) Chirality-related physicochemical properties may lead to a wide range of potential biological applications. There will be great opportunities to elucidate nanoscale chirality interactions with biological systems, such as chirality coupled biological processes and their evolution, chiral cascaded bioactivity levels, even for modulation of metabolic and immune processes. (v) The high degree of manipulating chiral-related spins, electrons, and photons is promising for real-time biological applications, such as light-responsive molecular machines or excellent candidates for better therapeutics or biomimetic nanocatalysts.

Jiangnan University
05 January 2022

*Chuanlai Xu*
*Hua Kuang*
*Liguang Xu*

# 1

# Introduction

*Chuanlai Xu, Hua Kuang, and Liguang Xu*

*Jiangnan University, International Joint Research Laboratory for Biointerface and Biodetection, State Key Lab of Food Science & Technology, and School of Food Science and Technology, 1800 Lihu Road, Wuxi, Jiangsu Province, 214122, China*

Chirality is a widespread phenomenon in nature and plays an important role in chemistry, pharmacology, biology, and medicine. If a molecule cannot be superimposed on its mirror image, it is called a chiral molecule. The structure is mirrored into symmetrical molecules and chiral enantiomers. As a ubiquitous phenomenon in nature, chirality has been extensively studied at the molecule level. Most biomolecules are chiral and exhibit single structure, such as DNA, amino acids, peptides, and sugars. Moreover, lots of important biological reactions involve chiral interactions [1, 2]. Since Louis Pasteur firstly separated the tartaric acid enantiomers in 1848, chirality has fascinated the chemistry world [3, 4]. In the pharmaceutical industry, the left-handed or right-handed isomers of some molecules show completely opposite pharmacological effects and toxicity. To avoid the adverse effects of chiral drugs, the selective synthesis of chiral drugs has been a research hot spot in the field of pharmacy. Therefore, it is of great scientific and practical significance to explore the role of material chirality in the origin of life, chemical reactions, and disease treatments. Until now, molecular-scale research is relatively mature, and chiral research and exploration promotes the progress of science and technology in different fields, especially in nanoscience.

## 1.1 Chirality and Asymmetry

Chirality is a geometric property that implies that a mirror image of an object cannot overlap with the original object [5, 6]. Notably, this attribute is related to the scale or the state of matter; thus, it can exist in any object, real or imaginary. The term "chirality" refers to unique geometric shapes of atoms, molecules, particles, etc., other than the resulting optical or chemical effects. In chemistry, two nonsuperimposable objects that have a mirror-symmetry relationship are called enantiomers. The simple physical properties of pure enantiomers of organic materials, such as melting temperature, evaporation temperature, and color, exhibit identical features.

*Chiral Nanoprobes for Biological Applications*, First Edition. Edited by Chuanlai Xu.

Chiral molecules with the same chemical composition and have no mirror symmetry are diastereomers. Diastereomers have different physical properties. Because of many choices for the geometric positions of atoms in nanoparticles (NPs), diastereomers with small differences are the most common in NPs, such as phase transition temperatures or color. Furthermore, the secondary properties of chiral assemblies (including optical, chemical and biological performances) are the derivatives of the geometry of a particular chemical object at one or all the scales mentioned above. Of course, these properties are the main reasons for formation of chiral molecular and nanoscale structures in most cases and determine the functionality and practicality of chiral nanomaterials.

Moreover, the difference between individual objects and a collection of a mass of these objects also require the secondary effects of chiral nanostructures. The collection of chiral objects is usually in the solution of chiral molecules or dispersions of individual chiral NPs. The relative number of respective enantiomers determines the possibility of observing the secondary properties of chiral chemical structures. Notably, individual objects, such as molecules, nanoparticles (NPs), or higher order structures, can be chiral. However, positive and negative polarization rotations of different enantiomers may cancel out each other. Thus, the spectral detection of dispersions or solutions may display no rotatory optical activity. Moreover, ensembles of chiral molecules, NPs, and their nanostructures can usually exhibit equivalent enantiomers. Thus, these ensembles are called racemic mixtures. When the amount of one enantiomer is higher than that of the other one, the mixture is called non-racemic. Typically, when the molar preference of one enantiomer is relative to the other one, it is usually expressed as an enantiomeric excess. The enantiomeric excess is measured as a percentage (%), and typically abbreviated as ee. When a solution or dispersion is made up of a single enantiomer, it is called homochiral with an ee of 100%. As pointed out by Kelvin [7], Pasteur [8], and Piutti [9], homochiral compounds are common in bioorganic materials. However, they are not applicable in many inorganic nanostructures, as discussed in this book. Through chiral separation [10], e-beam lithography, and NP self-organization [11], nearly all homochiral chemical systems have been obtained from inorganic nanostructures.

## 1.2 Chiral Origin of Nanomaterials

### 1.2.1 Chiral Geometries of Nanoparticles

The typical pictorial representation of NPs as perfect spheres, cubes, ellipsoids, etc. may not be the best way to keep in mind when it comes to understand their chirality. Asymmetric NP cores, ridges, apexes, adatoms, vacancies, dislocations, etc. conduce to chiral geometric forms of these chemical species [12]. In addition, edges, apexes, and faces of an inorganic material cannot be coated with surface ligands uniformly or in a widely accepted radial manner. Therefore, the surface layer contains asymmetry, except for core defects, and causes the chiral geometric elements. The latter can happen, as exemplified by the epitaxial lattice-to-lattice coalescence of NPs, also

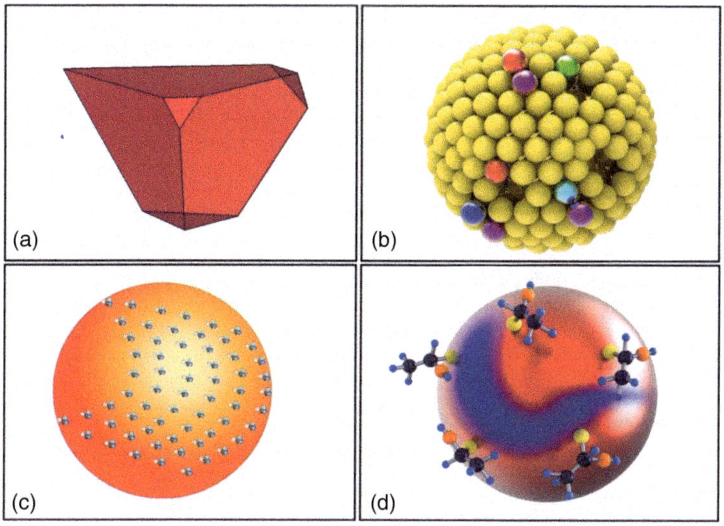


**Figure 1.1** (a) Type 1 chirality: the inorganic core with chiral shape. (b) Type 2 chirality: chiral surface of the inorganic core. (c) Type 3 chirality: chiral pattern of the surface ligands. (d) Type 4 chirality: polarization effects [1]. Source: Ma et al. 2017, Reproduced with permission of American Chemical Society.

known as oriented connection [13]. Even if the overall shape of the NP is spherical, or the inorganic core crystallinity is perfect, or the stabilizers evenly modify, the NP may still be chiral. Under the external bias conditions of a given electric or magnetic field, the high polarizability of the inorganic core makes the electron density asymmetrical, resulting in chirality. Therefore, there may be multiple sources of chirality in NPs, and it may be difficult as well as interesting to know which of them is/are responsible for specific secondary effects.

The first type of chirality is related to the asymmetry of the nanostructure's inorganic core. The object here is a continuous part of an inorganic substance, which defines the nanostructure's inorganic properties (Figure 1.1a). NPs with different shapes have been included in this group, such as tetrahedrons with variable truncations [8], twisted gold clusters [14, 15], and twisted nanorods (NRs) [16]. Apart from the NPs, nanohelices [17], twisted ribbons [18], and asymmetric objects with less traditional shapes (such as nanopillar caps [19]) are also included to this part.

In most situations, instead of packing a collection of atoms in a specific crystal lattice, the geometry of the NP core can be taken into account as a whole, especially for nanocolloids via bottom-up techniques. Overall shape of NPs should be given for the materials with chiral lattices, exemplified by macroscale crystals of tartaric acid separated by Pasteur. Assuming that an inorganic material was fabricated with a chiral crystal lattice by the top-bottom method, such as lithography, a mirror-symmetric cube is realistic. However, a non-racemic ensemble of such NPs and a chiral bias at the atomic scale during the growth will be required to explore chirality in such materials.

From the synthetic point of view, though the organization of the inorganic core is often strongly affected by the surface ligands, Type 1 chirality does not consider surface ligands. In fact, it is difficult to distinguish whether the surface layer chirality is resulted from the chiral packing of atoms in the small cores (0.5–1.5 nm) or vice versa. Thus, from this perspective, one may consider it as a controversial issue to distinguish the ligand shell chirality and core chirality; however, this is not the case. Actually, it is possible to decouple the chirality of the NP cores and the organic shells. Chiral geometries of inorganic cores with several nanometers have been reserved in the case of altered composition of the surface layer, such as twisted rods and helices [20–22].

Moreover, from the thermodynamic view, geometrically distorted structures will be more stable than perfectly packed symmetric crystal lattices. Therefore, organic ligands are not essential for the occurrence of the NP cores chirality. For example, as observed in gold nanoclusters, the energy preferred to the chiral structure instead of the symmetric structure, which resulted in the transition from the symmetric cubic to the twisted lattice [23]. As a point to illustrate the importance of understanding the chirality of ligand-free NPs, chiral inorganic NPs with a limited amount of stabilizers might be characteristics of interstellar matter.

The driving force of the chiral configurations of NP cores is different from that of surface ligands, including the thermodynamically favorable distortion of the nanoscale inorganic substance lattice. For example, a perfect tetrahedron can be prepared merely by one atomic packing; however, a tetrahedron with four different apexes possesses degenerate ways to arrange these apexes. Compared with an achiral perfect tetrahedron, such degeneracy makes the defects stable with an entropy of c. 8 kJ $mol^{-1}$. Therefore, according to the enthalpy contributions of each defect, a clear distribution of chiral NPs can be obtained. Nevertheless, not all NPs can be explained by such a purely geometric illustration. It is essential to explore the amount of entropy and enthalpy caused from the defects of chirality.

Other types of chiral NP geometries are associated with the existence of surface ligands. These surface ligands make adequate for experimental observations under certain conditions. Type 2 chirality of NPs is related to asymmetries of the NP core surface. When the chiral surface ligands result in such asymmetries, it can be called chiral footprint [24]. More generally, it is inorganic surface chirality. In the presence of stabilizers or other factors, the NP core atoms exhibit distortions and displacements, such as addition of atoms, erosion of atoms, or atomic reconstruction responding to chiral media (Figure 1.1b). Different from Type 1 chirality, in this case, the shape of the inorganic core may be achiral. The surface ligands display intense interactions with the core of inorganic materials. Usually, two anchoring groups are carried along with the inorganic core, and the two oxygen atoms are from two carboxylates, or carboxylate groups and thiolates. Particularly, the anchoring groups may lead to the distortions easier than the weak ligands with single-point attachment. For example, the L-cysteine methylester, *N*-isobutyryl-cysteine, and D/L-penicillamine ligands possess carboxylates and thiol groups, which may result in chiral distortions [25–27].

In most cases, Type 2 chirality is related to surface ligands. Note that surface ligands can be chiral, racemic, or achiral. Type 2 chirality exhibits a significant difference from Type 1 and Type 3 (discussed later). Type 2 chirality is affected and related to the specific surface states of inorganic matter, while the electronic states related to the inorganic core (plasmonic, excitonic, etc.) are not affected [24]. To further illustrate the importance and universality of Type 2 chirality of individual NPs, it will be discussed in the context of classical researches of surface chirality, such as inorganic surfaces associated with the origins of life [28]. This type of chirality may happen in a lot of cases, as exemplified by surface reconstruction. For a detailed example, when all mirror symmetry planes are destroyed, the surface atoms shift in unusual configurations can appear after the adsorption of tartaric acid on Cu [29, 30] or Ni surfaces [24] (Figure 1.1b). It seems that the double or higher coordination of the ligands is essential; otherwise, the ligands might easily rotate freely around their unique NP bonds, and such thermal agitation may result in full racemization ($ee = 0\%$).

Type 3 chirality in individual NPs is related to chirality of the stabilizer shell, as exemplified by the molecules with chiral organic shell and achiral for their core and surface [10, 31]. Importantly, achiral surface ligands may also result in chirality. Here, the stabilizer molecules packing on their surface are related to the symmetry breaking (Figure 1.1c). In other words, the geometry of the object considered here is a collection of the surface-bound organic constituents. Type 3 chirality of chiral NPs shows obvious different biological and chemical properties, since it is the main determinant factor for NP–protein interactions, NP assembly into the superlattices, as well as chiral separations.

The bonding mode of stabilizer molecules might not follow the underlying inorganic crystal lattice symmetry and may lack the necessary symmetry elements. This chirality can be produced by the chiral arrangement of Au–thiolate ligands, such as a staggered arrangement. Taking $Au_{38}(SCH_2CH_2Ph)_{24}$ NP as an example, it displays a prolate shape and has a face-fused biicosahedral $Au_{23}$ core protected by three short $Au(SR)_2$ and six long $Au_2(SR)_3$ staples. The $Au_{23}$ core is a perfect $D_{3h}$ symmetry [10]. In particular, the long staples are organized in a chiral staggered configuration, which consists of two triblade fans with three staples. They can rotate either in a clockwise or anticlockwise. In the case of short staples, they are slightly tilted, to follow the handedness of the long staples. Similarly, $Au_{40}(2\text{-PET})_{24}$, $Au_{38}(2\text{-PET})_{24}$, and $Au_{38}(SCH_3)_{24}$ clusters were reported to show chirality analogous to Type 3 chirality, which are arranged around the achiral metal core by the chirality of the Au-thiolate ligand [32, 33]. Some of the organic shells may show a patchy-like geometry, which may cause a nonsuperimposable mirror image, particularly for the three-dimensional (3D) surface layer [34]. It is worth to mention that when the temperature rises to the activation energy threshold for the translation of molecules, as exemplified by thiols along the underlying inorganic surface, mobility of the organic molecules on the surface of NPs may give rise to the transformation of one NP enantiomer into another one.

Type 4 chirality in individual NPs can be described as chiral field effect (Figure 1.1d) [35]. This type of chirality, unique to NPs, is related to high polarizability of some inorganic nanomaterials. Similar to Type 3 chirality, the atoms are achiral distributed in the NP core; however, the electrical field of the surface ligands or other source may affect the distribution of electrons. In achiral ligands with chiral adsorption patterns, it may produce asymmetric polarization patterns. The asymmetric polarization patterns can also occur in chiral ligands with achiral adsorption patterns, as well as in the conditions for chiral ligands with chiral adsorption patterns. Chirality can be produced near enantiomers by high polarizability of the inorganic core typical for plasmonic NPs. It is normal for chiral ligands with achiral adsorption patterns to show Type 4 chirality. The modified NPs often show a chiroptical activity in the ultraviolet (UV) range (200–400 nm). Because of the polarizability of the NPs, the circular dichroism (CD) spectrum during the UV part may change in comparison to their original chiral ligands. Moreover, chiral bands can also be produced in the plasmonic region, such as ~405 nm for Ag NPs, 520 nm for Au NPs. These NP systems can include peptides [36], peptide nanotubes [37], or proteins [38], which are physically or chemically adsorbed on surfaces.

Under the incident electromagnetic radiation, the high polarizability of metal NPs can give rise to localized areas high electrical field around them, which are usually called "hot spots." And then, it can lead to intense electronic transition enhancement for organic ligands. In addition to Raman scattering enhancement, these hot spots can also enhance chiral activities [39]. This strong effect derives from the polarization of electron density in NPs, which reflects the existence of the organic matter chirality in the hot spots. Therefore, it has been classified as Type 4. Notably, because of the resonance between the adjacent plasmon polarization oscillations, chiral band is enhanced by the hot spots in NP assemblies. Due to their popularity in the academic literatures, it can be attractive to attribute the chiral activities of the NP assemblies to the plasmonic enhancement effects. As a matter of fact, the diverse contributions to the CD activity requires to be evaluated [22].

In general, it is essential to understand the atomic and molecular structures to distinguish the different types of chirality (Type 1, Type 2, Type 3, and Type 4) for NPs. In fact, it is still quite difficult to decouple the chiral effects from all those asymmetric geometric features based merely on the results of CD or optical ration dispersion (ORD). Nevertheless, this is feasible to illustrate the molecular dynamics of NPs and their interactions with the incident photons based on multiscale computational methods. It can be simply assumed that there is Type 4 chirality in plasmonic NPs with a chiral surface layer.

### 1.2.2 Structural Origins of Chirality

Type 1 chirality in NPs was distinguished in experiments [40] and simulations for Au nanoclusters [41]. Taking advantage of genetic algorithms and many-body potentials, the density functional theory (DFT) calculations method serves as a powerful tool for simulations of chiral nanostructures among the computational tools. Only relatively small atomic systems of the above-mentioned chiral nanoclusters can be studied at present. DFT studies have shown that methyl thiol ligands on a truncated cuboctahedron with a face-centered cubic structure can result in chiral atomic packing of the inorganic core, such as $Au_{38}$ cluster coated with 24 $HSCH_3$ molecules resulting in chiral distortions on the $Au_{38}(SCH_3)_{24}$ cluster [42]. The Hausdorff chirality measure (HCM) shows a quantitative measure of asymmetry for any geometric figure [43], the HCM value increases with the number of thiol molecules. HCM is equal to 0.160 and 0.121 for methyl thiol-stabilized $Au_{28}(SCH_3)_{16}$ and $Au_{38}(SCH_3)_{24}$ clusters, respectively [44]. Similarly, some theoretical studies have shown that many of the asymmetric or asymmetric topological structures have energy close to or lower than the lowest energy ordered isomers for bare Au nanoclusters in the range of 10–75 atoms [45].

Type 2 chirality was distinguished in NPs capped by *N*-isobutyrylcysteine [25]. These molecules were attached to the Au surface in the form of two points and distorted the underlying inorganic core, resulting in intense CD activity (Figure 1.2A,B) [26]. Another representative research was about $Au_{102}(p\text{-MBA})_{44}$ clusters. A chiral stabilizer shell appeared due to the hydrogen bonding between carboxylic acids; the achiral *p*-MBA gave rise to a chiral displacement of atoms on the Au surface [47]. The ligand exchange following the synthesis experiments demonstrated the chiral memory of NPs. Taking CdTe NPs as an example, the surface memory was also found for NPs with Type 2 chirality [27]. The surface of thiol-coated CdTe NPs was distorted with the addition of a chiral ligand. The achiral thiol-coated CdTe still showed a symmetrical mirror chiroptical band, which was similar to the original chiral L-cysteine methyl ester hydrochloride-coated CdTe, after the ligand exchange reaction in the presence of the achiral thiol. Such effects were related to the "stored" chirality on the NPs surface instead of that in the NP core.

Type 3 chirality was distinguished in NPs with a symmetric Au core. The ligand stabilizers were attached to the "staples" motif and created chiral patterns on the surface of NPs. In the case of $Au_{38}(SR)_{24}$ protected by three RS(AuSR) and six $RS(AuSR)_2$ units, both of the crystallographic nanostructures and the optimally simulated nanostructures exhibited an achiral gold core with 23 atoms (Figure 1.2C) [46]. However, the geometric arrangement of the nine Au-S units indicated chiral $D_3$ symmetry. Such nanoclusters demonstrated intense chiral optical activity corresponding to the metal–metal or ligand–metal charge transfer (Figure 1.2D). Similarly, $Au_{40}(2\text{-PET})_{24}$ generated enantiomers with opposite CD signals following the separation on a chiral cellulose-mediated high performance liquid chromatography (HPLC) column, in spite of the achiral molecule of the thiolate stabilizer. The 2-positron emission tomography (PET) staples on the $Au_{40}$ cluster surface were believed to cause the enantiomeric preference in NP adsorption to cellulose [32]. Notably, other types of chirality also existed in the above-discussed NPs. It may be worth to point out that rarely NPs only displayed one type of chirality. Due to the involvement of different scales and different matter (organic or inorganic), the secondary effects of all types were obvious different.

Type 4 chirality in NPs is suitable for an achiral metal core placed in a chiral environment. Charge-perturbed particle-in-a-box calculations demonstrated that chiral stabilizers placed in achiral adsorption patterns on a metal NP surface resulted in chiral image charges in inorganic crystal lattices. The asymmetry of the adsorption mode and the influence of the ligand on the electrostatic perturbation have a significant inducing effect on the chirality of the symmetrically polarizable core

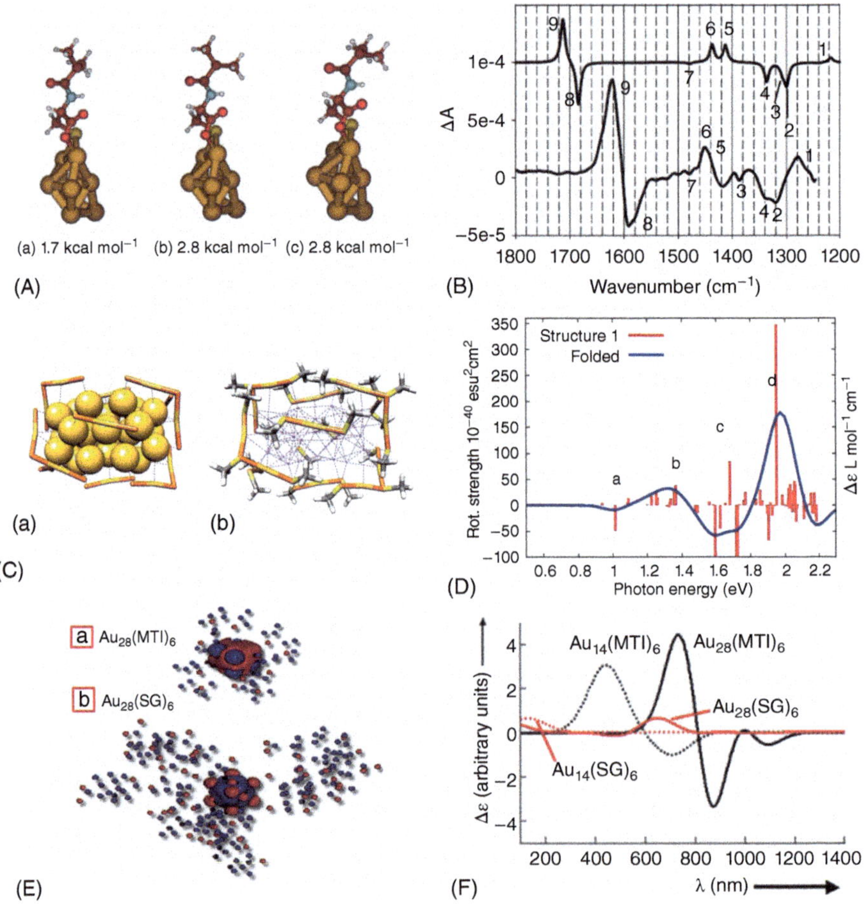


**Figure 1.2** (A) Calculated conformers of deprotonated *N*-isobutyryl-L-cysteine on $Au_8$ nanocluster. (B) Comparison between simulated and experimental VCD spectra of *N*-isobutyryl-L-cysteine on Au nanoclusters [26]. (C) Chiral $D_3$ arrangement of the Au-S atoms (a). Optimal $SCH_3$ arrangement on structures (b). (D) Simulated rotatory strength and chiral spectra [46]. (E) $Au_{28}$(R-methylthiirane)$_6$ (a) and $Au_{28}$(glutathione)$_6$ (b) using partial charges for the molecules and their image charges. (F) The simulated CD spectra [35]. Source: (A–C) Reproduced with permission from Refs [26, 46], respectively. Copyright 2006 and 2010, respectively, American Chemical Society; (E, F) Reproduced with permission from Ref [35]. Copyright 2006 Royal Society of Chemistry.

(Figure 1.2E,F) [35]. Some researches ascribed the observed optical and chemical effects to this type of chirality. Typically, their dispersions show chiral optical activity of the stabilizers (such as L-glutathione) and inorganic core (such as Au) in the spectral window. This observation indicated that the chirality of surface ligands was transferred to the plasmonic oscillation structure of the metal core through a variety of possible mechanisms.

The near-field effect often includes the description of various plasmonic NPs and has been extensively treated theoretically [48] and experimentally [49]. As a matter

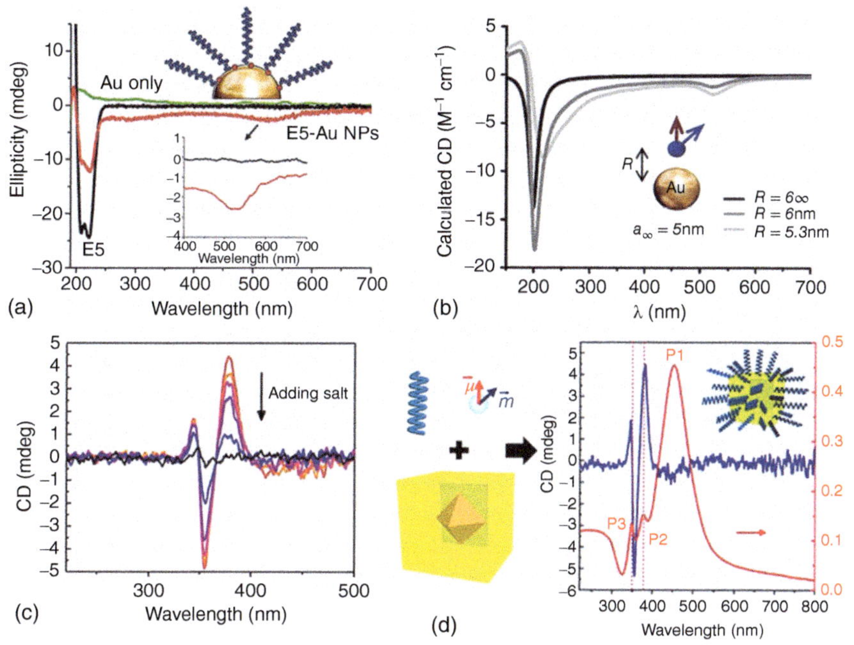


**Figure 1.3** (a) CD spectra of Au NPs with or without E5 peptide. (b) The calculated CD spectra for a dipole of chiral molecule and for a dipole–nanoparticle complex with two separations [51]. (c) Salt-dependent CD spectrum evolution of ssDNA-modified Ag NPs with ionic strength. (d) The design of "individual plasmonic chiral NPs" using Au/Ag core–shell nanocube. The UV–vis absorption and CD spectra of ssDNA-coated Ag nanocubes [52]. Source: (a–c) Reproduced with permission from Refs [51, 52], respectively. Copyright 2011 and 2013 American Chemical Society.

of fact, an explanation of the CD activity of the plasmonic NPs typically involves the effect of the NP field on chiral stabilizers and/or the field of stabilizers on the NP core, such as Au NPs with chiral 1,3-disubstituted diaminocalix arene ligands [50], peptide-coated Au NPs (Figure 1.3a,b) [51], and photo-synthetic proteins [38]. High enhanced rotatory optical activity is ascribed to the field effects (Figure 1.3c,d) [52], which displayed substantial practical value.

The far-field electromagnetic coupling between a localized plasmon of an achiral structure and the surrounding chiral molecular layer may also give rise to CD activity. Far-field electromagnetic coupling effect caused the unexpected chiral distortions of the electron gas. It has been confirmed by comparing experimental results with a simple electromagnetic model, which included a plasmonic object coupled with a chiral medium [49]. So far, even a chiral NP may be considered a complex chiral system with multiple chiral geometric structures simultaneously. In addition, due to other chirality triggered by the chiral arrangements of the stabilizer shell, it is quite difficult to point out the thermodynamic origin of chirality in NPs. Noted that the secondary properties-optical, chemical, biological properties could be derived from one configuration rather than the other enantiomer configuration, resulting in the understanding of different contributions to chirality in NPs. Therefore, it is still challenging to determine the origin of secondary effects of chirality in many nanomaterials.

Therefore, the ligand exchange of chiral ligands on the NP surface, such as thiolate to phosphine, often gives rise to the chiral band modification. However, it remained unclear as to the NP structure alterations in this process and remaining chiral/achiral structure of the metal core [53]. The chiral ligand (*N*-isobutyrylcysteine) was exchanged and caused the chiral inversion by stabilizers, but the atomic structure of the NPs remained unchanged. Even if the ligand was exchanged with achiral ligands, the induced surface chiral distortion could be retained, especially for multiply anchored ligands, such as carboxylate groups and thiolate ligands. Importantly, both the stabilizer shell and the metal core may change by the ligand exchange, including the average size. The ligand exchange can give rise to smaller or larger size clusters. For example, if the ligands are exchanged with hexanethiol, $Au_{75}$ clusters can be obtained [54]; however, $Au_{13}$ icosahedron can be formed from $Au_{25}$ by etching the core [55]. The ligand exchange chirality causes alterations in the metal core, which requires further studies.

## 1.3 Relationship with Known Chiral Materials

### 1.3.1 Similarity to Chiral Inorganic and Organic Molecules

Considering the diversity of chiral structure origins of NPs, it should be emphasized that these geometric shapes also exist in many other chiral materials based on the traditional chemistries and compounds. The structural design of different scales and multifunctional "building blocks" may be extended from various branches of classical chemistry to nanomaterials of "nano"chemistry [56]. For example, the NP assemblies resemble the well-known supramolecular assemblies of organic molecules [57]. Besides, as the similar in the scales and geometries between the NPs and the liquid crystal (LC) phases [58], the similarity has led to the emergence of hybrid nanoparticle–liquid crystal (NP–LC) materials, which is a promising direction in the field [59]. The cholesteric LCs may be a preferentially selected system for future study. It may be an attractive research direction to study the chiral induction of achiral LCs in the nematic phase.

The twisted NR pairs from Au or tellurium (Te) exhibit obvious similarities to the well-known binaphthyl compound (BINAP) with the geometry analogous to that of open scissors [60]. Chiral compounds assembled through metal center coordination bonds also have many similarities, although their scale usually is smaller than individual NPs or their assemblies [61]. Moreover, propeller-like coordination complexes, as one of the earliest chiral inorganic compounds, have been achieved in inorganic NP assemblies with a DNA-bridged NR assembly around an NP [62]. Large-scale coordination compounds, such as metal–organic frameworks (MOFs) and polyoxometalates (POMs) [63], have also been extended to the nanoscale. Importantly, the chiral geometries and chirality transfer mechanisms of chiroptical POMs resemble those of individual NPs. Therefore, it has paved a flexible and feasible way for the construction of chiral POM–NP hybrid nanostructures [64].

### 1.3.2 Similarity to Biological Nano- and Mesoscale Compounds

As in biology, a great majority of biological molecular and supramolecular compounds are chiral, which can be attributed to homochirality of life on Earth. There are multiple considerations for the similarities between the chirality of NPs and biomacromolecules. One common phenomenon is that chiral biomolecules are generally used as ligands at the surfaces of NPs, which partly contributes to chirality of individual NPs. Moreover, the size of the NPs is similar to the observed biological macromolecules. The characteristic size of numerous proteins is a few nanometers, while that of amino acids and nucleotides is usually smaller than 1 nm. Some proteins are even larger, with a size of tens of nanometers. Furthermore, some anisotropic protein molecules reach mesoscale dimensions. Therefore, a 300-nm-long triple helix of collagen molecule assembles in over 20 μm fibrils. Almost all soluble globular proteins, their oligomers, and complexes display intricate molecular shapes. These nanoshapes show obvious chiral activity and belong to a number of chiral point groups. In this case, the analogy to NPs with Type I chirality is very clear. To further expand this idea, the surface chemistry of water-soluble NPs shows an obvious similarity to the chemistry at the interface of protein molecules and aqueous media due to the dependence on the same chemical groups: –OH, –COOH, $-NH_2$, etc. Therefore, the stabilizer shell chirality can be compared with the atomic-scale packing of similar chemical groups and segments on the outer surface of protein or DNA globules. Similar parallel processing can also be carried out between the "staples" of the surface ligands on NPs and packaging of peptides, such as chaperonin molecular units [65].

### 1.3.3 Hierarchical Chirality

The chiral geometric structures of inorganic and biological nanostructures show obvious similarities. New insights may be brought into NPs and other nanostructures by the juxtaposition of biological and inorganic nanostructures, one of which is hierarchical chirality. For example, in proteins, cooperative behavior of hydrogen bonds and other intermolecular interactions can give rise to secondary nanostructure patterns from amino acid chirality at the atomic scale, as exemplified by α-helices and β-sheets. The secondary chiral structures may further generate the tertiary and quaternary structures of proteins. As common in protein complexes and rodlike viruses, helicoids can be produced with the self-assembly and aggregation of proteins [66]. Due to the complementarity of the chiral motifs and cooperativity of the interactions, atomic-scale chirality of NP stabilizers and NP cores can result in a similar hierarchy of chiral inorganic NPs structures. In the arrangements of stabilizers on the NP surface, cooperativity interactions are evident [67]. Hierarchical chirality can be observed in NP assemblies, such as the chiral supraparticles from Au NRs with a twisted rod arrangement originated from the CdTe NPs shell with amino-acid surface ligands. More complex hierarchical assemblies can also be prepared. The assembly of chiral NPs is similar to that of tobacco mosaic virus [68].

Other research fields associated with chiral NPs may also be inspired by the relationship between the chirality of individual NPs and biological molecules, which are based on the difference in biological activity of different enantiomers. The interactions between chiral graphene quantum dots and mammalian cell membranes show the difference, which have been observed experimentally and computationally. Different enantiomers of NPs may selectively interact with chiral biomolecules, such as enzymes, membranes, and even DNA. Therefore, they may bring distinct environmental and health hazards. These chiral NPs have not been observed in experiments and will be resolved in the future.

## 1.4 Characterization Methods

### 1.4.1 Optical Spectra of Chiral Nanostructures

Optical characterizations are the core secondary properties of chiral molecules and nanostructures in the field of physical phenomena. Therefore, it may be useful to study the origin of chiral NPs, their relationship with geometric structures, and chiroptical information from different types of NPs. Most of the chiroptical activity of individual NPs is associated with two quantum mechanical phenomena at the nanoscale, which are plasmons in metallic NPs and excitons in semiconducting NPs. The delocalization of charge carriers over the NPs can be characterized by these electronic states and therefore mainly depend on the crystal lattice of the NP core. In addition, metal-to-ligand charge transfer (MLCT) transitions may occur at the interface of the stabilizers and metal core [69]. All these excited states are influenced by shapes, sizes, surface ligands, and the surrounding dielectric environment of the NPs. When nanostructures exhibit a chiral shape, their plasmons and excitons may display chirality. Moreover, any other optoelectronic properties may also show chirality, which depend on the symmetry of the inorganic core and the overall electronic structure related to the excited states of the NP core. Certainly, the MLCT transitions depend on the chirality of ligands. They also depend on the symmetry of the NP core because of Type 2 chirality in the NPs and the contribution of other chirality.

In general, CD spectra and optical rotation dispersion (OR or ORD) spectra are the most commonly used spectroscopy tools in this field. They are generally used to describe the electronic transitions of chiral NPs from the visible (vis) to the near ultraviolet (UV) photons, including electronic and plasmonic transitions. The circular polarization of the incident photons can determine their vibrational components and can be measured by vibrational circular dichroism (VCD) and the Raman optical activity (ROA) methods [70]. Similarly, in the case of inorganic structures, CD and ORD spectroscopy are the most commonly used tools, and other spectroscopies can be applicable as well [26].

The ORD spectrum describes that optical rotation of a substance changes with the wavelength of light that passes through it. In this case, smaller wavelengths lead to greater optical rotation with the same path traveled by the light. The ORD spectrum can describe polarization plane rotation when a linearly polarized light beam passes through a material. Equation (1.1) describes the plane rotation per unit length, which expresses the circular birefringence of the chiral media. When the left circular polarization and right circular polarization components of linearly polarized light propagate in the chiral media, the difference between the refractive indices $n^L$ and $n^R$ can be positive or negative, related to the common chemical names of right- and left-handed media, respectively.

$$\alpha = \frac{\pi}{\lambda}(n^L - n^R) \tag{1.1}$$

ORD spectrum can offer responsible experimental signals for chiral NPs, which can also be calculated by theory. However, these spectra are inconvenient for answering the fundamental question on the chiral activities of electronic states in NPs due to their intense effect by scattering. In the meanwhile, ORD spectroscopy can also illustrate the practicality of chiral nanomaterials in polarized optoelectronics and other fields. As in the case of CD spectroscopy, it can detect the differential absorption of left- and right-handed circularly polarized light (electromagnetic radiation) and gives rise to their ellipticity alteration when the radiation beam passes through the sample. The chiral CD bands appear in the same spectral windows as the standard absorption bands of non-polarized light. Either positive or negative CD bands can occur, depending on the relative absorption of each circularly polarized component of the incident radiation. As a differential method, complex CD spectra may occur.

It is essential to clearly understand the interaction between light and matter in asymmetric nanostructures for the determination of the chiral geometric elements and the sign and intensity rationalization of multiple CD "waves." It may seem too complicated and convoluted. However, in fact, it can be quite simple to display commercial software packages (such as Lumerical, Gaussian, COMSOL) when one can identify the chirality type of the individual NPs, as exampled by Type 1, Type 2, Type 3, Type 4 or their combination thereof. It can also be applied to NP assemblies. Usually, the chemical characterization and separation of the nanoscale enantiomer mixtures determine the complexity of CD or ORD spectrum calculations [1]. As expected with improved computational power, it can be a standard part of the NP researches.

From the perspective of computational aspects, the ORD and CD spectra can be calculated based on semiclassical electrodynamic models with quantum mechanical perturbation theory. And special equations can be derived, which can address polarized light scattering caused by the oscillating electric dipole moment (linear oscillation) and magnetic dipole moment (circular oscillation). Both linear oscillation and circular oscillation are caused by the incident electromagnetic radiation. In the case of CD, the transition moment of the electric dipole interferes with the transition moment of the magnetic dipole, and the rotatory strength $R_j$ for the $j^{th}$ transition is obtained according to Eq. (1.2) [71].

$$R_j = Im(\langle \psi_0 \,|\hat{d}|\, \psi_j \rangle \cdot \langle \psi_j \,|\hat{m}|\, \psi_0 \rangle) \tag{1.2}$$

From Eq. (1.2), basic physical insights about the origin of quantum mechanics of CD activity can be obtained. Notably, only the symmetry requires to be considered when one predicts the intensity and algebraic sign of any transition. If the transition is to be observed, the two integrals of the ground state wave function $\psi_0$, the $j^{\text{th}}$ excited state wave function $\psi_j$, the electric dipole moment operator $\widehat{d}$, and the magnetic dipole moment operator $\widehat{m}$ cannot disappear at the same time. In other words, the electric and magnetic dipole moments must allow this transition at the same time. This strict selection rule limits the point groups where two transition moments do not disappear at the same time. The only possible groups for the CD spectrum are $C_n$, $D_n$, O, T, and I. In fact, for these point groups, chirality is usually possible not only the CD spectrum of chiral media. Notably, there are many chiral biological, organic, and nanoscale systems that contain the point groups of $C_n$ and $D_n$. However, in the case of chiral point groups O, T, and I, they are usually less observed. Nevertheless, their chirality can be detected when the latter point groups are observed.

As shown in Eq. (1.2), transition moments are vectors, and they are combined into a dot product to produce the scalar rotatory strength of each $j^{\text{th}}$ state. The rotatory intensity is defined as a dot product, which also means its null magnitude in the case of two transition moments perpendicular to each other. The greater contribution may origin from the parallel transition moments (positive rotatory strength) or antiparallel transition moments (negative rotatory strength). Interestingly, for any CD transition, the ground state wave function $\psi_0$ is the same (Eq. (1.2)). It means that the amplitude differences and the algebraic sign differences of various transitions mainly come from the alteration in the excited state wave functions $\psi_j$.

It is also easy to see the abnormally high intensity of CD spectra and high anisotropy g-factor value of the inorganic chiral nanoscale structures from Eq. (1.2). The electrical and magnetic susceptibility of metals and semiconductors are higher than that of organics, which typically represents the inorganic materials. Therefore, the instantaneous dipole moment and dot product of the interaction between the incident photon and the nanoscale inorganic matter are larger than those of organic objects of similar sizes. The CD spectrum calculation may depend on some different methods in Eq. (1.2) to calculate the wave function of the excited state. Generally speaking, time-dependent density functional theory (TD-DFT) is the selected method for calculating the CD spectrum of individual NPs [1].

### 1.4.2 Methodological Aspects for Computational Studies

The evaluation of the chiral properties of the individual NPs and their higher order structures can be done through by computer simulations. As shown in Eq. (1.2), the CD spectrum calculation can depend on some various methods to achieve ground wave functions and excited state wave functions. In addition, the classical models use the polarizability of individual atoms as input information. It can calculate the CD spectra from local dipole moment polarization as a static response to the electric field of other dipoles to generate a dipole vector system and the incoming electromagnetic radiation. Taking into account the development of high-performance computing facilities and efficient computer codes in the past 20 years, the DFT method can handle small NPs at the atomistic scale, while classical electrodynamic calculations, such as finite difference time-domain method and finite element integration, may continue to be used in larger assemblies containing many coarse-grained NPs [72].

Accurate knowledge of their shape is required for the simulation of CD bands of the individual NPs and their assemblies. An atomic-level detail for individual NPs can be offered from high-resolution imaging techniques [73]. It is essential to confirm the inherent asymmetry (Type 1 chirality) of inorganic cores. However, it can only describe NPs with sufficient high stability and a high-temperature phase transition at the atomic level for the electron tomography. In addition, chirality of NPs is determined by the average geometric shapes in a large ensemble. When realizing tomographic geometric shapes in simulation, the variability of their geometric shapes should be carefully considered. The organic molecules on the surface of NPs are transparent to the electron beam, and their geometric positions require to be determined by various techniques; for example, it may be firstly purified by electrophoresis or solvent extraction, and then characterized by nuclear magnetic resonance (NMR), infrared (IR), and VCD spectra, finally simulated the chirality activity (VCD, CD) based on the geometrically placed ligand on the surface of the NPs [26]. The orientation of the ligand on the surface of the NPs can be determined by the transition dipole moment orientation under different vibrations. X-ray diffraction (XRD) may indicate the lowest energy conformation pattern, which can be used as candidate methods to characterize the geometric ligand placement on the surface of the NPs [46].

In the case of sole computation, one should ideally start with a high-level and ab initio quantum chemistry model, which is composed of inorganic cores, organic shells, and solvents. However, currently most of the NPs of interest cannot afford the computational cost, and these restrictions will most likely be relaxed in the next two to five years. In practice, the choice of the theoretical level largely depends on the NP sizes, which range from classical molecular modeling and coarse-grained models to large nanostructures with dozens or hundreds of NPs, as well as the most advanced quantum chemical modeling for small clusters with tens of atoms. It is notable to mention that the trend of using DFT should be pointed out instead of wave-function-based methods, because the former provides a better compromise between accuracy and efficiency in systems with transition metal atoms than the latter [74].

Models based on DFT and wave functions need to choose a basis set to describe the electronic density. These basis sets are composed of more mathematical functions for better description of the electron density and other properties that depend on them. However, in practice, the individual NPs and their assemblies are too large to fully describe all the electrons. To overcome this inherent computational limitation, most of the DFT calculations depend on the effective core potential (ECP), which is a simplified empirical expression for the average electrostatic potential generated by the closed shell of the heavier atoms and core electrons, which are usually starting from the third row on the periodic table. In fact, the calculated electrostatic shielding internal electrons are in the atomic nucleus of an atom, which requires complete quantum chemical processing of the electrons. We add an empirical function to correctly parameterize the element and get the same shielding effect. This approximation is usually feasible because most molecular structures, bonding patterns, and electronic properties rely entirely on valence electrons, not core electrons. Finally, the description of valence electrons may also require some approximation. The most commonly used basis set describes the single-electron atomic orbit as the empirical sum of Gaussian functions. Each function has an exponent $\zeta$ defining the distance between the decay rate of the function and the nucleus. Generally, the larger the number of Gaussian functions describing valence electrons, the better the result, but when we move from a single-$\zeta$ basis (one Gaussian function per valence electron) to a double-$\zeta$ basis (two Gaussian functions), the performance will rapidly decline. In practice, recent DFT calculations depend on a double-$\zeta$ basis, with ECP as a compromise between efficiency and accuracy. Obviously, with the improvement of the computing power, larger and better basis sets may become feasible.

It is a difficult task to carry out the quantum chemical computation even used with described methodological trade-offs. Most computational implementations use some form of linear response theory to deal with the TD-DFT problem and achieve a description of the excitation energy and excited state orbit. We assume that it is equivalent to the excited state wave functions in Eq. (1.2). The TD-DFT formalism is accurate in principle, but the approximate form of the exchange correlation function may introduce errors. In practice, it is necessary to verify the calculation results with experimental data. Another inaccuracy source is the requirement of the narrow orbital range around the bandgap, which can be included in the TD-DFT modeling of electronic transitions. In a precise theory, all Kohn–Sham one-electron orbitals have entered an iterative process of mixing them, resulting in multi-electron orbitals for the holes and charges associated with each transition. In practice, an NP with hundreds of atoms may generate thousands of orbitals, even if small ECP basis set is used to reduce the electron number in the model; this is a prohibitively large number of one-electron Kohn–Sham orbitals. In the description of the hole and charge pairs for the electronic transition, the choice of the number of occupied orbitals and virtual/unoccupied orbitals to be mixed is a matter of trial and error, and a compromise is sought between the computational cost and the accuracy of results.

The root number in the iterative process also affects the calculation performance and the quality of results. Each root corresponds to an electronic transition, starting from the energy band transition to a higher energy transition. There is no rule of thumb to define how many root calculations are required to describe the energy in the NP's UV–vis spectrum range. It is usually best to gradually increase the root number until the spectrum matches the energy range of interest. In this incremental search, two problems are required to be solved before obtaining reliable results. First, as increasing the number of the excited state, it needed to evaluate the range of orbits entering into the iterative mixing, as low-lying occupied orbitals and highly unoccupied orbitals might become important, that is, their weight in describing the average orbit of the excited state can no longer be ignored. Secondly, the model calculation needs to go deeper than the last band of interest in the high-energy region of the spectrum, because the last band is usually not clearly defined, and this additional number of excited states significantly increases the computational cost of the modeling. Autschbach discussed in detail the trade-off between accuracy and cost in a review of similar tutorials [1, 71].

### 1.4.3 Selected Computer Modeling Study

Researchers investigated CdS NPs stabilized with penicillamine via DFT calculations with the BP86 functional and a double-ζ ECP basis set for the Cd atoms. Although a detailed optimized structure was calculated by means of a series of angles and the distances between atoms to evaluate the chirality of the structure, we must consider this method as qualitative, because of these raw data do not offer a chirality measure of NP, which requires calculation of CD spectrum or chirality index. The electronic structure is also qualitatively analyzed by the method of Kohn–Sham orbitals (sometimes referred to as molecular orbitals), but these orbitals are single-electron properties, while CD and other electronic excitation phenomena are essentially multi-electron processes, requiring the actual evaluation of the state of the excited state orbitals. A very attractive qualitative finding from their work is the discovery of multiple binding ligand molecules NP. The key observation is that a single self-coordinating ligand may have a free rotating surface, and this degree of freedom will result in thermal breakdown of the arrangement of chiral ligands.

For L-cysteine- and D-cysteine-stabilized CdS NPs, there is a quantitative match between the CD spectra in dispersions and TD-DFT calculations (Figure 1.4A,B) [75]. Although the NP in the simulation is much smaller than that in the experiment, all the main spectral characteristics and the sign of polarization rotation are almost consistent with the experimental spectrum. The calculation is based on 16 representative transitions of an electromagnetic frequency characteristic window of a nuclear reactor. Interestingly, the TD-DFT model shows that the chiroptical band around 390 nm may not only come from the bandgap transition but also from the p → s type excitation of sulfur atoms. Based on the quantum state of semiconductor nanowires with screw dislocations, a chiral semiconductor nanowire or nanorod model is calculated. Two-dimensional or three-dimensional periodic lattices can produce optically active quantum supercrystals. Due to the quantization of the size of the nanowire energy spectrum, its activity can change in many ways. The model shows that optical chirality is caused by screw dislocations in cylindrical nanowires that are much longer than the diameter (Figure 1.4C–E) [12]. The model proposes that inherent chirality of CdSe/ZnS nanocrystals is caused by naturally occurring chiral distortions (Figure 1.4F) [76]. For large-scale semiconductor nanomaterials, the finite-difference time-domain Maxwell solver based on the interaction of circularly polarized photons with the model structure can calculate a medium-scale semiconductor spiral [11].

A ring was formed from such double coordination between the ligand and the metal cluster, which contains the stereogenic center, the α-carbon atom. Here, the cluster is composed of only eight atoms, and the two types of chirality (Type 1 and

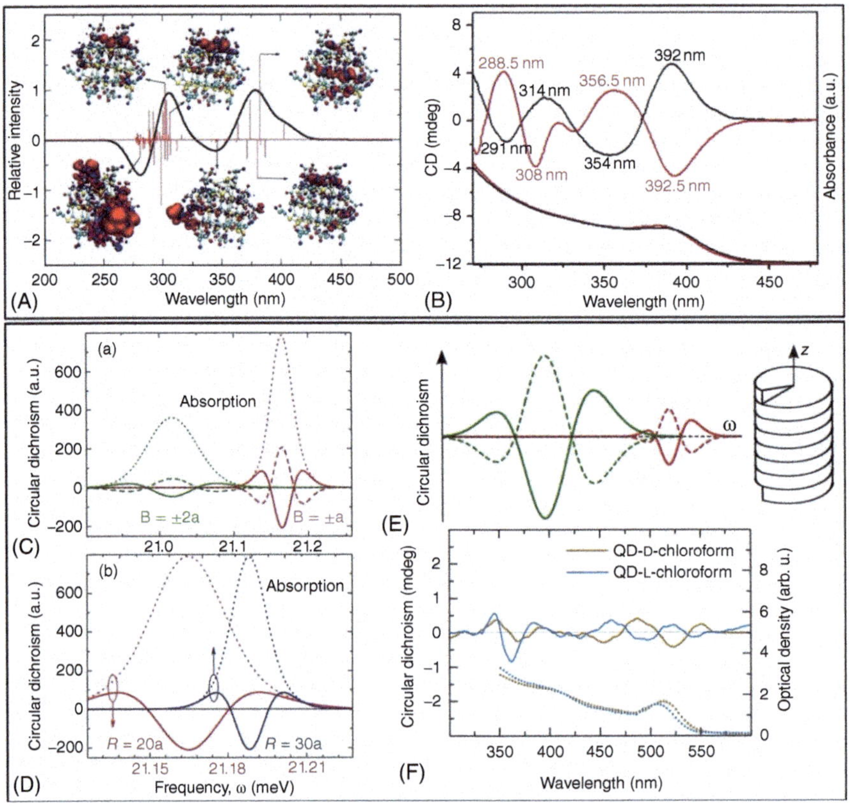


**Figure 1.4** (A) TD-DFT CD spectra for D-CdS [75]. (B) UV–vis and corresponding CD spectra of CdS-D and CdS-L NPs from experimental results. (C, D) Theoretically simulated CD spectra upon intraband transitions (1, ±2) → (1, ±3) in ZnS nanowires with (a) right- and left-handed screw dislocations [12]. (E) The right-handed screw dislocation and the general calculated CD spectra [12]. (F) The intrinsic CD of the L- and D-CdSe/ZnS semiconductor NRs [76]. Source: (A,B) Reproduced with permission from Ref. [75]. Copyright 2016 Royal Society of Chemistry; (C–F) Reproduced with permission from Refs [12, 76]. Copyright 2015 American Chemical Society.

Type 2) are actually the same. The ligand in the cluster induces a twisted structure (Type 2), but the cluster is too small, and then affects the whole. The structure of the cluster changes, not only its surface atoms (Type 1). The degree of chirality depends on the degree of the asymmetry of the stereocenter and, therefore, on the local conformation around it. The authors calculated six different conformations of VCD and found that the intensity, position, and algebraic sign of the vibration band are related to the conformation around the chiral carbon atom, but not to the side chain conformation. One thing to pay attention is the sampling of the conformational space of the ligand, because the flexible chain always has a large number of local minima related to the rotation of the dihedron. These degrees of freedom are not properly sampled by standard geometric optimization techniques, which are usually based on the energy gradient, which is not enough to bring the molecule from one local minimum to another, due to the energy gradient close to zero at the minimum. This is a problem for this small model with only one ligand, while for a larger and more realistic system, there are dozens of ligands on the surface of a large NP, which may become a serious flaw.

To sample the conformation of Au NPs from $Au_{28}$ to $Au_{75}$, a genetic–symbiotic algorithm was used [44]. This optimization technique is not based on the potential energy gradient, so in principle, it can effectively sample the local minimum in the construction space. For efficiency reasons, this initial optimization was performed using classical potentials. Only the representative local minimums were further optimized at the quantum chemical level, using local density approximation (LDA) and generalized gradient approximation (GGA) using pseudo potential (a form of ECP) and double-ζ basis functions. If there are high-quality classical parameters available for the conformation search, this method should be considered as an acceptable choice, because a large number of independent structures need to be considered, and the quantum chemical potential requires a too high thorough sampling. The authors discovered two populations of $Au_{28}$ NP, one is formed by nearly degenerate amorphous clusters and the other formed by ordered NPs belonging to the T point group, in which the T isomers lying c. 60 kJ $mol^{-1}$ above the disordered nanostructure. Based on the HCM, the degree of chirality of each population was evaluated. Both the disordered and the T isomers presented values larger than zero, with a slightly larger value for the T isomer. They also evaluated the effect of adding $SCH_3$ thiol groups on the surface of NPs. The general trend was either the induction of chirality in the achiral NPs or the increase in the chirality degree of the chiral NPs. The analysis of the optimized structures showed that thiol passivation tends to induce a disordered structure in the NP core, leading to the elimination of the symmetry plane, otherwise it will make the NP achiral. The $Au_{28}$ NPs were subsequently used to test a semi-analytical model and used the dipole moment approximation to calculate the CD spectra, which was first proposed by DeVoe and Applequist. Unfortunately, the authors did not use quantum chemistry methods to calculate the CD spectra of NPs, nor did they compare the calculated spectra with the experimental results. Therefore, we are unable to assess the quality of the results.

The TD-DFT calculations used the same theoretical level, sampling the first excited states of each NP. In the case of determining the excited states number, the calculated energy range of the UV–vis and CD spectra varied according to the NP size, from 1.5–3.5 eV for the smallest NP to 2.2–3.3 eV for the largest NP. The spectra included a Gaussian broadening of 0.1 eV to produce the line shapes of the rotation intensity calculated by TD-DFT. The result is a more complex spectrum compared to the UV–vis line shape of the same broadening in most research systems. For the CD spectra of these calculated capped NPs, the most important and common observation is that there is a single transition with a negative or positive rotatory strength for any broadened band, because the overall sign of any CD band comes from the sum of many individual transitions. This is important for the experimental data analysis of the changes in either the intensity or the algebraic sign of CD bands. It cannot be attributed to a single transition but must be regarded as the collective result of many changes in individual transitions. It may be shifted, damped, or even suppressed. In any case, multiple transitions must be evaluated to describe a single-frequency band [1].

An important problem is the effect of solvation on chirality of NPs. Most calculations were performed on a single NP in a vacuum, whereas most experimental data were carried out in suspensions. Although we do not have specific literatures on the influence of solvents on the chiral properties of individual NPs, there is general literature dedicated to this topic. We can point out that two main contributions are expected to play an important role in the evaluation of general chiral properties. First, because the relative population of different minimum energy conformations will change significantly under different solvation conditions, the solvent effect should be considered from the beginning of the geometric optimization step. Second, the phase transition is also affected by solvation. These two effects can be observed even with simplified continuum models [77].

In summary, a review of the theoretical research on chirality of NPs proposes quantum chemistry and electrodynamic calculations as natural choices for understanding the chiral activity origin of inorganic nanostructures and calculation methods suitable for different chirality type. It is worth noting that single NPs have greater correlation with quantum mechanical methods, and electrodynamic calculations are more suitable for larger nanostructures. There is still room for improvement in these two methods, and the widespread use of high-performance computing equipment and software will be conducive to the increasing research on rival computers in the next few years. Prospectively, the combination of experimental and computational research will provide a basis for the design and fine-tuning of materials based on chiral NPs [2, 78, 79].

## References


1 Ma, W., Xu, L.G., de Moura, A.F. et al. (2017). Chiral inorganic nanostructures. *Chemical Reviews* 117 (12): 8041–8093.

2 Hao, C., Xu, L., Kuang, H., and Xu, C. (2020). Artificial chiral probes and bioapplications. *Advanced Materials* 32 (41): 1802075.

3 Ma, W., Hao, C.L., Sun, M.Z. et al. (2018). Tuning of chiral construction, structural diversity, scale transformation and chiroptical applications. *Materials Horizons* 5 (2): 141–161.

4 Crassous, J. (2009). Chiral transfer in coordination complexes: towards molecular materials. *Chemical Society Reviews* 38 (3): 830–845.

5 Kuang, H., Xu, C., and Tang, Z. (2020). Emerging chiral materials. *Advanced Materials* 32 (41), 2005110.

6 Aouani, H., Rahmani, M., Navarro-Cía, M., and Maier, S.A. (2014). Third-harmonic-upconversion enhancement from a single semiconductor nanoparticle coupled to a plasmonic antenna. *Nature Nanotechnology* 9 (4): 290–294.

7 Ganesh, N., Zhang, W., Mathias, P.C. et al. (2007). Enhanced fluorescence emission from quantum dots on a photonic crystal surface. *Nature Nanotechnology* 2 (8): 515–520.

**8** Yeom, J., Yeom, B., Chan, H. et al. (2015). Chiral templating of self-assembling nanostructures by circularly polarized light. *Nature Materials* 14 (1): 66–72.

**9** Zhu, Z., Guo, J., Liu, W. et al. (2013). Controllable optical activity of gold nanorod and chiral quantum dot assemblies. *Angewandte Chemie International Edition* 52 (51): 13571–13575.

**10** Dolamic, I., Knoppe, S., Dass, A., and Bürgi, T. (2012). First enantioseparation and circular dichroism spectra of Au38 clusters protected by achiral ligands. *Nature Communications* 3 (1): 798.

**11** Feng, W., Kim, J.-Y., Wang, X. et al. (2017). Assembly of mesoscale helices with near-unity enantiomeric excess and light-matter interactions for chiral semiconductors. *Science Advances* 3 (3): e1601159.

**12** Baimuratov, A.S., Rukhlenko, I.D., Gun'ko, Y.K. et al. (2015). Dislocation-induced chirality of semiconductor nanocrystals. *Nano Letters* 15 (3): 1710–1715.

**13** Niederberger, M. and Cölfen, H. (2006). Oriented attachment and mesocrystals: non-classical crystallization mechanisms based on nanoparticle assembly. *Physical Chemistry Chemical Physics* 8 (28): 3271–3287.

**14** Lechtken, A., Schooss, D., Stairs, J.R. et al. (2007). $Au_{34}^-$: a chiral gold cluster? *Angewandte Chemie International Edition* 46 (16): 2944–2948.

**15** Gu, X., Bulusu, S., Li, X. et al. (2007). $Au_{34}^-$: a fluxional core–shell cluster. *The Journal of Physical Chemistry C* 111 (23): 8228–8232.

**16** Ben-Moshe, A., Wolf, S.G., Sadan, M.B. et al. (2014). Enantioselective control of lattice and shape chirality in inorganic nanostructures using chiral biomolecules. *Nature Communications* 5: 4302.

**17** Mark, A.G., Gibbs, J.G., Lee, T.-C., and Fischer, P. (2013). Hybrid nanocolloids with programmed three-dimensional shape and material composition. *Nature Materials* 12 (9): 802–807.

**18** Srivastava, S., Santos, A., Critchley, K. et al. (2010). Light-controlled self-assembly of semiconductor nanoparticles into twisted ribbons. *Science* 327 (5971): 1355–1359.

**19** Yeom, B., Zhang, H., Zhang, H. et al. (2013). Chiral plasmonic nanostructures on achiral nanopillars. *Nano Letters* 13: 5277–5283.

**20** Zhang, F., Wu, J., Wang, R. et al. (2014). Portable pH-inspired electrochemical detection of DNA amplification. *Chemical Communications* 50 (61): 8416–8419.

**21** Zhao, Y., Xu, L., Ma, W. et al. (2014). Shell-engineered chiroplasmonic assemblies of nanoparticles for zeptomolar DNA detection. *Nano Letters* 14 (7): 3908–3913.

**22** Ma, W., Kuang, H., Xu, L. et al. (2013). Attomolar DNA detection with chiral nanorod assemblies. *Nature Communications* 4: 2689.

**23** Karttunen, A.J., Linnolahti, M., Pakkanen, T.A., and Pyykkö, P. (2008). Icosahedral Au72: a predicted chiral and spherically aromatic golden fullerene. *Chemical Communications* 4: 465–467.

**24** Humblot, V., Haq, S., Muryn, C. et al. (2002). From local adsorption stresses to chiral surfaces: (R,R)-tartaric acid on Ni(110). *Journal of the American Chemical Society* 124 (3): 503–510.

**25** Gautier, C. and Burgi, T. (2008). Chiral inversion of gold nanoparticles. *Journal of the American Chemical Society* 130 (22): 7077–7084.

**26** Gautier, C. and Burgi, T. (2006). Chiral N-isobutyryl-cysteine protected gold nanoparticles: preparation, size selection, and optical activity in the UV-vis and infrared. *Journal of American Chemical Society* 128 (34): 11079–11087.

**27** Nakashima, T., Kobayashi, Y., and Kawai, T. (2009). Optical activity and chiral memory of thiol-capped CdTe nanocrystals. *Journal of American Chemical Society* 131 (30): 10342–10345.

**28** Hazen, R.M. and Sverjensky, D.A. (2010). Mineral surfaces, geochemical complexities, and the origins of life. *Cold Spring Harbor perspectives in biology* 2 (5): a002162.

**29** Ortega Lorenzo, M., Baddeley, C.J., Muryn, C., and Raval, R. (2000). Extended surface chirality from supramolecular assemblies of adsorbed chiral molecules. *Nature* 404 (6776): 376–379.

**30** Stepanow, S., Lin, N., Vidal, F. et al. (2005). Programming supramolecular assembly and chirality in two-dimensional dicarboxylate networks on a Cu(100) surface. *Nano Letters* 5 (5): 901–904.

**31** Li, Y., Yu, D., Dai, L. et al. (2011). Organo-soluble chiral thiol-monolayer-protected gold nanorods. *Langmuir* 27 (1): 98–103.

**32** Knoppe, S., Dolamic, I., Dass, A., and Bürgi, T. (2012). Separation of enantiomers and CD spectra of $Au_{40}(SCH_2CH_2Ph)_{24}$: spectroscopic evidence for intrinsic chirality. *Angewandte Chemie International Edition* 51 (30): 7589–7591.

**33** Qian, H., Eckenhoff, W.T., Zhu, Y. et al. (2010). Total structure determination of thiolate-protected $Au_{38}$ nanoparticles. *Journal of the American Chemical Society* 132 (24): 8280–8281.

**34** Singh, C., Ghorai, P.K., Horsch, M.A. et al. (2007). Entropy-mediated patterning of surfactant-coated nanoparticles and surfaces. *Physical Review Letters* 99 (22): 226106.

**35** Goldsmith, M.-R., George, C.B., Zuber, G. et al. (2006). The chiroptical signature of achiral metal clusters induced by dissymmetric adsorbates. *Physical Chemistry Chemical Physics* 8 (1): 63–67.

**36** You, C.C., Agasti, S.S., and Rotello, V.M. (2008). Isomeric control of protein recognition with amino acid- and dipeptide-functionalized gold nanoparticles. *Chemistry (Weinheim an der Bergstrasse, Germany)* 14 (1): 143–150.

**37** George, J. and Thomas, K.G. (2010). Surface plasmon coupled circular dichroism of Au nanoparticles on peptide nanotubes. *Journal of the American Chemical Society* 132 (8): 2502–2503.

**38** Carmeli, I., Lieberman, I., Kraversky, L. et al. (2010). Broad band enhancement of light absorption in photosystem I by metal nanoparticle antennas. *Nano Letters* 10 (6): 2069–2074.

**39** Lieberman, I., Shemer, G., Fried, T. et al. (2008). Plasmon-resonance-enhanced absorption and circular dichroism. *Angewandte Chemie International Edition* 47 (26): 4855–4857.

**40** Reilly, S.M., Krick, T., and Dass, A. (2010). Surfactant-free synthesis of ultra-small gold nanoclusters. *The Journal of Physical Chemistry C* 114 (2): 741–745.

**41** Hidalgo, F., Sánchez-Castillo, A., Garzón, I.L., and Noguez, C. (2009). First-principles calculations of circular dichroism of ligand-protected gold nanoparticles. *European Physical Journal D* 52 (1): 179–182.

**42** Tlahuice, A. and Garzón, I.L. (2012). Structural, electronic, optical, and chiroptical properties of small thiolated gold clusters: the case of $Au_6$ and $Au_8$ cores protected with dimer $[Au_2(SR)_3]$ and trimer $[Au_3(SR)_4]$ motifs. *Physical Chemistry Chemical Physics* 14 (20): 7321–7329.

**43** Buda, A.B. and Mislow, K. (1992). A Hausdorff chirality measure. *Journal of the American Chemical Society* 114 (15): 6006–6012.

**44** Garzón, I.L., Beltrán, M.R., González, G. et al. (2003). Chirality, defects, and disorder ingold clusters. *The European Physical Journal D – Atomic, Molecular, Optical and Plasma Physics* 24 (1): 105–109.

**45** Soler, J.M., Beltrán, M.R., Michaelian, K. et al. (2000). Metallic bonding and cluster structure. *Physical Review B* 61 (8): 5771–5780.

**46** Lopez-Acevedo, O., Tsunoyama, H., Tsukuda, T. et al. (2010). Chirality and electronic structure of the thiolate-protected $Au_{38}$ nanocluster. *Journal of the American Chemical Society* 132 (23): 8210–8218.

**47** Jadzinsky, P.D., Calero, G., Ackerson, C.J. et al. (2007). Structure of a thiol monolayer-protected gold nanoparticle at 1.1 A resolution. *Science* 318 (5849): 430–433.

**48** Govorov, A.O. (2011). Plasmon-induced circular dichroism of a chiral molecule in the vicinity of metal nanocrystals. application to various geometries. *The Journal of Physical Chemistry C* 115 (16): 7914–7923.

**49** Abdulrahman, N.A., Fan, Z., Tonooka, T. et al. (2012). Induced chirality through electromagnetic coupling between chiral molecular layers and plasmonic nanostructures. *Nano Letters* 12 (2): 977–983.

**50** Ha, J.-M., Solovyov, A., and Katz, A. (2009). Postsynthetic modification of gold nanoparticles with Calix[4]arene enantiomers: origin of chiral surface plasmon resonance. *Langmuir* 25 (1): 153–158.

**51** Slocik, J.M., Govorov, A.O., and Naik, R.R. (2011). Plasmonic circular dichroism of peptide-functionalized gold nanoparticles. *Nano Letters* 11 (2): 701–705.

**52** Lu, F., Tian, Y., Liu, M. et al. (2013). Discrete nanocubes as plasmonic reporters of molecular chirality. *Nano Letters* 13 (7): 3145–3151.

**53** Yao, H., Miki, K., Nishida, N. et al. (2005). Large optical activity of gold nanocluster enantiomers induced by a pair of optically active penicillamines. *Journal of the American Chemical Society* 127 (44): 15536–15543.

**54** Balasubramanian, R., Guo, R., Mills, A.J., and Murray, R.W. (2005). Reaction of $Au_{55}(PPh_3)_{12}C_{16}$ with thiols yields thiolate monolayer protected $Au_{75}$ clusters. *Journal of the American Chemical Society* 127 (22): 8126–8132.

**55** Heaven, M.W., Dass, A., White, P.S. et al. (2008). Crystal structure of the gold nanoparticle $[N(C_8H_{17})_4][Au_{25}(SCH_2CH_2Ph)_{18}]$. *Journal of the American Chemical Society* 130 (12): 3754–3755.

**56** Wang, Y., Qi, W., Huang, R. et al. (2015). Rational design of chiral nanostructures from self-assembly of a ferrocene-modified dipeptide. *Journal of the American Chemical Society* 137 (24): 7869–7880.

57 Hoeben, F.J.M., Jonkheijm, P., Meijer, E.W., and Schenning, A.P.H.J. (2005). About supramolecular assemblies of $\pi$-conjugated systems. *Chemical Reviews* 105 (4): 1491–1546.

58 Borshch, V., Kim, Y.K., Xiang, J. et al. (2013). Nematic twist-bend phase with nanoscale modulation of molecular orientation. *Nature Communications* 4: 2635.

59 Hegmann, T., Qi, H., and Marx, V.M. (2007). Nanoparticles in liquid crystals: synthesis, self-assembly, defect formation and potential applications. *Journal of Inorganic and Organometallic Polymers and Materials* 17 (3): 483–508.

60 Auguié, B., Alonso-Gómez, J.L., Guerrero-Martínez, A., and Liz-Marzán, L.M. (2011). Fingers crossed: optical activity of a chiral dimer of plasmonic nanorods. *The Journal of Physical Chemistry Letters* 2 (8): 846–851.

61 Arias, S., Freire, F., Quiñoá, E., and Riguera, R. (2014). Nanospheres, nanotubes, toroids, and gels with controlled macroscopic chirality. *Angewandte Chemie International Edition* 53 (50): 13720–13724.

62 Wu, X., Xu, L., Ma, W. et al. (2016). Propeller-like nanorod-upconversion nanoparticle assemblies with intense chiroptical activity and luminescence enhancement in aqueous phase. *Advance Materials* 28 (28): 5907–5915.

63 Zhang, J., Hao, J., Wei, Y. et al. (2010). Nanoscale chiral rod-like molecular triads assembled from achiral polyoxometalates. *Journal of the American Chemical Society* 132 (1): 14–15.

64 Zhang, B., Guan, W., Zhang, S. et al. (2016). Controlled chiral electrochromism of polyoxometalates incorporated in supramolecular complexes. *Chemical Communications* 52 (30): 5308–5311.

65 Liu, G., Chen, H., Peng, H. et al. (2011). A carbon nanotube-based high-sensitivity electrochemical immunosensor for rapid and portable detection of clenbuterol. *Biosensors and Bioelectronics* 28 (1): 308–313.

66 Gibaud, T., Barry, E., Zakhary, M.J. et al. (2012). Reconfigurable self-assembly through chiral control of interfacial tension. *Nature* 481 (7381): 348–351.

67 Knoppe, S. and Bürgi, T. (2014). Chirality in thiolate-protected gold clusters. *Accounts of Chemical Research* 47 (4): 1318–1326.

68 Zhou, Y., Marson, R.L., van Anders, G. et al. (2016). Biomimetic hierarchical assembly of helical supraparticles from chiral nanoparticles. *ACS Nano* 10 (3): 3248–3256.

69 Yao, H. and Ishikawa, Y. (2015). Finite size effect on magneto-optical responses of chemically modified $Fe_3O_4$ nanoparticles studied by MCD spectroscopy. *The Journal of Physical Chemistry C* 119 (23): 13224–13230.

70 Barron, L.D. (2004). *Molecular Light Scattering and Optical Activity*, 2e. Cambridge: Cambridge University Press.

71 Autschbach, J. (2009). Computing chiroptical properties with first-principles theoretical methods: background and illustrative examples. *Chirality* 21 (1E): E116–E152.

72 Guerrero-Martínez, A., Alonso-Gómez, J.L., Auguié, B. et al. (2011). From individual to collective chirality in metal nanoparticles. *Nano Today* 6 (4): 381–400.

73 Van Aert, S., Batenburg, K.J., Rossell, M.D. et al. (2011). Three-dimensional atomic imaging of crystalline nanoparticles. *Nature* 470 (7334): 374–377.

**74** Cramer, C.J. and Truhlar, D.G. (2009). Density functional theory for transition metals and transition metal chemistry. *Physical chemistry chemical physics : PCCP* 11 (46): 10757–10816.

**75** Liang, Z., Bernardino, K., Han, J. et al. (2016). Optical anisotropy and sign reversal in layer-by-layer assembled films from chiral nanoparticles. *Faraday Discussions* 191 (0): 141–157.

**76** Mukhina, M.V., Maslov, V.G., Baranov, A.V. et al. (2015). Intrinsic chirality of CdSe/ZnS quantum dots and quantum rods. *Nano Letters* 15 (5): 2844–2851.

**77** Mennucci, B., Cappelli, C., Cammi, R., and Tomasi, J. (2011). Modeling solvent effects on chiroptical properties. *Chirality* 23 (9): 717–729.

**78** Hua Kuang, C.X. and Tang, Z. (2020). Emerging chiral materials. *Advanced Materials* 32 (41): 2070304.

**79** Ma, W., Xu, L., Wang, L. et al. (2019). Chirality-based biosensors. *Advanced Functional Materials* 29 (1): 1805512.

# 2

# Chiral Nanocrystals

*Chuanlai Xu*[1,2], *Liguang Xu*[1,2], *and Si Li*[3]

[1]*Jiangnan University, International Joint Research Laboratory for Biointerface and Biodetection, School of Food Science and Technology, Wuxi, Jiangsu, 214122, China*
[2]*Jiangnan University, State Key Laboratory of Food Science and Technology, Wuxi, Jiangsu, 214122, China*
[3]*Jiangnan University, International Joint Research Center for Photo-responsive Molecules and Materials, School of Chemical and Material Engineering, Wuxi, Jiangsu, 214122, China*

Chiroptical properties of colloidal nanocrystals (NCs) and their nanostructures have received huge attention in various areas including fundamental science and technological applications. Intense chiroptical inorganic nanocrystals can provide powerful technologies for construction of novel chiral nanomaterials. Moreover, a series of ultrasensitive biosensing detection methods have been developed based on chiral NCs, which include small molecules, micro-ribonucleic acids, DNA segments, and cancer antigens. In addition to the experimental studies, NCs have also been a chiral model for theoretical research. Due to the developments in NCs preparation and assembly techniques, numerous inspiring chiral phenomena have been observed, which include chiral memory, chiral inversion, and chiral-coupled electron spin in semiconductor NCs. Intense chiroptical activity has been successfully applied in efficient chiral recognition and separation, chiral-induced spintronic devices, circularly polarized luminescence, and so on [1–3].

In the past decades, chiral NCs were greatly developed and expanded. Thus, it is necessary to systematize a mass of inorganic nanomaterials into different categories according to their geometric elements and circular dichroism (CD) activities. Such classification methods have been originated from physical geometries of chiral nanostructures. It is still challenging and usually uncertain to build up the relationship between the model structures of NP chiral dispersions and chiroptical or other secondary properties, even on the computer simulation [4, 5]. Indeed, most of the inorganic NCs possess chiral symmetry with multiple geometric elements and distinct scales in the meantime. Generally, chiral inorganic nanomaterials cover semiconductor nanoparticles (NPs) [6], metal NPs [7], magnetic NPs [8], silica nanostructures [9], carbon nanomaterials [10], and others [11]. Herein, in this chapter, we simply classify the chiral nanocrystals into semiconductor NPs, metal NPs and metal compound NPs. This classification of chiral NCs is actually based on

*Chiral Nanoprobes for Biological Applications*, First Edition. Edited by Chuanlai Xu.

the band structure of solids, which is simple and meaningful. Also, it offers a direct transition to chemical, optical, and other properties. Notably, other classification methods either by chiral geometries/configurations or materials/constituent elements are not contradictory to each other but can be complementary.

## 2.1 Synthetic Methods

The chiral NP synthetic methods have been first discussed for better understanding of the geometry of NPs at atomic, molecular, and nanometer scales. Although chiral surface-ligand-coated nanoparticles have been established for quite a long time [4, 12], the synthesis of sol-based single colloidal nanoparticles with a chiral shape for inorganic nuclei remains unknown [13]. It is worth to mention that the question here is not whether it is possible to prepare chiral NPs, rather whether it is possible to perform such synthesis selectively or whether it is possible to separate enantiomers from masses of diastereomers. Generally, the common method for chiral NP preparation is dependent on the nucleation and cessation of growth in a solution medium with chiral surface ligands, which act as stabilizers. These molecules cover the surface of nanoparticles, giving thermodynamic stable nanoparticle colloids, and are fundamental to understand all aspects of the chemical, physical, and biological behavior of nanoparticles [14].

This synthesis method is actually a well-known chiral transfer reaction from chiral biological components to products. To obtain chiral nanoparticles, chiral biomolecules were used as surface ligands, such as nucleotides, amino acids, peptides, lipids, and proteins. Chiral molecules were often chemically bonded on the surface of NPs, which can be also physically adsorbed [15]. Even staying free of chiral molecules in the solution without coating on NPs, it is possible that a certain degree of chirality can be occurred by the chiral environment created surrounding the growing NPs, particularly in the case of intense interaction between chiral solute and solvent [16].

Usually, the chiroptical activity of the dispersed NPs exhibits chiral signals in the visible light or near-infrared light regions, which are at the characteristic absorption position of inorganic NP cores. For example, when the L-type of the amino acid cysteine was replaced by its D-type, the rotatory optical activity of the dispersed NPs usually switched to the opposite optical activity [17]. The stabilizer enantiomer added during the precipitation process determined the sign of the CD band corresponding to the interband or intraband transition of inorganic materials. And it implied the chiral transfer to the NP core from the surface ligands. Various geometric objects of distinct scales and chirality measurements can be produced based on a large number of chemical and physical effects. Actually, little has been known about the real geometric organization of atoms. The geometric arrangement of these atoms resulted in the chiroptical activity of NPs. However, molecular-scale chirality of the surface ligand may be propagated to the geometric structure at a higher level through the collective effect of NP assembly [18], along with the assembled individual NPs into a larger structure [14].

## 2.2 Chiral Semiconductors

### 2.2.1 Semiconductor Nanoparticles

Semiconductor nanoparticles (also known as quantum dots, QDs) have been extensively studied in the recent few years [19–21]. QDs have the advantages such as long fluorescence lifetime, photoelectric effects, extensive biological coupling potential, and photochemical stability [22]. Because of their unique photoelectric properties, chiral QDs have broad application prospects in chiral and fluorescence detection, etc., they have gained widespread attention. In recent years, functionalized QDs coupled with chiral molecules have received huge attention. Highly selective enantiomeric reactions and other novel optical properties can be induced by the chiral effect, and those optical characteristics are extremely important.

The typical chiral QDs synthesis method was the same as the method of preparing particle growth in solution. One of the earliest reported studies of chiral QDs was the synthesis of chiral CdS NPs with penicillamine as a surface ligand [23]. To prepare such particles, basic solutions of cadmium perchlorate and D-, L-, or *rac*-penicillamine were mixed together. Subsequently, by adding thioacetamide, weak luminescent CdS nanoclusters were achieved. Following the annealing process, high-luminescent chiral CdS NPs were obtained [6]. Following similar methods, (D,L)-cysteine (Cys) or (D,L)-glutathione (GSH) coated chiral cadmium telluride (CdTe) NPs were also prepared [24, 25]. Chiral graphene quantum dots (GQDs) were also made based on similar technologies [17]. Under the circumstances, the chiral surface ligands were bound on the edges of GQDs by covalent bonds. Due to the chiral interactions at the "crowded" edges induced by chiral ligands, GQDs were caused helical buckling to produce the chiroptical activities. However, scientists are still devoted to similar studies in II–VI semiconductor NPs. According to the latest data of lattice dislocations, the chiroptical activity of the inorganic core was induced and related to the overall chiral shape of NPs [2].

The chirality of QDs may be ascribed to those similar structural effects to metal clusters. Generally, they can be summarized into three types. Type 1, their chirality can be from the chiral distortion from surface atoms as shown in Figure 2.1A [6]. Type 2, chirality can be attributed to chirality transfer between the surface of chiral organic shells and inorganic cores (Figure 2.1B) [26]. Type 3, it can be attributed to the atomic stacking resulted from inherent chirality (Figure 2.1C) [27].

Gun'ko and colleagues demonstrated the chiroptical activity in QDs for the first time, and later they also constructed chiral tetrapods (Figure 2.2a) [23, 28]. Among all the above researches, the CD signals were detected in the excitonic spectral range. Especially, when the CdS core was replaced by CdSe, they observed a redshift of the CD band (Figure 2.2b) [29]. In consideration of their emission characteristics, the fluorescence of these QD NPs mainly come from the defect state rather than the NP exciton state. The chiroptical activity was attributed to the chiral distortion of surface atoms. This conclusion was later further confirmed by calculating the surface structure of NPs ab initio (Figure 2.2c) [23]. Subsequently, Kotov and colleagues observed

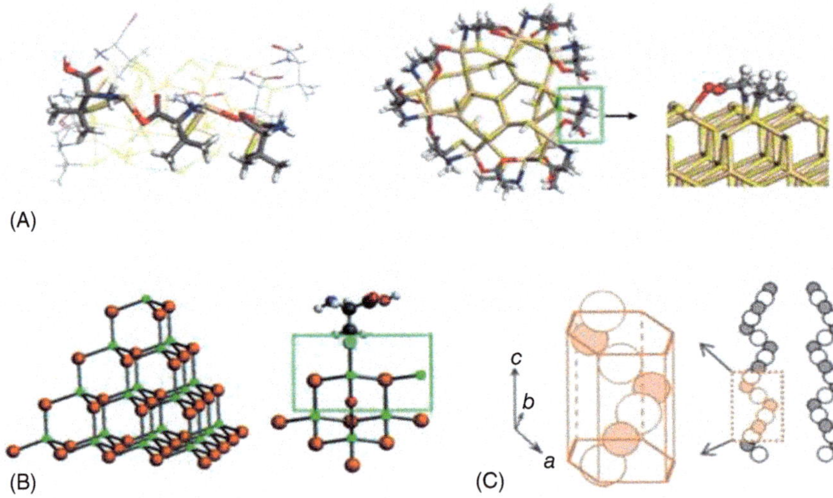


**Figure 2.1** (A) Top and side views of a (1010) face of the CdS cluster model [6]. (B) Ideal tetrahedron of CdTe NPs in calculations (left) and model of a chiral tetrahedral apex (right) [26]. (C) A scheme for reconstructing the opposite helix of a small fraction of atoms within a hexagon cell frame [27]. Source: (A, B) Reproduced with permission from Refs. [7, 26], respectively. Copyright 2008 and 2010 American Chemical Society; (C) Reproduced with permission from Ben-Moshe et al. [27]. Copyright 2013 Wiley-VCH.

a clear blueshift of the CD band from the chiral CdTe NPs relative to the band-edge transitions, which were ascribed to the surface states (Figure 2.2d) [26].

In the work of Tang and colleagues, broad CD absorption was observed in the visible light region for chiral NPs stabilized by chiral GSH and Cys, which can be regulated and controlled through their size modulation [25]. Moreover, the dependence of CD intensities on different materials and differently sized NPs was also studied [30]. Nakashima et al. have demonstrated corresponding chirality effects. For example, even if the chiral ligand was replaced by an achiral ligand, the chirality activity of CdTe NPs could still be maintained, and the prepared CdS NPs with a ferritin template produced circularly polarized emission. According to the prediction of the theoretical calculations, dependent on the chiral ligand–protected NPs model, broad chiral absorption bands should be attributed to the Coulomb dipole–dipole interactions between chiral molecules and NPs [31].

Due to the wide size distribution of QDs synthesized in aqueous phase, it is difficult to interpret the induced CD band in chiral QDs. The chiral amino acid, exchanged with the ligand, was used to prepare organic-phase semiconductor nanorods (SNRs); thus the synthetic nanorods (NRs) showed much clearer characteristic bands. Unfortunately, most of the early studies were mainly devoted to the chirality induction mechanism, while there were few studies on chirality enhancement. On the other hand, the ligand exchange for the prepared chiral QDs,

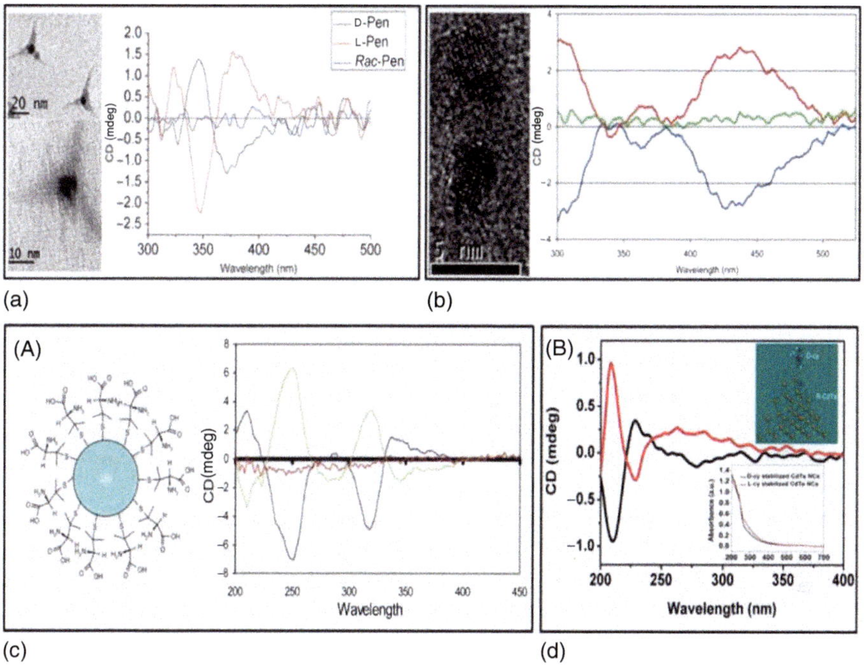


**Figure 2.2** (a) Chiroptical activity of penicillamine (Pen)-capped CdS NPs and relevant transmission electron microscopy (TEM) image [28]. (b) CD spectra of CdSe NPs capped with D-penicillamine, L-penicillamine, and *rac*-penicillamine and relevant TEM image [29]. (c) Scheme illustration of the D-penicillamine-capped CdS NP [23] (A) and CD spectra of the CdS NPs capped with D-penicillamine, L-penicillamine, and *rac*-penicillamine (B). (d) L-cysteine-capped CdTe NPs and D-cystine-stabilized CdTe NPs after 16 hours [26]. Source: (a) Govan et al. [28]/with permission of The Royal Society of Chemistry, (b) Gallagher et al. [29]/with permission of The Royal Society of Chemistry, (c) Reproduced with permission from Moloney et al. [23], respectively. Copyright 2010, 2010, and 2007 The Royal Society of Chemistry; (d) Reproduced with permission from Ref. [26]. Copyright 2010 American Chemical Society.

give an opportunity to study the interactions between the chiral NCs and chiral ligand.

In general, semiconductors with strong chiroptical activity can promote the study of preparation of chiral materials, photovoltaics, stereoselective catalysis and photocatalysis. To achieve intense chiroptical signals, the post-synthesis ligand exchange method offered an attractive way to achieve strong chiral properties and quantum size effects for the chiral organic-shell-capped NPs. And the diversity of amino acids further opened up new possibilities and methods for the preparation of strong chiral-active semiconductor nanomaterials. The electron transfer effects in nanoscale chirality coupling to photoelectrochemical, such as in water splitting, provided new strategies and application potentials for chiral research [1, 21, 32–34].

In a recent review from Tang's group, the opportunities and challenges of chiral QDs have been systematically overviewed [35]. In their previous studies, they found that the chiroptical signals of chiral CdSe nanorods (NRs) varied with size, and the characteristic bands of NRs gradually redshifted with the increasing length of NRs [36]. Chiroptical activity is affected by the crystal structure of semiconductors. For example, chiral nanoplate CdSe semiconductors with various crystal structures, such as wurtzite (WZ) and sphalerite (ZB), can even display obviously different chiral signals when coated with the same chiral ligands. All these researches comprehensively confirmed the inherent chirality origin mechanism of semiconductor systems. Further, it built the preliminary foundation for the preparation of chiral-active semiconductors.

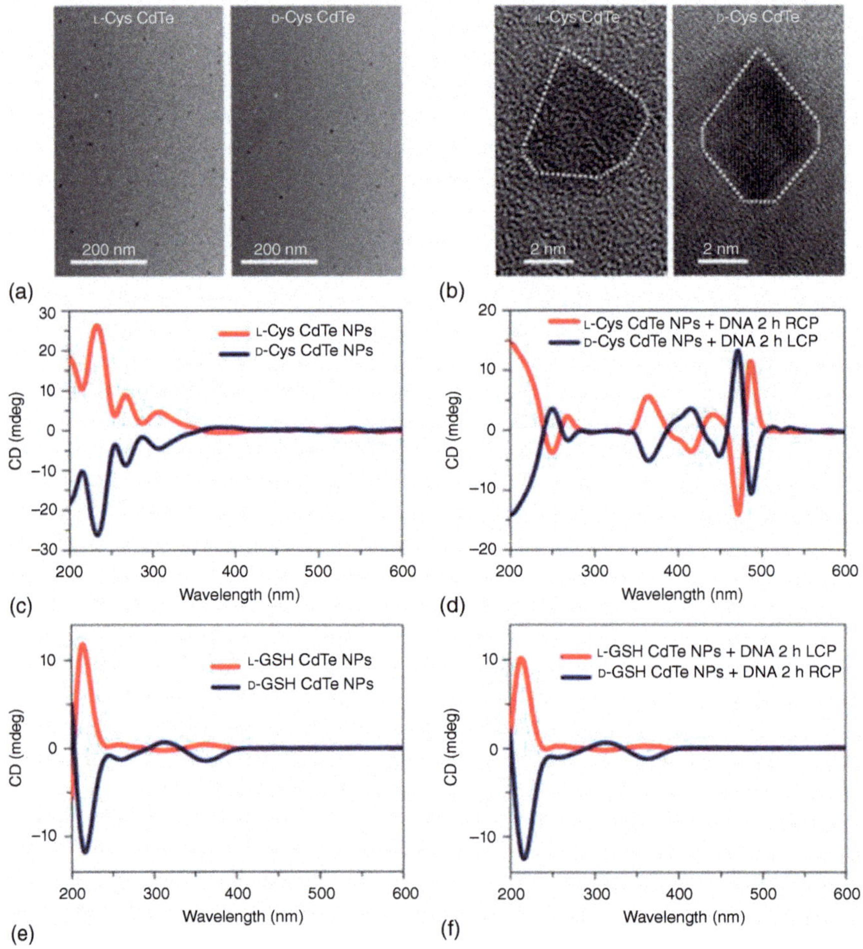


**Figure 2.3** Chiral CdTe NPs preparation and characterization [37]. TEM (a) and high-resolution TEM (HRTEM) (b) images of truncated tetrahedron-shaped L-Cys CdTe NPs and D-Cys CdTe NPs. (c) CD bands of L-Cys CdTe NPs and D-Cys CdTe NPs. (d) CD bands of L-Cys CdTe NPs and D-Cys CdTe NPs mixed with 1839 bp DNA and under two-hour illumination at 405 nm RCP/left-handed circularly polarized light (LCP). (e,f) CD bands of L-GSH CdTe NPs and D-GSH CdTe NPs before (e) and after (f) the mixture of 1839 bp DNA and two-hour illumination with 405 nm LCP/RCP. Source: Sun et al. [37]/with permission of Nature Publishing group.

In a latest research report by Kuang's group, truncated tetrahedral chiral CdTe NPs were synthesized with a diameter of 4.5 ± 0.3 nm (Figure 2.3a,b). Such truncated tetrahedral chiral CdTe NPs were composed of D-Cys or L-Cys [37]. The truncated tetrahedral CdTe QDs displayed a CD signal at 200–320 nm (Figure 2.3c) [38]. Mirror and broad CD features were detected for the D-Cys- and L-Cys-capped CdTe NPs between 350 and 410 nm (Figure 2.3c). The L-Cys CdTe NPs exhibited a negative signal, while the D-Cys CdTe QDs band was positive at these wavelengths. After mixing the CdTe nanoparticles with a natural double-stranded DNA of approximately 1839 bp (Salmon sperm DNA), no obvious CD bands of the CdTe nanoparticles changed in the visible region. Interestingly, under the two-hour right-handed circularly polarized light (RCP) illumination of the L-Cys CdTe–DNA mixture, the CD curves of the L-Cys CdTe displayed various CD bands at 365, 415, 470, and 488 nm within the absorption bandgap transitions (Figure 2.3d). They also further explored other kinds of ligands for the CdTe nanoparticle preparations in their study, such as GSH (Figure 2.3e,f), *N*-isobutyryl-L-cysteine, Cys-Phe dipeptides, *N*-acetyl-L-cysteine and L-cysteine methyl ester hydrochloride as well as achiral 2-mercapto acetic acid.

Based on the above study, Xu and colleagues further prepared Cys-decorated CdTe NPs for subsequent biomimetic nanozyme activity [39]. The prepared water-soluble L-Cys CdTe NPs were characterized by TEM and HRTEM. As shown in Figure 2.4a, the size of L-Cys CdTe was about 4.21 ± 0.4 nm. Similarly, the truncated tetrahedral shape and crystal lattice were observed as displayed in the HRTEM images (Figure 2.4b). A broad CD band was detected between 200 and 450 nm, after the L-Cys modification (Figure 2.4e). And the CD band at 200–310 nm was originated from the chiral molecule of L-Cys. Notably, compared to pure L-Cys, the L-Cys CdTe NPs exhibited intense CD signal and ultraviolet-visible (UV–vis) spectra between 350 and 410 nm. Interestingly, as shown in the TEM (Figure 2.4c) and HRTEM (Figure 2.4d) images, the morphology of the L-Cys CdTe was transformed from NPs to nanorods (NRs), under irradiation by 405 nm RCP for two hours after mixing with the DNA. The NR length was measured to be 22 ± 1.8 nm and the width was 5 ± 0.18 nm. Additionally, the UV–vis absorption shifted from 410 to 450 nm, and the fluorescence peak changed from 610 to 617 nm with a decrease in intensity (Figure 2.4f). After illumination, CD signals were more complicated. Five characteristic peaks emerged at 365, 415, 448, 471, and 488 nm, which corresponded to Te-rich CdTe NRs.

By the construction of complex architectures from an abundance of components in a single structure enabled to study the combinatorial properties, especially for manipulation and enhancement of optical activities. There were many reports on the QD materials, such as CdS rods, CdSe@CdS dot rods, end-site doped CdS-Pt, CdSe@CdS-Au/Pt/Pd, and multicomponent heterogeneous ZnS-CdS/Au/Pt/Pd nanowires; however, most of these studies focused on the photocatalysis research. The plasmonic effects are beneficial to the adjustment of the signal-to-noise ratio of electronic structures and can promote the rapid transfer and separation of electrons. Furthermore, the plasmonic effects can induce optical enhancement, especially for the adjustment and enhancement of CD signal [40–44].

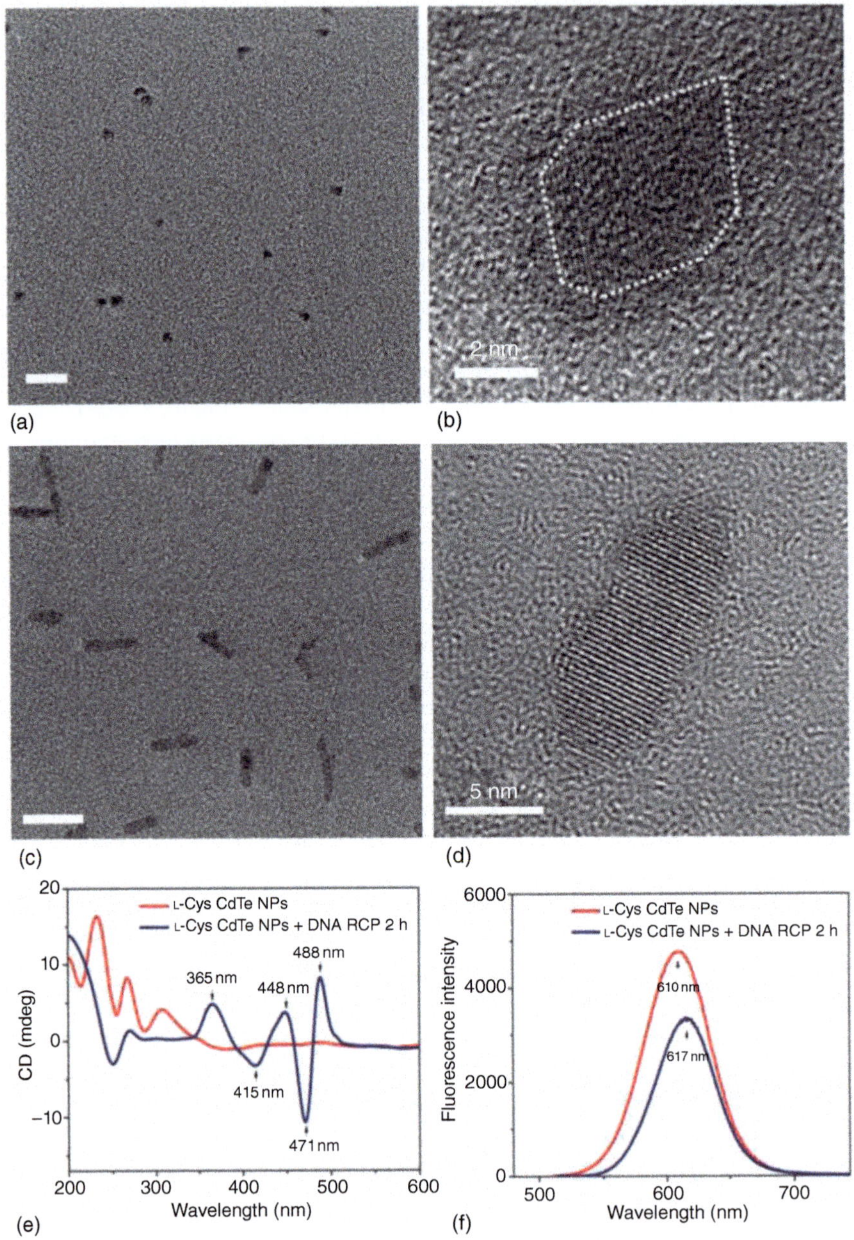


**Figure 2.4** Geometric and optical characterization of the L-Cys CdTe NPs. (a) TEM and (b) HRTEM images of the L-Cys CdTe NPs. (c) TEM and (d) HRTEM images of the L-Cys CdTe NPs treated with DNA under two-hour RCP (100 mW cm$^{-2}$, 405 nm). (e) CD spectra of the L-Cys CdTe NPs with and without DNA and two-hour RCP (100 mW cm$^{-2}$, 405 nm) treatment. (f) Fluorescence spectra of L-Cys CdTe NPs with and without DNA and two-hour RCP (405 nm, 100 mW cm$^{-2}$) treatment. Source: Wang et al. [39]/with permission of Wiley.

A recent work by Ma and colleagues reported intense chiroptical response of CdSe@CdS SNR based on phase transfer mediated by two amino acids (cysteine and histidine), which was further enhanced by the post growth of plasmonic metals (such as Au, Pt) [1]. Based on the classical hot-injection technique with slight modifications, wurtzite CdSe and CdSe@CdS SNRs were prepared in trioctylphosphine oxide (TOPO). The amount of the CdSe seed, concentrations of hexylphosphonic acid and octadecylphosphonic acid were optimized to tune the aspect ratio of the CdSe NR and CdSe@CdS. Well-grown rod shapes of synthesized SNRs (CdSe, CdSe@CdS) were observed with good dispersibility and exhibited typical band-edge transitions of semiconductor NCs. With strong quantum confinement, the complex energy band of the QDs produced multiple peaks of CD spectra with stronger signals at 539–558 nm or in the shorter wavelength region of the 445–461 nm bands. CdSe@CdS $SNR_1$ showed a much stronger anisotropy factor (g-factor, $2 \times 10^{-4}$) than that of CdSe nanorods ($7.9 \times 10^{-5}$), where the maximum g-factor of CdSe nanorods was close to the aspect ratio of 1.8.

When the mixtures of racemic cysteine and histidine mixtures (*rac*-Cys, *rac*-His) were used for phase transfer, the SNR showed no chiroptical signals, which was mainly due to the elimination of overall chirality at the semiconductor interface by the racemic mixture of L-Cys and D-Cys. Different from CdSe QDs, non-thiol chiral ligands used for phase transfer, such as tartaric acid or a combination with an achiral thiol ligand (3-mercaptopropionic acid, MPA), cannot cause CD signal in the visible light range. Meanwhile, only the tartaric acid induced CD bands at 240 nm. $SNR_1$ phase-transferred with two amino acids (cysteine and histidine) exhibited more intense chiroptical activity than that of single cysteine-phase-transferred $SNR_1$ in the range of 200–350 nm; however, CdSe displayed weak chiroptical activity.

The combination of D-Cys, D-His (or L-Cys, L-His) produced more intense chiroptical activity; meanwhile, the SNR maintained strong luminescence. Longer SNR rods gave rise to more intense CD signals (461 nm, 1120.1 $M^{-1}$ $cm^{-1}$), as well as larger g-factors ($1.2 \times 10^{-3}$ of $SNR_1$; $8.4 \times 10^{-4}$ of $SNR_2$; $5.9 \times 10^{-4}$ of $SNR_3$).

With the increasing aspect ratio, the increased CD intensity was probably attributed to the arrangement of Cd—S bonds or the combination of carboxyl groups in the long axis direction of the SNR. Under the same concentrations of the SNR, along with the increasing number of cysteine or histidine molecules and more bound or coupled amino acids, longer SNRs showed stronger CD intensity (Figure 2.5). Besides, an obvious redshift (11 nm) was observed from the CD bands for larger sizes of the SNR, which may be attributed to the larger dimension of the SNR. Moreover, $SNR_1$ phase-transferred with two amino acids possessed a much higher g-factor value of $1.2 \times 10^{-3}$ than that of the single D-cys or L-cys $SNR_1$ ($5.2 \times 10^{-4}$), indicating as the potential and efficient chiral construction method for organic-phase-prepared nanostructures.

The CD intensity can be further enhanced by the multisite post growth of platinum (Pt) and gold (Au), which can also induce the energy band transition

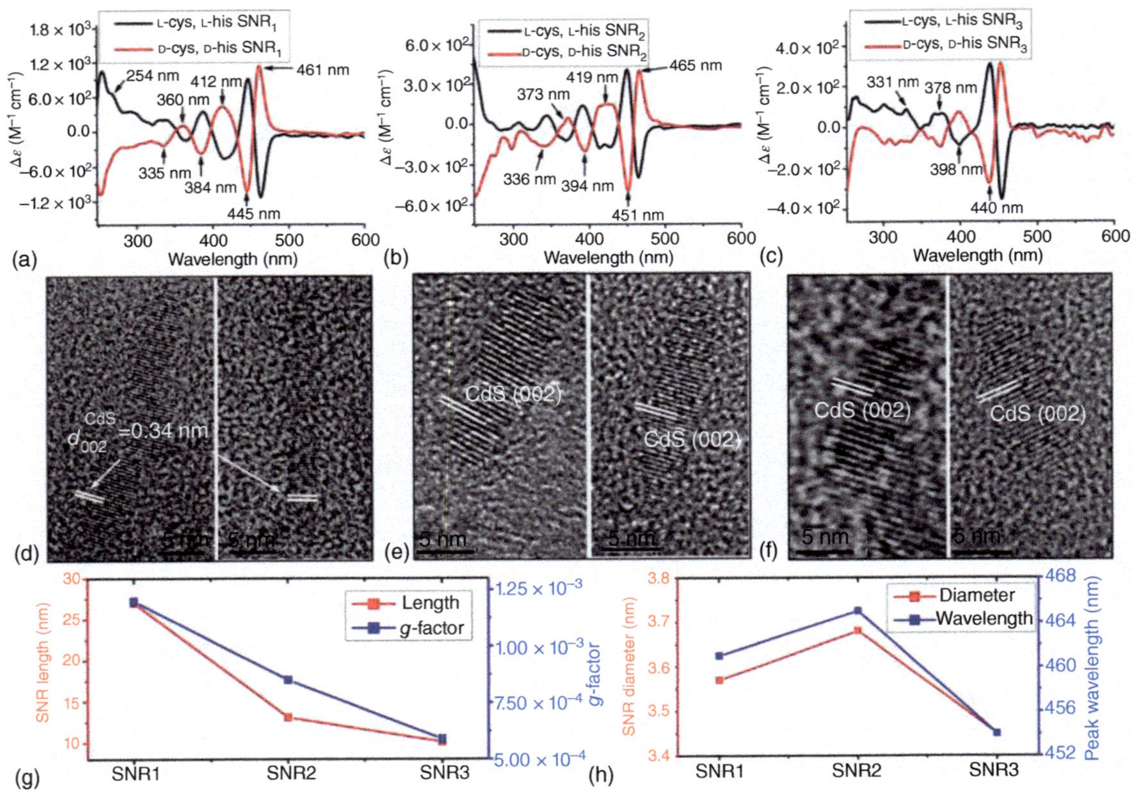


**Figure 2.5** (a–c) CD activity of the D-Cys, D-His (and L-Cys, L-His) phase-transferred $SNR_1$ (aspect ratio of 7.2) (a), $SNR_2$ (aspect ratio of 3.6) (b), and $SNR_3$ (aspect ratio of 3.1) (c); (d–f) HRTEM images for the D-Cys, D-His $SNR_1$ and L-Cys, L-His $SNR_1$ (d); (e) D-Cys, D-His $SNR_2$ and L-Cys, L-His $SNR_2$; (f) D-Cys, D-His $SNR_3$ and L-Cys, L-His $SNR_3$; (g, h) *g*-factor (g) and peak wavelength (h) for the length and diameter of $SNR_1$, $SNR_2$, and $SNR_3$. Source: Ma et al. [1]/with permission of Elsevier.

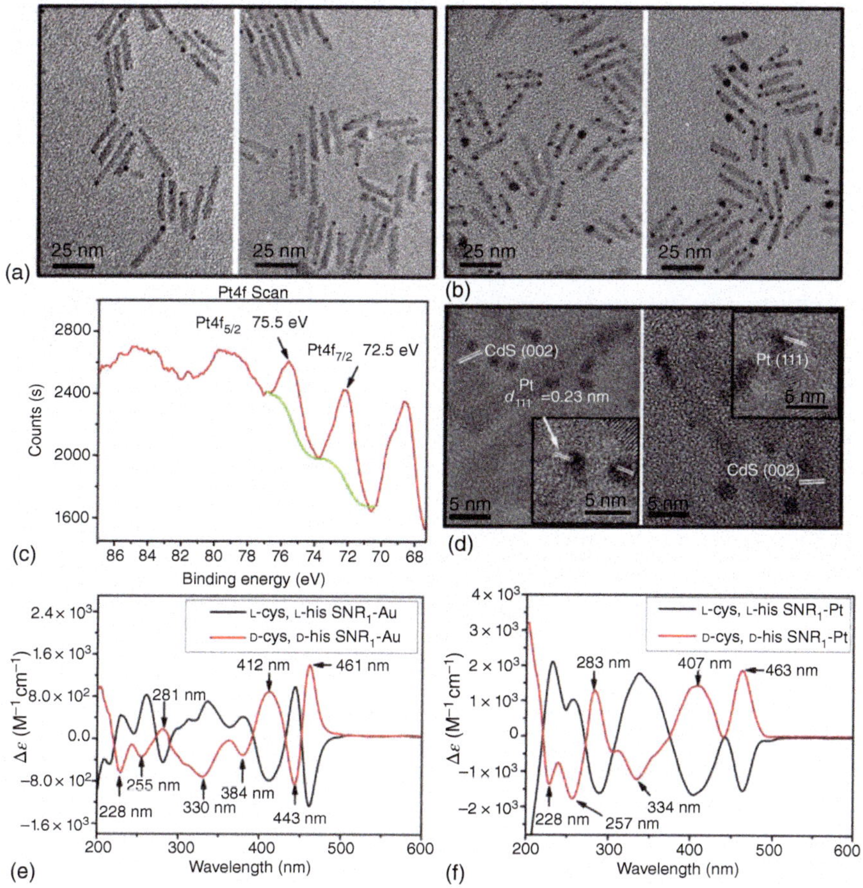


**Figure 2.6** (a) Representative TEM images of D-Cys, D-His $SNR_1$-Au nanorods (left), L-Cys, L-His $SNR_1$-Au nanorods (right); (b) representative TEM images of D-Cys, D-His $SNR_1$-$Pt_2$ nanorods (left) and L-Cys, L-Cys $SNR_1$-$Pt_2$ nanorods (right); (c) XPS spectrum for D-Cys, D-His $SNR_1$-$Pt_2$. (d) Representative HRTEM images for D-Cys, D-His $SNR_1$-$Pt_2$ (left) and L-Cys, L-His $SNR_1$-$Pt_2$ (right). (e) CD spectrum for D-Cys, D-His- and L-Cys, L-His $SNR_1$-Au nanorods; (f) CD spectrum for D-Cys, D-His- and L-Cys, L-His $SNR_1$-$Pt_2$ nanorods. Source: Ma et al. [1]/with permission of Elsevier.

of CD. The strongest CD intensity band of 461 nm increased from 1120.1 to 1413.7 $M^{-1}$ $cm^{-1}$, while other CD bands (such as 443, 412, 384, and 330 nm) maintained the intensity close to that of the original SNR. Because of the Au NCs on the SNR, the CD intensity of the CD bands (at 228, 255, and 281 nm) were amplified and distinguished. They found that Pt post growth of $SNR_1$ occurred at multiple sites on the sides and ends of $SNR_1$ (Figure 2.6). Particularly, the CD signal at 461–463 nm increased from 1120.1 to 1858.5 $M^{-1}$ $cm^{-1}$, meanwhile the CD intensity at 407–412 and 334–335 nm increased significantly (454.5–1525.5 $M^{-1}$ $cm^{-1}$, and 229.5–1404.1 $M^{-1}$ $cm^{-1}$, respectively). The size of metallic platinum was the key to enhance the electronic-state hybridization and achiral SNR and chiral ligands.

During the growth process, the high Pt content can destruct the combination of surface amino acids and cause release, while, the low content of Pt cannot produce the maximum enhancement effect.

From the electromagnetic calculation results, intense electromagnetic field enhancement was produced with metal post growth. And the enhancement from platinum was stronger than that of Au. The higher photocatalytic activity was displayed by the optimized chiral multisite platinum of the SNR, indicating that the deposited metal imparted energy transfer and enhancement of the separation for photoexcited electrons and holes. Therefore, these findings demonstrated the multisite-deposited metals on the NRs, which offer a way to generate highly optically active nanostructures mediated by the morphology of semiconductor nanorods.

Moreover, selenium, tellurium, and α-HgS have been demonstrated to show inherent chiral crystal lattices, which can be influenced by chiral molecules [2]. By using chiral ligands (penicillamine), a chiral crystal phase was observed from the α-HgS NPs with a large enantiomeric excess [27]. Similarly, they have also explored colloidal tellurium and selenium structures with crystal lattice and structural chirality based on the thio-chiral biomolecules (glutathione, penicillamine, and cysteine) [2]. Thus, the chiral molecules can trigger the enantioselective nucleation and growth and eventually form a chiral lattice conformation of NPs.

In summary, different chiral models have been proposed for semiconductor NPs. Note that these models were based on the limited experimental results, and further experiments are needed to verify them. The relationship between the surface coating or stabilizer type and the CD signals is informative and intriguing, but their chiral bands are generally less stronger than that of metal nanostructures.

### 2.2.2 Semiconductor Helicoids

Many methods for synthesizing spiral or twisted semiconductor nanomaterials have been used through the racemization process. For instance, spirally rotating epitaxial branched nanowires of PbS "pine trees" were prepared via chemical vapor deposition. The probability of occurrence of right- or left-handed "pine tree" is roughly equal, and the experimental ee ratio is 107 : 126.47 [45]. Biomolecular templates are often used in the formation of nanostructures of inorganic materials, which can regulate the geometry of the nanostructures. This strategy can overcome the difficulty of enantioselectivity. The polymer assembly templates were often used to synthesize helical nanostructures, which showed the right- and left-handed helices of growth CdS helices (Figure 2.7a) [46, 48]. The pitch of the inorganic helix was about 40–60 nm (Figure 2.7b,c). However, this process is not enantioselective. Another method used for ZnS showed clearly right-handed spiral nanoribbons in metal cholate (Figure 2.3d–g) [47]. Because of the addition of chiral molecules of cholate, this process was enantioselectively right-handed. The incorporation of metal ions can provide a variety of functionalities and merits to other nanohelices such as $SiO_2$.

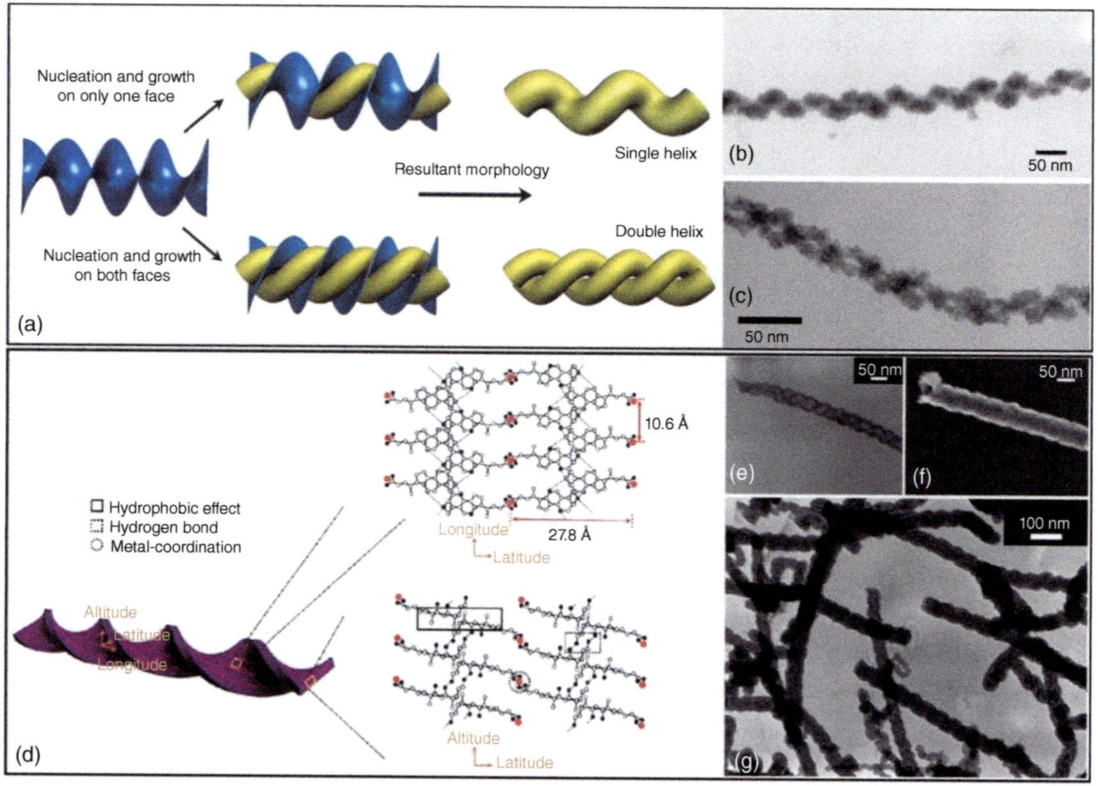


**Figure 2.7** (a) Molecular graphic model of a twisted ribbon in ethyl methacrylate. Nucleation and growth on one side of the twisted ribbons result in single CdS helices, while nucleation and growth on both sides of the ribbon lead to double helices. (b) A mature CdS single helix. (c) Representative TEM images of CdS double helices [46]. (d) Schematic illustration of twisted nanoribbon consisting of parallel longitudinal stripes; (top right) top view and (bottom right) cross-sectional view of the molecular model. (e–g) Representative TEM and SEM images of ZnS semiconductor helical nanotubes [47]. Source: (a–c) Sone et al. [46]/with permission of Wiley-VCH; (d–g) Reproduced with permission from Qiao et al. [47]/with permission of American Chemical Society.

## 2.3 Chiral Metal Nanostructures

Chiral metal NPs have more structural information than chiral semiconductor NPs. Among all the nanostructures of similar sizes, metal nanoparticles have the strongest chiroptical activity. Their potential applications in metaoptics [49, 50] and enantioselective catalysis have also received special attention [7].

Several general methods have been reported for preparing metal NPs for organic surface ligands. These methods differ depending on the nature of the chemical reaction, resulting in metal formation [51–53], including chemical reduction of metal salts, electrochemical reduction, replacement from organometallic complexes, laser ablation, and thermal, photochemical, or sonochemical decomposition of metal–organic precursors. Most of the researches synthesized the metallic NPs with chiroptical activity through the chemical reduction method [51]. Gold (III) tetrachloride acid and silver nitrate were the typical precursors, and the most frequently employed reduction reagents were sodium borohydride and sodium citrate. In the case of surface ligands, amino acids, peptides, and DNA oligomers were often used for the synthesis of chiral metal NPs. Ligand exchange reaction was rarely used to perform post-synthesis modification of metal NPs with chiral molecules or clusters [54]. Compared with other nanoscale materials, the combination of stabilizers and the cores of NPs (containing heavy metals) can be very intense. Therefore, the growth of metal nanoparticles were often terminated at a relatively early stage, forming Au and silver (Ag) NPs of about 1–2 nm. Such NPs can be also called nanoclusters.

Chiral plasmonic NPs can display much more intense chiroptical activity compared with the relevant molecular response. The first study of chiral Au NPs has been reported by Schaaff et al. [55]. Since then, numerous chiral metal NPs have been prepared. Au NPs with certain polarization rotation activity have been synthesized by using a variety of chiral ligands based on the chemical reduction method. By the aforementioned chemical reduction method, glutathione, *N*-isobutyryl-L-cysteine [56], D-/L-captopril [57], 1,1′-binaphthyl-2,2′-dithiol [58], and DNA oligomers were prepared [59]. More importantly, achiral molecules, such as 2-phenylethanethiolate [60], *p*-mercaptobenzoic acid [61], and others [62], can also induce chiral metal NPs by their chiral arrangements on metal surface. These studies demonstrated the complexity of chiral geometric recognition, which is responsible for the optical and other secondary characteristics of individual NPs. Chiral separation techniques, including electrophoresis, chromatography, and chiral biomolecules selective precipitation, may be beneficial to the yield enhancement of the specific sizes [63], and better understanding of chiral nanoparticle structures [64]. Furthermore, rotatory optically active dispersions can be separated from the initial racemic NP mixtures.

### 2.3.1 Metal Nanoparticles/Nanorods

Au and other metallic NPs are the representative chiral nanoscale systems that have been best studied either theoretically or experimentally. In general, Au NPs grown in contact with chiral organic molecules were chiral, which included the D- and L-isomers of diphenylalanine [65]. By employing enantiomers as surface ligands, the chiral Au NPs usually exhibited opposite polarity with bisignate signal at their plasmonic bands. Au nanorods (NRs) coated with chiral molecules also showed intense chiroptical activity. Because of their stronger transition moment, Au NRs exhibited stronger CD bands compared to those of Au NPs. Along with the imperceptible variations of the atomic and nanoscale geometries, the transition moment intensity or their relative directions may make a difference. Similarly to amino acids made up of proteins, the chiroptical signals in the UV range can be employed to quantify the collective conformations of surface ligands surrounding the NPs and NRs. These ligands can be comparable to the α-helical components of folded peptides [66].

Chiral mesoporous silica (CMS) layer-coated Au NRs showed chiroptical bands after porous shells functionalized with the D/L-cysteine [67]. The silica shell enabled the storage of chiral molecules for the chiroptical activity enhancement of amino acids in the UV–vis and near-infrared (near-IR) ranges. The metal–silica composites showed intense plasmonic CD signal and a clear chiral discrimination using the surface-enhanced Raman scattering (SERS) with stretching band at 730 $cm^{-1}$. Fan and Govorov used the electromagnetic theory to calculate chiroptical activity in the case of individual chiral NPs with twisters, antitwisters, asymmetrical pyramids, and chiral tetrahedron-like structures [13]. From the experimental aspects, they have synthesized metal helices with similar dimensions through stepwise glancing angle deposition. And the CD signal in the plasmonic absorption part was observed, which partially corresponded to theoretical calculation carried by Hou et al. [68] and Mark et al. [50].

Enantiomeric ligand-functionalized silver NPs also displayed high chiroptical activity [69]. Surface ligands covered captopril, glutathione, as well as a mixture of captopril and glutathione (Figure 2.8a) [70]. In the plasmonic part of the spectra, the researchers found that opposite handedness stabilizers induced signal signs for both positive and negative CD bands [71]. However, at the moment, one should be careful to ascribe it to concrete electronic transitions. It is rational that the CD signal of the silver NPs coated with D/L-penicillamine was assigned to metal-based electronic transitions (Figure 2.8b) [72], resembling the identified metal-based electronic transitions for chiroptical activity of $Au_{25}$ clusters (Figure 2.8c) [73].

Taking into account the different types of chirality, the highly degenerated molecular electronic state of $Ag_{55}SCH_3$ and $Au_{55}SCH_3$ can be chiral or achiral, indicated by time-perturbed density functional theory calculations, and they are dependent on the adsorption position and direction of the molecules [74]. The results showed that intense optical activity *n* can be observed in the case of thiols adsorbed between

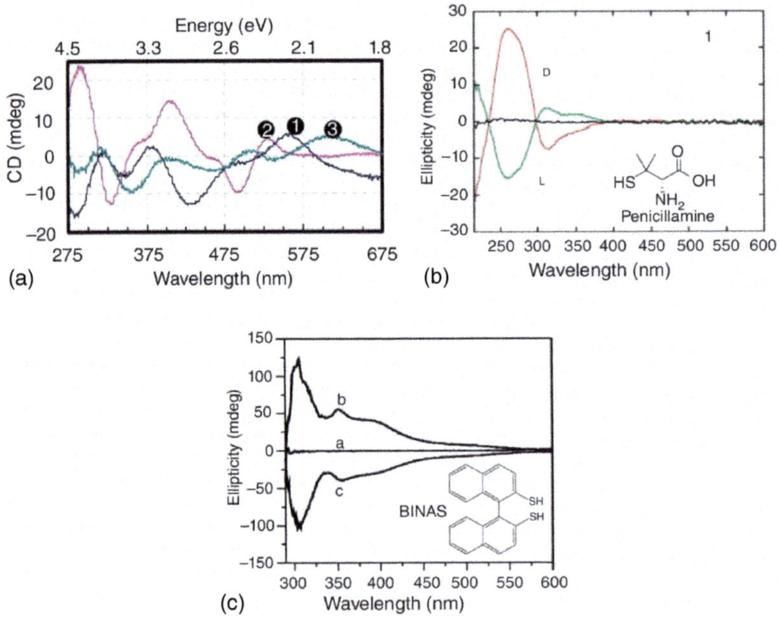


**Figure 2.8** (a) CD spectra of chiral Ag NPs coated with various enantiomeric thiol ligands [70]. (b) CD spectra of silver nanocluster compound [72]. (c) CD spectra of $Au_{25}$ coated with R/S-1,1′-binaphthyl-2,2′-dithiol (BINAS) [73]. Source: (a–c) Reproduced with permission from Refs [70, 72, 73] respectively. Copyright 2009, 2007, and 2009 American Chemical Society, respectively.

two atoms with different coordination numbers. Compared with Ag NPs, a single ligand can more effectively convert the Au NPs into plasmonic chirality, while the chirality of Ag may depend on the direction of the thiol molecule. In fact, the chirality of these NPs displayed two distinct mechanisms. The Au NPs metallic core was actually chiral, while the Ag NPs chirality was due to the misalignment between the plane formed by the ligand and the atomic planes within the metallic core. According to our classification, Au NPs belonged to Type 1 chirality. And even naked Au NPs may belong to the same type [75]. Due to the relative orientation of the single-ligand molecule for the overall chiral structure of the system, the Ag NPs should be classified as Type 3. It should be noted here that singly coordinated ligands show a large degree of rotation freedom. Even at room temperature, they may be randomized by thermal agitation. In addition, it was not possible to experimentally observe a ligand for each NP.

Larger silver NPs were wrapped by different chiral molecules, including DNA, glutathione, penicillamine, L-GS-bimane [76], peptides [77], cysteine and its derivatives [78]. NPs with different stabilizer shells showed tremendously various CD spectra. It indicated that different types of chirality contributed to the optical properties to a certain extent, including collective chiral-packing-induced Type 2 and Type 3 chirality. Therefore, glutathione and penicillamine exhibited CD signals; however, lysine- and glutamine-stabilized NPs did not show CD bands. Markovich and his colleagues reported that chiral-molecule-stabilized larger

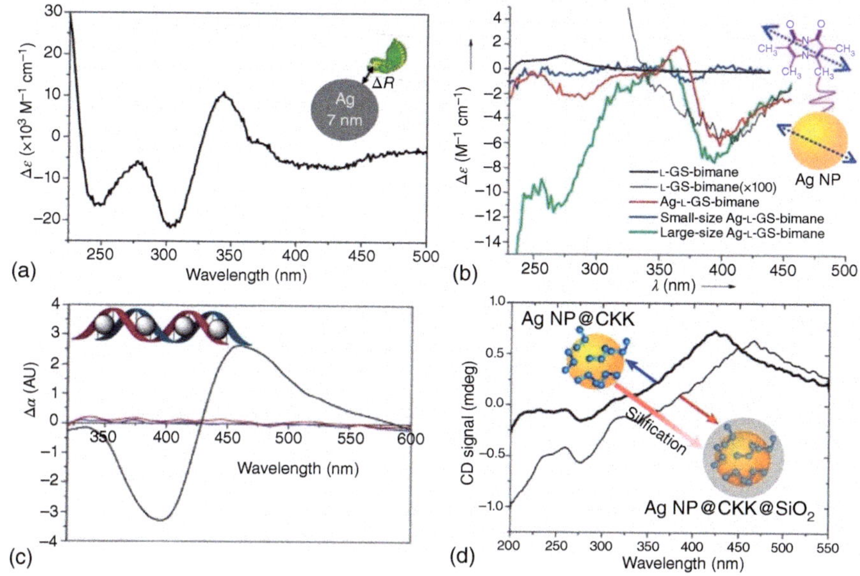


**Figure 2.9** (a) CD spectra of chiral molecular ligand-coated silver NPs [68]. (b) CD spectra of different samples [76]. (c) CD spectra of Ag NPs grown directly on DNA, Ag NPs grown without DNA, and Ag NPs adsorbed on the DNA [79]. (d) CD spectra of peptide-coated silver NPs and Ag/peptide@$SiO_2$ NPs [77]. (a, c, d) Reproduced with permission from Refs [69, 77, 80], respectively. Copyright 2012, 2006, and 2011 American Chemical Society, respectively; (b) Reproduced with permission from Graf et al. [77]. Copyright 2008 Wiley-VCH.

silver NPs displayed CD bands, including molecular and plasmonic CD bands (Figure 2.9a) [68]. Similarly, L-GS-bimane was attached to the Ag NP surface, and led to an increase in the CD spectra. And it can be explained by the increase in plasmon-induced resonant absorption caused by the larger particles (Figure 2.9b) [76]. The optical activity for silver NPs coated with cysteine and its derivatives at different concentrations and pH values also showed characteristic CD bands. In other studies, glutamine and cysteine cannot give rise to plasmonic CD signal. These reported studies only showed chiroptical activity in the adsorbate absorption part of the spectra. Therefore, it can be considered that Type 3 chirality does not necessarily lead to Type 1 and Type 2 chirality [78]. The result of NP synthesis in the presence of Ag NP poly(dG) -poly (dC) dual-stranded (ds) DNA showed that NPs displayed a chiral plasmonic CD band (Figure 2.9c) [79]. A silica shell was coated on the peptide-modified NPs to protect the NPs and make them more biocompatible (Figure 2.9d) [77].

Several groups reported the NP synthesis based on the DNA [71], peptides [80], and biomolecular fibrils [81]. For instance, Markovich and his colleagues reported that the Cotton effect on the plasmon part of the spectra of silver NPs (g~ 0.001) was much larger compared to other methods [79]. It was assumed that the atomic structure of chiral metal core was affected by the characteristics of Type 1 chirality. Silver NPs were also grown on closed supercoiled dsDNA, which generated changed conformation and chiral NP cores. When Ag NPs were prepared based on a chiral template, such as insulin helical fibrils (the outer diameter of 4–6 nm, the length of a few microns, and the helical pitch of 65 nm [80]), their CD showed an obvious chiral band in the plasmonic range. Chiral fibers with a diameter of 50–60 nm and lengths of several micrometers were prepared, after the gentle reduction of the enantiomeric silver coordination compound [81]. The organogel with a particle size of 10–20 nm displayed strong CD activity and a large anisotropy factor of ~0.02. In short, further experimental evidence was needed to determine the molecular geometry of these NPs, which may help establish accurate geometric models. Particularly, it will be crucial to clarify whether the chirality is confined to the surface, ligand shell, or metal core. The chiral memory effect indicative of surface distortions is unlikely in metal clusters. The chirality of the metal cores can be evaluated by CD or ORD spectra, which are useful for the construction of Type 1 and Type 2 chirality in metal clusters.

Wu et al. reported chiral Au/(DNA-Ag) core–shell NPs based on a DNA-template growth process [82]. They modified the Au NPs with $C_{30}$ thiolated single-stranded (ss) DNA (Au NPs-DNA). Then by adding silver nitrate ($AgNO_3$) to the solution, silver ions ($Ag^+$) were strongly bound to the $C_{30}$ ssDNA scaffolds, forming metal complexes (Au/DNA-$Ag^+$). Through the chemical reducing agent ascorbic acid, these $Ag^+$- bound molecules were finally transformed into a dense Ag shell on the surface of each Au NP. By adjusting the amount of $AgNO_3$ solution, Ag shells with different thicknesses were formed on the surface of Au NP (Figure 2.10a). As the thickness of the shell varied, the optical activity of the structure changed significantly (Figure 2.10b). When the thickness of the Ag shell increased, an absorption peak at 400 nm (the peak of the Ag plasmonic resonance band) appeared and was dominant. At the same time, as their absorption characteristics varied, a substantial increase at 413 nm in chiroptical activity was observed, and there was a slight redshift of 7 nm (Figure 2.10b). When the Ag shell increased to $8.0 \pm 0.6$ nm, the observed CD signal at 420 nm was $105 \pm 2.3$ mdeg, and the anisotropy factor, g, of the NPs reached $1.93 \times 10^{-2}$. It was very impressive for individual NPs, and their g factor value was about an order of magnitude higher than that of plasmonic or non-plasmonic nanoclusters or NPs.

As shown in Figure 2.11a, the starting cores of Au NPs were achiral with no chiral features. The DNA-bound Au cores (Au NPs-DNA) exhibited weak bands ($0.46 \pm 0.1$ mdeg) in the range of 500–550 nm, which was caused by the electronic "printing" of the DNA bound to the Au cores. Au NPs-DNA (Au/DNA-$Ag^+$) bound to $Ag^+$ showed a chiral morphology similar to Au NPs-DNA in the plasmonic region of the CD spectrum. Before the final growth of the silver shell, all samples had no plasmonic optical activity or weak CD bands at 500–550 nm. From the results in Figure 2.11a, the naked Au@Ag NPs had no chiral absorption at all. Secondly, Au@Ag NPs were modified by DNA, and the CD signal around 400 nm was $0.9 \pm 0.1$ mdeg. Thirdly, prepared DNA-Ag NPs, Ag NPs grown on DNA templates without Au core, displayed weak chiroptical activity ($2.5 \pm 0.2$ mdeg) at 428 nm.

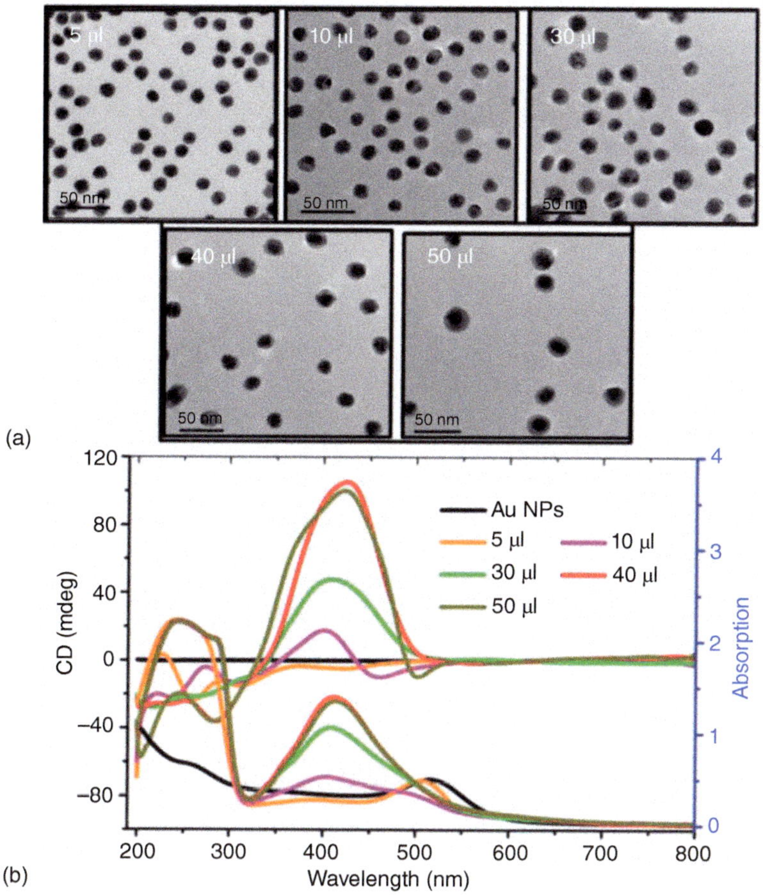


**Figure 2.10** Growth of Au/(DNA-Ag) core–shell NPs with different shell thickness. (a) Representative TEM images of Au/(DNA-Ag) core–shell NPs with $AgNO_3$ solution of 5, 10, 30, 40, 50 μl, respectively. (b) CD and UV–vis absorption spectra of Au/(DNA-Ag) core–shell NPs. Source: Wu et al. [82]/with permission of Wiley-VCH.

The Ag shell of Au@Ag NPs seemed to be epitaxially deposited on the surface of Au NPs and exhibited a uniform single crystal structure and (111) bare surface. In contrast, Au/(DNA-Ag) core–shell NPs presented a polycrystalline structure, mainly composed of (111) planes with a lattice spacing of 0.23 nm (Figure 2.11d). During the growth process, chiral DNAs were embedded in the Ag shell, strongly interfering with the epitaxial growth of the Ag shell. Thus, by growing a silver shell on a gold NP core with a cytosine–silver–cytosine–mismatched silver shell, a strong chiral particle with a size of 40 nm was achieved. And its g-factor value was up to $1.93 \times 10^{-2}$, which was similar to the anisotropy factor of chiral composite nanomaterials.

Recently, on the basis of previous researches, Hao et al. used cysteine to grow the silver shell on the surface of the gold core, and prepared enantiomeric chiral gold–silver nanostructures (GGS NSs) with interior nanobridged gaps [83]. GGS

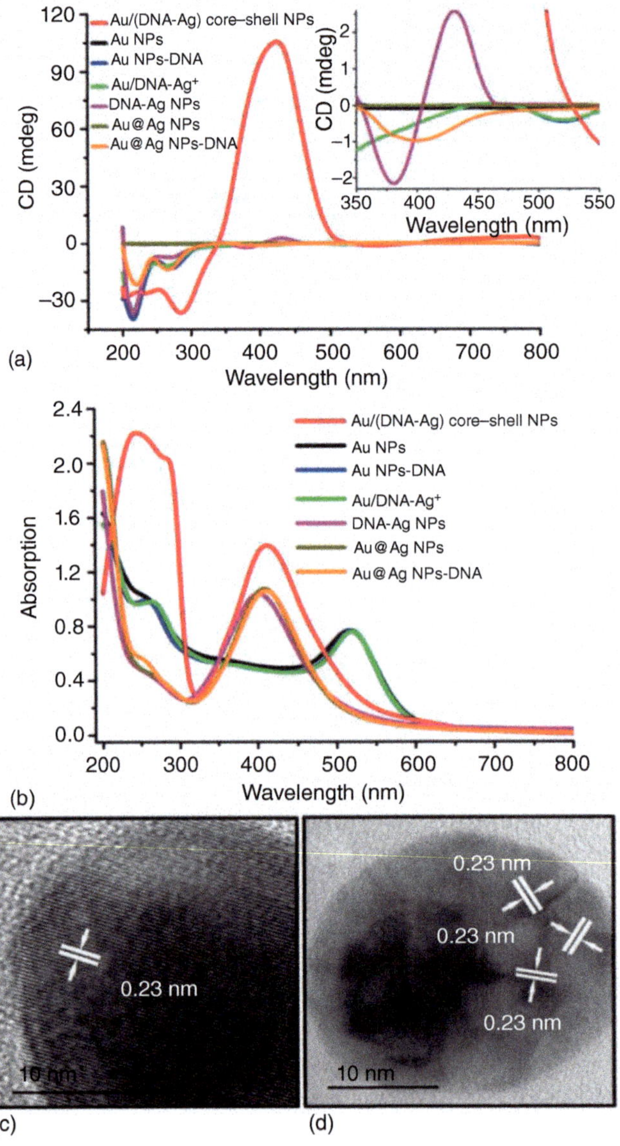


**Figure 2.11** Chiroptical activity and geometric properties of Au/(DNA-Ag) core–shell NPs. (a) CD spectra and (b) UV–vis absorption spectra of Au NPs, Au NPs-DNA, Au/DNA-$Ag^+$, Au/(DNA-Ag) core–shell NPs, DNA-Ag NPs, Au@Ag NPs, and Au@Ag NPs-DNA, respectively. (c, d) Representative HRTEM images of achiral Au@Ag NPs. Source: Wu et al. [82]/with permission of Wiley-VCH.

NPs were synthesized from the citrate-modified Au NPs. Methoxy-polyethylene glycol-thiol (PEG-SH) and L-cys ($0.01 \times 10^{-3}$ M) was coated on the AuNP-cit surface to prepare AuNP-L-cys/PEG, which were used as seeds for subsequent GGS-L NSs preparations. PEG-SH was employed to prevent the formation of disordered AuNP aggregates by cysteine molecules. Subsequently, precursor ($AgNO_3$),

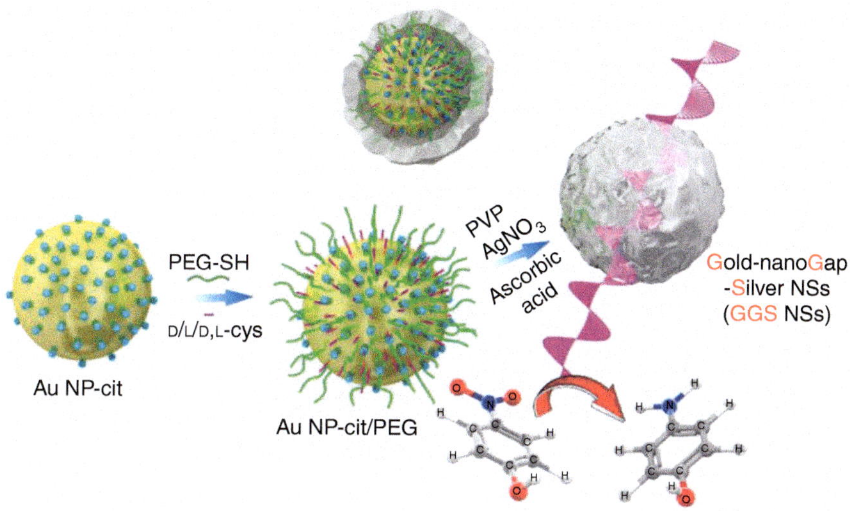


**Figure 2.12** Scheme of the chiral GGS NSs preparation [83]. Source: Reproduced with permission of Wiley-VCH.

reductant (L-ascorbic acid), and stabilizer (polyvinylpyrrolidone, PVP) were added sequentially for the GGS-L-0.01 NS preparations (Figure 2.12).

In the initial preparation stage, Au-Ag heterodimers were prepared by the growth of Ag structure budding on the surface of the Au core (Figure 2.13a). During the first 5–20 minutes, the budding NSs further grew and connected to each other, forming a complete silver shell and interior nanogaps (Figure 2.13b–e). The nanobridges connected the Au core, and the outer Ag shell was clearly observed by TEM images (Figure 2.13f). Along with the completely grown Ag shell, longitudinal bands gradually showed blueshift until disappearing, while redshift of Ag surface plasmonic resonance was observed, along with gradually increasing intensity. Only one absorption band of the final product at 430 nm was observed; no longitudinal plasmonic absorbance was found. As for the CD spectra, increased chiroptical activity at the Ag plasmonic part occurred after the Ag shell formation. Actually, Ag plasmonic CD signal increased from $23 \pm 1.2$ mdeg to $62 \pm 3.1$ mdeg during the 5–20 minutes, and an obvious redshift was observed (405–425 nm). In the case of Au plasmonic bands, the CD signal decreased from $6.2 \pm 0.4$ mdeg to $2.5 \pm 0.1$ mdeg and shifted from 570 to 535 nm. After 15 minutes, the Au plasmonic CD band disappeared, and similar conditions were observed in the UV–vis absorption spectrum. Notably, GGS-L-0.01 and GGS-D-0.01 NSs exhibited opposite CD line shapes in the 200–700 region window (Figure 2.13g). The opposite CD signal demonstrated the formation of opposite crystal enantiomers.

Importantly, the interior nanogaps sizes were corresponding to the intensity of CD signal; thus, they carried out further studies to explore their relationship. Generally, the varied amount of L-/D-/D,L-cys can manipulate the nanogap size of GGS NSs (Figure 2.14). A series of GGS with varied interior nanogaps can be obtained. For example, $0.5 \pm 0.1$ nm interior nanogaps of GGS-L/−0.01 NSs were

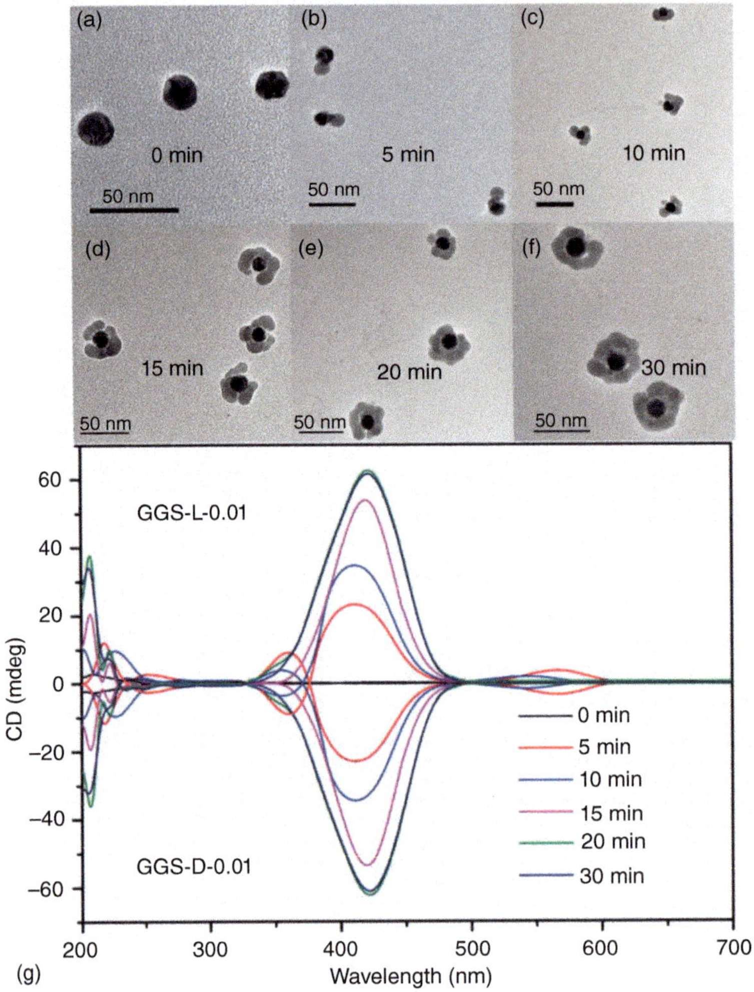


**Figure 2.13** (a–f) Representative TEM images of GGS-L-0.01 NSs for various reaction times. (g) CD bands for GGS-L-0.01 NSs and GGS-D-0.01 NSs. Source: Hao et al. [83]/with permission of Wiley-VCH.

observed at $0.01 \times 10^{-3}$ M of L-cys (Figure 2.14a–c). Meanwhile, as for using the same concentration of D-cys or D,L-cys, the GGS-D-0.01 NSs and GGS-D,L-0.01 NSs showed $0.5 \pm 0.2$, and $0.4 \pm 0.2$ nm, respectively. Larger interior nanogaps could be achieved with increased L-/D-cys concentration of $0.05 \times 10^{-3}$ M and the nanogap of $1.1 \pm 0.5$ and $1.2 \pm 0.4$ nm, respectively (Figure 2.14d,e). When the concentration of Cys further increased to $0.1 \times 10^{-3}$ M, large multi-nanohole-like NSs were observed; however, the GGS-L-0.1 NSs were not stable and even aggregated for a short period of time. Importantly, no interior nanogaps of Au@Ag NPs can be detected without the addition of cysteine in the synthetic process (Figure 2.14f).

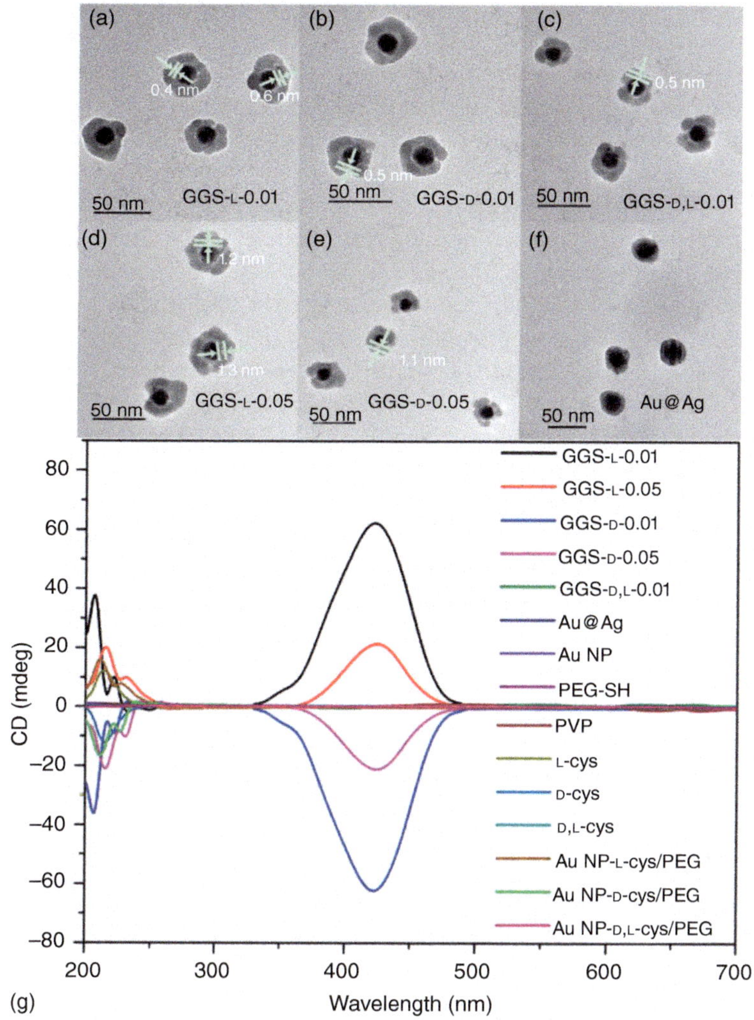


**Figure 2.14** TEM images of (a) GGS-L-0.01 NS, (b) GGS-D-0.01 NS, (c) GGS-D,L-0.01 NS, (d) GGS-L-0.05 NS, (e) GGS-D-0.05 NS, and (f) Au@Ag NP. (g) CD spectra. Source: Hao et al. [83]/with permission of Wiley-VCH.

Importantly, this work demonstrated the distinctive plasmon-induced chiroptical performance in the visible region and made progress for the enantioselective fabrication of chiral plasmonic NSs with tunable optical chirality. By simply altering the chiral cysteine amount, the nanogap of GGS NSs was tailored, as well as their corresponding CD signal (Figure 2.14g). It was believed that the dipole interaction between chiral cysteine and meal core-shell gave rise to plasmonic chirality and further enhanced by electromagnetic coupling. Moreover, they found unusual circularly polarized photocatalytic activity of the chiral GGS NSs under circular polarized light irradiation. It was believed that the preparation strategy could be extended to enantioselective synthesis of other high g-factor chiral NSs, and showed promising potential for chiral catalysis and biosensing.

Another representative study of chiral nanostructures based on wet-chemistry method was the fabrication of AuCuAu heterogeneous NRs reported by Xu group [84]. The chiral AuCuAu heterogeneous NRs were prepared in the presence of hexadecylamine (HDA), based on the selective growth of crystalline copper (Cu) along the Au NRs sides. Then Cu on the pre-synthesized can be etched based on the galvanic replacement reaction and resulted in crystalline Au-Cu alloy metal on one side of the Au NRs. Importantly, the added dipeptide (cysteine-phenylalanine, CF) ligands enabled the alloy heteronanorods AuCuAu (AuCuAu HNRs) to display intense CD in the 400–1000 nm wavelength range. A bisignate shape with typical positive (620 nm) and negative (838 nm) Cotton effect was shown in the CD spectra. By increasing the length of Au NR length, their plasmonic chiroptical activities can be enhanced. The largest g-factor value can reach as high as $0.57 \times 10^{-2}$, due to the local electromagnetic field (Figure 2.15).

The general method of fabricating chiral-shaped nanomaterials was mainly through the chiral molecule–mediated nanomaterial synthesis. By using glutathione as a chiral inducer, Gil Markovich and coworkers reported chiral-shaped tellurium (Te) nanostructures, which showed significant chiroptical activity [2]. Furthermore, because of its weak reduction capacity, a reducing agent, such as hydrazine, can also act as an essential factor in efficiently tuning the morphology of chiral Te NPs. Following the combination of four chiral thiolate molecules, the Te (IV) atoms further influenced the final geometry and chemical stability of the NPs. Two opposite vertices of Te NPs were bridged by twisted ridges. The asymmetrical side faces of the trigonal prismatic Te NPs structure determined the chiroptical activity. It can be concluded that coefficient modules of structural and atomic-scale chirality caused their chiral optical properties based on the discrete dipole approximation (DDA) calculation data and experimental results. Furthermore, they have also expanded the chirality to silver telluride or gold telluride chiral nanostructures on the basis of Te trigonal prismatic nanostructure and their chemical reactivity with Te crystalline. It is an important synthesis method for chiral nanostructure shapes from achiral nanomaterials, and the seed growth method is expected as a universal nanostructure synthetic technique for more inherent chiral-shaped nanomaterial fabrications.

### 2.3.2 Metal Nanohelices

In general, top-down methods, such as lithography, have been developed for the preparation of chiral structures. Many reports achieved the synthesis of enantioselective structures. Though it was difficult to prepare chiral metal nanohelices based on the bottom-up methods, it was feasible [85]. It was difficult for the ligand-regulated seed growth process to exactly manipulate the geometry, such as length, width, pitch. It would be more difficult to achieve enantioselectivity. Therefore, racemic mixtures were usually obtained for helical NWs. Kondo and et al. reported gold nanowires (Au NWs) in an ultrahigh vacuum [86]. The Au NWs

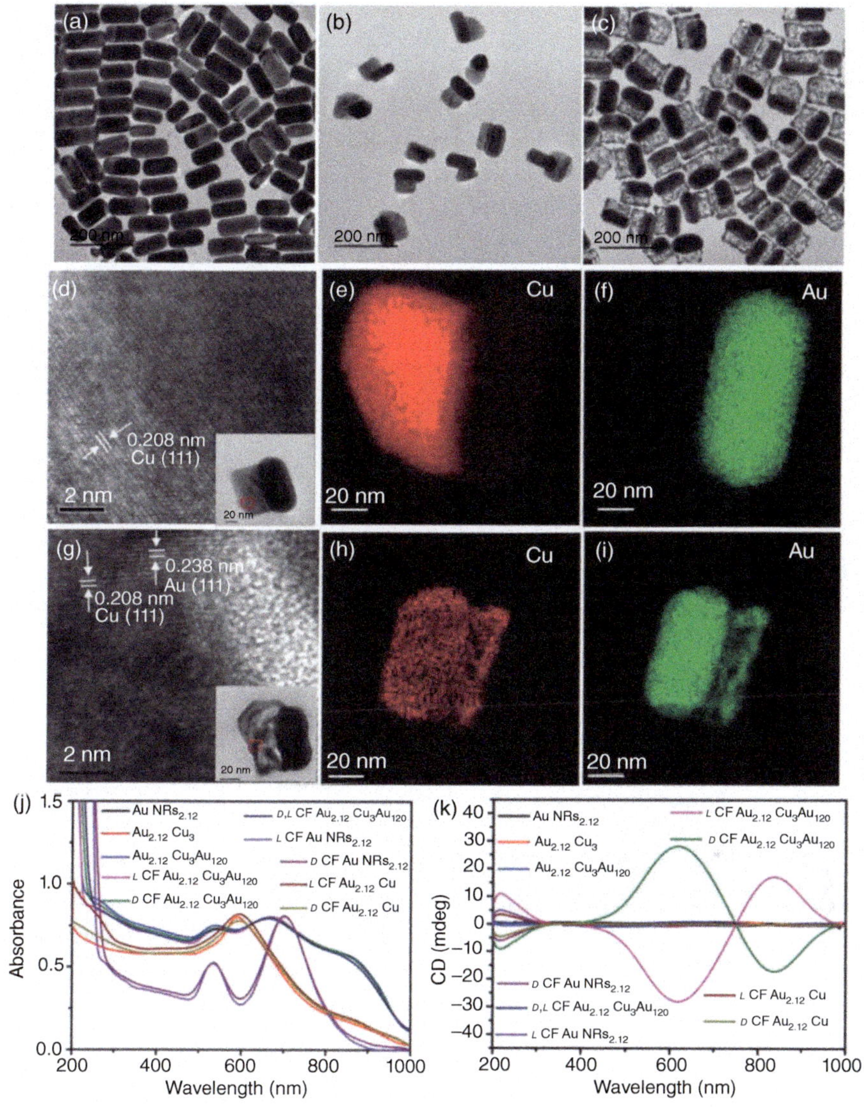


**Figure 2.15** TEM images of (a) Au $NRs_{2.12}$, (b) $Au_{2.12}Cu_3$, and (c) L CF $Au_{2.12}Cu_3Au_{120}$ HNRs. (d) HRTEM and (e, f) EDX mapping of $Au_{2.12}Cu_3$ HNRs. (g) HRTEM and (h, i) EDX mapping of L CF $Au_{2.12}Cu_3Au_{120}$ HNRs. (j) UV and (k) CD of L or D CF-modified Au $NRs_{2.12}$, L or D CF-modified $Au_{2.12}Cu_3$, D CF $Au_{2.12}Cu_3Au_{120}$ HNRs, L CF $Au_{2.12}Cu_3Au_{120}$ HNRs, and racemic D, L CF $Au_{2.12}Cu_3Au_{120}$ HNRs. Source: Wang et al. [84]/with permission of Wiley-VCH.

were composed of coaxial tubes with rows of atoms organized in a helical pattern. Interestingly, the shell-closing numbers enabled the preparation, where the number difference of helical atom rows in the outer and inner shells was 7. Subsequently, the helical NWs were achieved by Jose-Yacaman and coworkers based on the packing of interpenetrating icosahedral and decahedral units [87]. Those NWs were prepared as the Ag to Au was 3 : 1, by mixing the metal salts (Au and Ag) with oleylamine. The structure of icosahedral NWs was similar to those formed by two or three Boerdijk–Coxeter–Bernal helices roped on a single structure.

One representative study was the fabrication of double-helical Au nanowires with controllable chirality by Kawai and coworkers [88]. Chiral soft templates composed of two organogelators with active functions were employed. One of the organogelators was chiral D/L-12-hydroxystearic acid (HSA), which induced helicity into the template. The other organogelator was a long-chain amidoamine and served as a capping agent to control the Au shape. As shown in Figure 2.16, double-helical Au NWs obtained from the D- and L-HSA systems displayed right- and left-handed

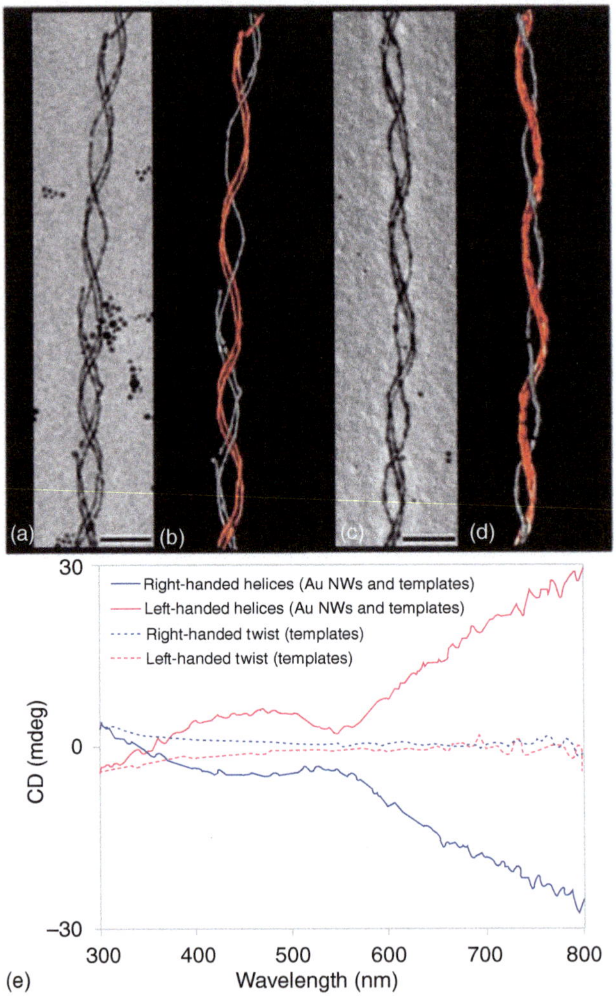


**Figure 2.16** (a, c) TEM images and (b, d) TEM–tomography images of Au NWs based on the right- and left-handed templates, respectively. (e) CD spectra of right- and left-handed double-helical Au NWs with twisted nanoribbons. Source: Nakagawa et al. [88]/with permission of American Chemical Society.

helicities, respectively. Following along both edges of the twisted nanoribbons, the Au NWs showed right- or left-handed helicities. Moreover, opposite CD bands were shown for the right- and left-handed double-helical Au NWs. The weak CD activity at the part of 300–350 nm was ascribed to soft templates with right- and left-handed twist. As a comparison, broad CD signals in the plasmonic part of 400–800 nm were derived from helical Au NWs. Therefore, by wet chemical synthesis method, chirality was successfully transcribed to the Au nanocrystals from the organic template.

In comparison to the above soft template method, amino-modified CMS can be used as a hard template to fabricate Ag NWs. Professor Che's group reported two-dimensional (2D) chiral silver nanowires (Ag NWs) based on CMS with a different multi-helical structure as a hard template [89]. Because of the coordination between the Ag ions and the amino group, silver ions could be easily introduced into the chiral pores. By the $AgNO_3$ impregnation with hydrogen reduction, the 2D chiral Ag NW arrays can be achieved within the channels of CMS. The 2D chiral Ag NWs can display two helical directions of the pores or the opposite orientation of the helical arrays. They may greatly influence the longitudinal and transversal chirality in the opposite direction.

As for diffuse-reflectance circular dichroism (DRCD), two obvious changes occurred. As the length of the CMS rod increased, the DRCD signals of Ag NW arrays within the antipodal L- or R-CMS also displayed bathochromic shift in consistent with longitudinal absorption (Figure 2.17). Secondly, a significant decrease in two enantiomer samples was observed in the characteristic CD absorption bands. This change can be explained by helicity. Interestingly, along with decreasing helicity, the CD intensity decreased, due to the weaken asymmetrical coupling between Ag NWs along the longitudinal direction with increased pitch lengths. The DRCD data demonstrated that the longitudinal pitch length can play a leading role in the CD activity of Ag NW array.

More recently, Fischer and coworkers reported chiroptical activity of single Au nanohelix. The nanoscale pattern of low-temperature shadow deposition caused the synthesis of helical nanocolloids with the size of 20 nm and a wide selection of materials, including Au, Au:Cu, and Cu (Figure 2.18a,b) [50]. As a top-down method, it provided helices with enantioselectivity as well as with tunable handedness and pitch. The experimental CD results were further confirmed by simulated gold helices (Figure 2.18c,d). Different Cotton effect relevant to plasmonic chiral NPs was observed from the left- and right-handed nanohelices. Moreover, the material composition variation can adjust the chiral signal of nanohelices in the visible range. Similar CD bands and redshift were shown from the CD spectra of the Ag:Cu (65 : 25), Au, and Cu nanohelices. Distinct compositions displayed switched CD activity (Figure 2.18e). This technology displayed promising potential to prepare anisotropic nanocolloids with magnetic, semiconducting, metallic, and insulating materials.

The Au-Ag alloy NW produced a metallic double helix by winding around itself in the growth of a thin metal (Pd, Pt, or Au) layer (Figure 2.18f) [85]. The metal deposition rate was demonstrated to be a crucial factor for the winding of nanowire formation. The helical shape formation may cause the chiroptical activity of Au-Ag

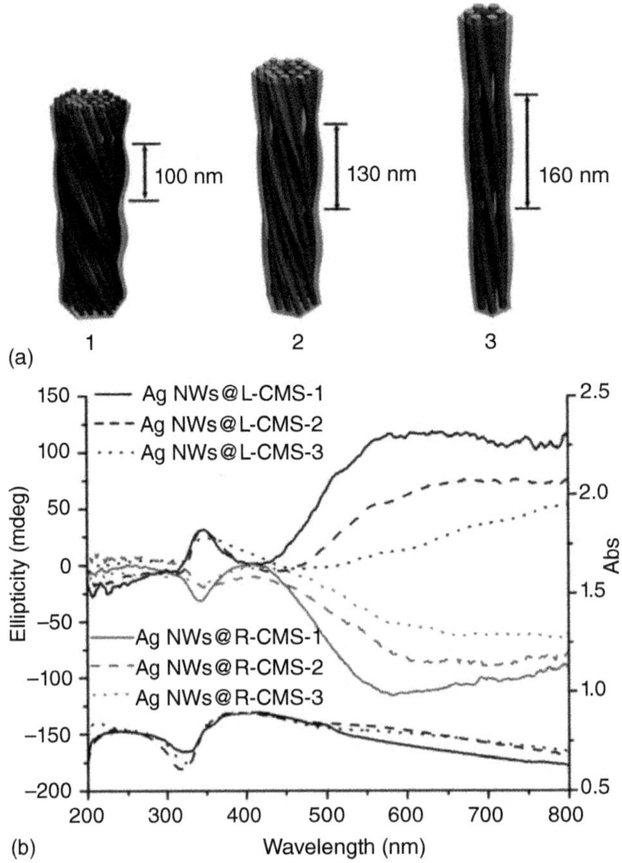


**Figure 2.17** (a) Schemes for varied pitch of Ag NWs@CMS. (b) DRCD and DR-UV/vis spectra of Ag NWs@CMS with distinct pitch lengths [89]. Source: Reproduced with permission of Wiley.

alloy NWs with a possible bcb-type twisted lattice. As observed in Figure 2.18g–i, one can clearly find left- and right-handed double helices. Importantly, the movement of a double helix formation under chemical stimuli may be used to develop smart nanodevices. At present, it is still a big challenge to synthesize enantioselective metal nanowires, but it is of great significance for obtaining metamaterials from the bottom-up methods.

### 2.3.3 Metal Nanorings and Superstructures

Taking advantage of the rapid development of wet chemistry nanomaterial preparation technologies, the morphology manipulation of nanomaterials has become more and more advanced. Generally, chiral-shaped nanomaterials were prepared through the chiral molecule–mediated nanomaterial method. Wang et al. fabricated chiral triangular Au nanorings with a platinum (Pt) framework (L/D-Pt@Au triangular nanorings, named L/D-Pt@Au TNRs) [90]. The hollow platinum (Pt) triangular

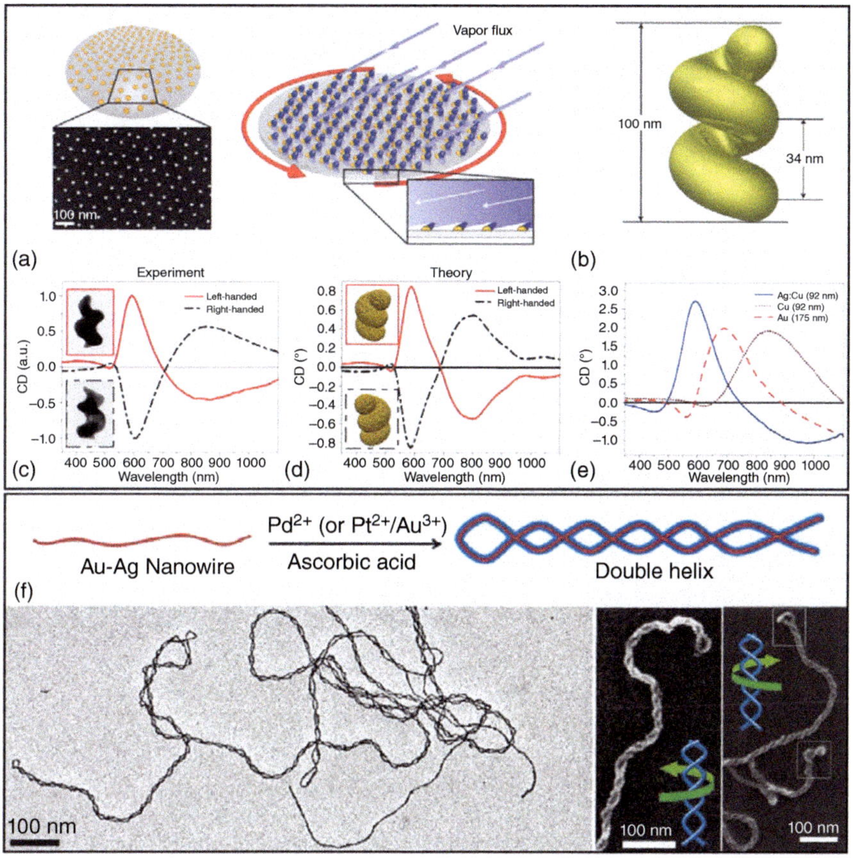


**Figure 2.18** (a) The glancing angle physical vapor deposition process. (b) Model two-turn gold helices. (c) Normalized CD activity of left- and right-handed helices. (d) Simulated CD spectrum. (e) CD signal of nanohelices with different metal compositions [50]. (f) Schemes of the double-helix formation by growing a metal layer on Au-Ag alloy NW. (g) TEM image of the AuAg NWs [85]. (h, i) SEM images of (h) right-handed (AuAg)@Pd double helices, and (i) left-handed (AuAg)@Pd double helices. Source: (a–e) Reproduced with permission from Mark et al. [50]. Copyright 2013 Nature Publishing group; (f–i) Kondo et al. [85]/with permission of American Chemical Society.

nanorings served as seeds, meanwhile L/D-glutathione (L/D-GSH) was used as chiral ligands. The Au was found to grow in an orderly arrangement on hollow Pt triangular rings. More importantly, intense CD response was detected for L/D-Pt@Au TNR, with an attractive and tremendous $g$-factor value of up to 0.023.

Generally, a series of synthesis processes were required for the preparation of chiral Pt@Au TNRs, which mainly included the rim-preferential Pt growth on Au nanoprisms, Au core etching, and gold regrowth on Pt nanorings. Firstly, the seeds of Au nanoprisms were synthesized; the edge length of Au nanoprisms was $83.7 \pm 2.2$ nm. Before the Pt deposition, a thin silver layer was coated on the surface of Au nanoprisms by electron shuttles. Then, $H_2PtCl_6$ was reduced by ascorbic

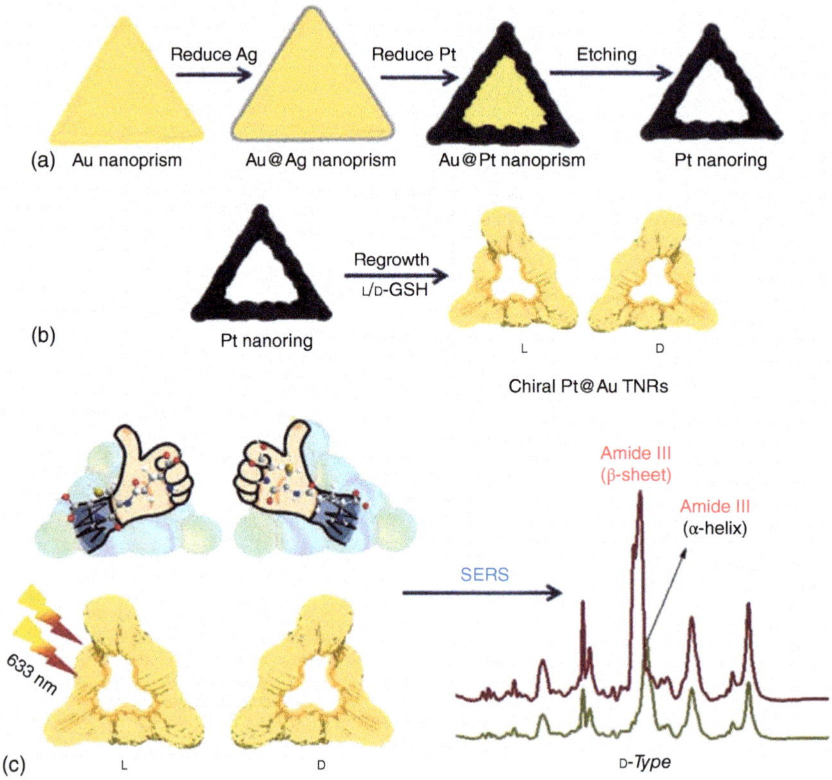


**Figure 2.19** Schemes of (a, b) chiral Pt@Au TNRs preparation and (c) bioapplication for $A\beta_{42}$ detection as a label-free SERS substrate [90]. Source: Reproduced with permission of Wiley.

acid. Due to the lower density of the hexadecylcetyltrimethylammonium bromide (CTAB) molecules and the presence of more undercoordinated metal atoms on the highly curved rim, Pt atoms were found preferentially deposited on the rim of Au nanoprisms, thus resulting in Au@Pt nanoprisms. After the addition of $HAuCl_4$ to selectively etch the gold cores in the Au@Pt nanoprisms, the Pt nanorings were obtained. Notably, GSH served as the chiral additives before etching (Figure 2.19).

As shown from the TEM images, Pt@Au TNRs with opposite vertex triangular orientations can be achieved in the case of distinct GSH enantiomers as chiral additives (Figure 2.20a–e). The nanorings were mainly made of Au and Pt, where the atomic ratio of Au and Pt was ~7.4, which was confirmed by high-magnification selected area electron diffraction (SAED) and energy-dispersive X-ray spectroscopy mapping. As shown from the high-resolution TEM images, the lattice distance of Pt@Au TNRs was 0.235 nm, corresponding to the bulk Au (111) spacing (0.236 nm). Another spacing of 0.204 nm corresponded to the (200) crystal plane, indicating the face-centered cube of the gold atoms. The side length of the Pt@Au TNRs with hollow triangle-like morphology was about $97 \pm 9.1$ nm; their framework thickness reached $20 \pm 2.8$ nm. Along the twisted tip of the triangle to one side, the GSH

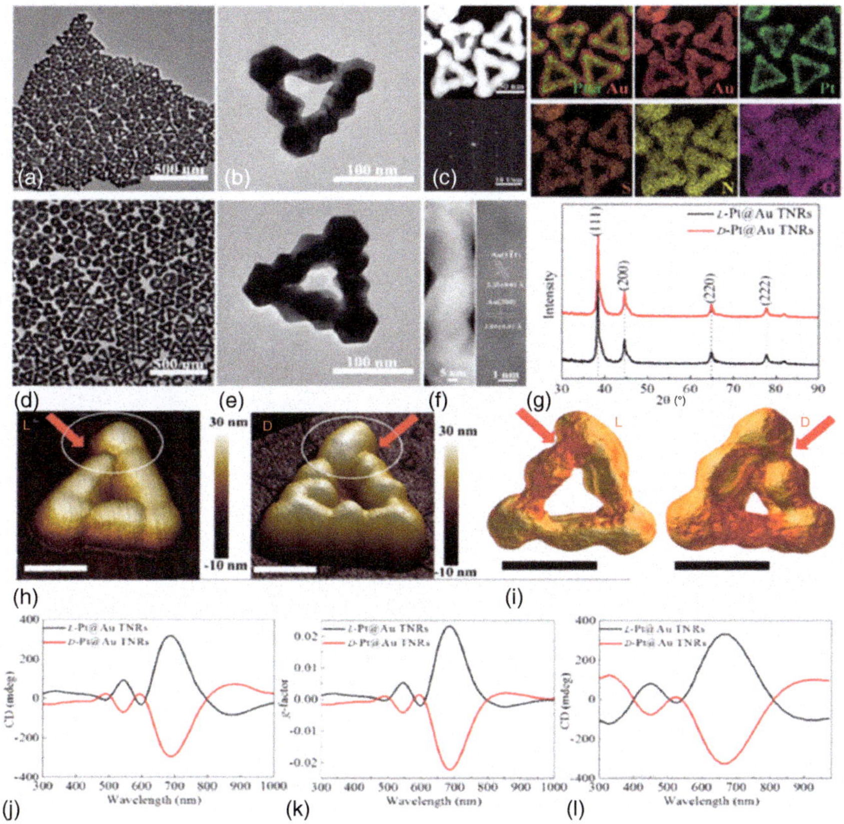


**Figure 2.20** (a, b) TEM images of L-Pt@Au TNRs (b), (c) HAADF, SAED image of L-Pt@Au TNRs and EDX-mapping image on L-Pt@Au TNRs. (d, e) TEM images of D-Pt@Au TNRs. (f) HRTEM image of L-Pt@Au TNRs. (g) Powder XRD patterns of chiral Pt@Au TNRs. (h) AFM of D-Pt@Au TNR and L-Pt@Au TNR. (i) 3D tomography images of D-Pt@Au TNR and L-Pt@Au TNR. (j) CD, (k) *g*-factor, and (l) Calculated CD spectrum of chiral Pt@Au TNRs. Source: Wang et al. [90]/with permission of Wiley.

enantiomers may cause opposite direction growth (Figure 2.20). The L-Pt@Au TNRs produced multiple CD bands at 489 nm (+), 543 nm (−), 594 nm (+), 689 nm (−), and 884 nm (+). The strongest CD band was at 689 nm. In the process of complete structural optimization, the L- and D-GSH enantiomers, L-Pt@Au TNRs, and D-Pt@Au TNRs showed mirror-image symmetries. Their maximum anisotropy factor ($g_{CD}$-factor) was calculated to be 0.023 at 686 nm. Using the final topology of the models, the frequency-domain finite integral method was used to calculate the CD activity of Pt@Au TNRs (Figure 2.20h–l).

With the increasing reaction time, the maximum CD intensity also enhanced (Figure 2.21a). Meanwhile, an obvious redshift was also observed, shown by 8.5 nm redshift from 682.9 to 691.4 nm during the reaction time of 10–50 minutes. Moreover, the $g_{CD}$ value can reach 0.023 at 50 minutes from 0.001 at 10 minutes (Figure 2.21b). As shown in Figure 2.21c,d, surface distortion were more evident

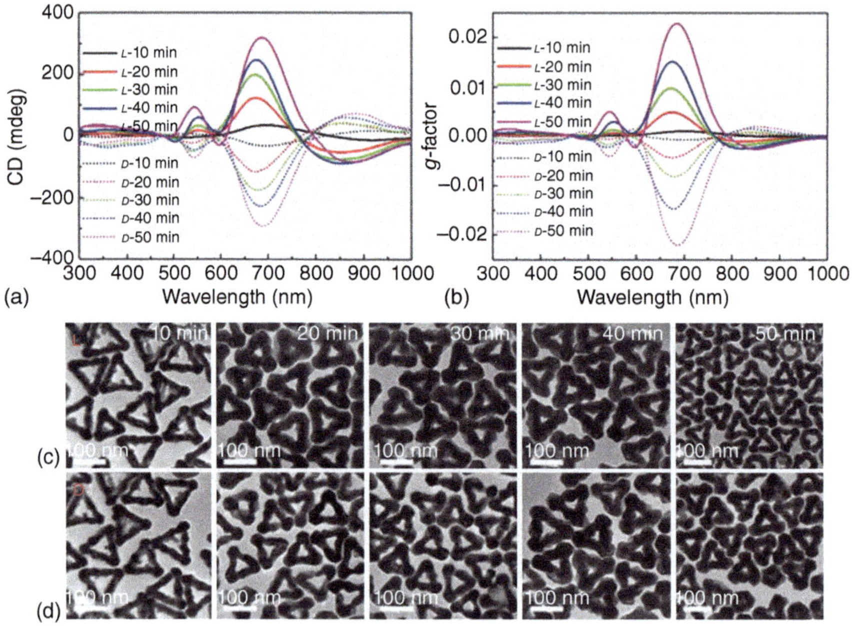


**Figure 2.21** (a) Chiroptical activity and (b) $g$-factor spectra of L/D-type Pt@Au TNRs under various reaction times. (c, d) Representative TEM images of L-type Pt@Au TNRs (c) and D-type Pt@Au TNRs (d) under different times. Source: Wang et al. [90]/with permission of Wiley.

for the Pt@Au TNRs with the increasing reaction times. Both L-Pt@Au TNRs and D-Pt@Au TNRs displayed a $g$-factor of 0.023. Moreover, due to the selective resonance coupling between the induced electric and magnetic dipoles, surface-enhanced Raman scattering-chiral anisotropy (SERS-ChA) effect was also found for the chiral plasmonic TNRs, which can be used effectively for the discrimination of enantiomers.

Neretina and colleagues used a light-driven solution to construct Au helices from substrate-bound seeds based on a plasmon-mediated growth mode [91]. The designed growth pattern depended on three separate symmetry-breaking events, which were involved by seeds lined with planar defects. The capping agent severely hindered early growth, and Coulombic repulsion occurred when they identically collided with the charge. As shown in Figure 2.22, by combining together, the planar growth was a preferential growth pathway, which originated from one side of the seed. Then the seed was encircled from both clockwise and counterclockwise directions. Following the collision of these two growth fronts, a self-perpetuating 3D spiral structure was achieved with the front one rising above the other. This study demonstrated a sequentially occurring symmetry-breaking-dependent spiral growth mode and a redirected growth pathway at critical junctures during the synthesis process.

More recently, a study on chiral-shaped nanostructure construction was carried out by Nam and coworkers (Figure 2.23) [92]. They proposed an efficient chirality

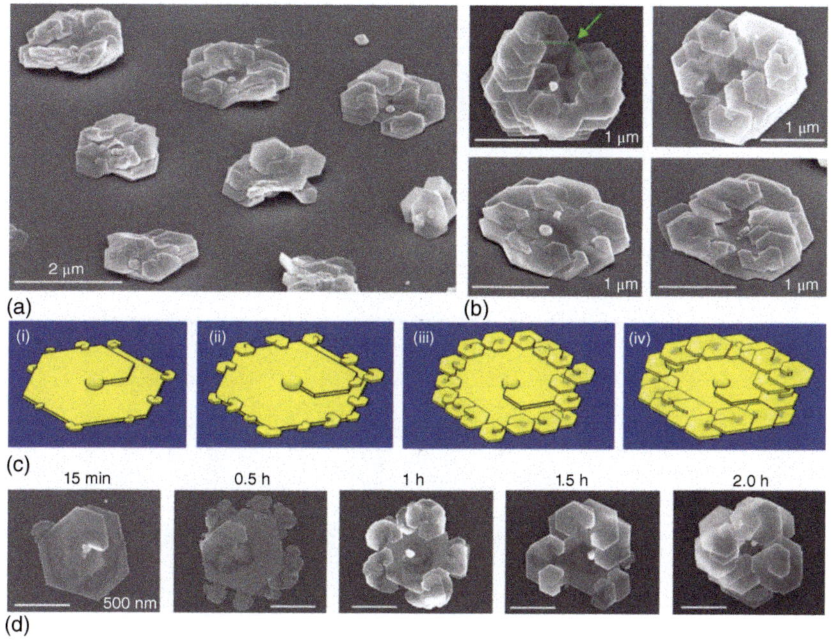


**Figure 2.22** (a) SEM images of Au spiral nanostructures via a two-step synthesis. (b) HRTEM images of individual nanostructures from top and tilted views. (c, d) Schemes and SEM images for the second-step growth progression occurring over a two-hour interval. Source: Golze et al. [91]/with permission of American Chemical Society.

transfer method to control the morphology of Au NPs in the presence of amino acids or peptides during the crystal growth process; amino acid-oriented interactions between the chiral ligand and achiral seeds manipulated the chiral nanostructure with a high-index-plane transition from a low-index plane. The resulting anisotropic chiral nanostructure was probably due to the enantioselective interaction between the chiral molecules and different Au NP surfaces, which also led to conformational asymmetry. When the octahedral seeds were replaced by cubic seeds and glutathione was used as chiral ligand, pinwheel-like chiral nanostructures showed excellent chiroptical responses with a good *g*-factor value of approximately 0.2 at 622 nm. In addition, such an intense anisotropy factor value made it rational and simple to adjust the wavelength of transmitted light for polarization-resolved color modulation.

## 2.4 Chiral Metallic Compound Nanoparticles

### 2.4.1 Chiral Oxide Nanoparticles

Recently, there were also some studies on chiral oxide NPs. One of the representative chiral oxide NPs was magnetic oxide nanoparticles, which may provide new insight into the chiral origin of NPs and can be used in homogeneous

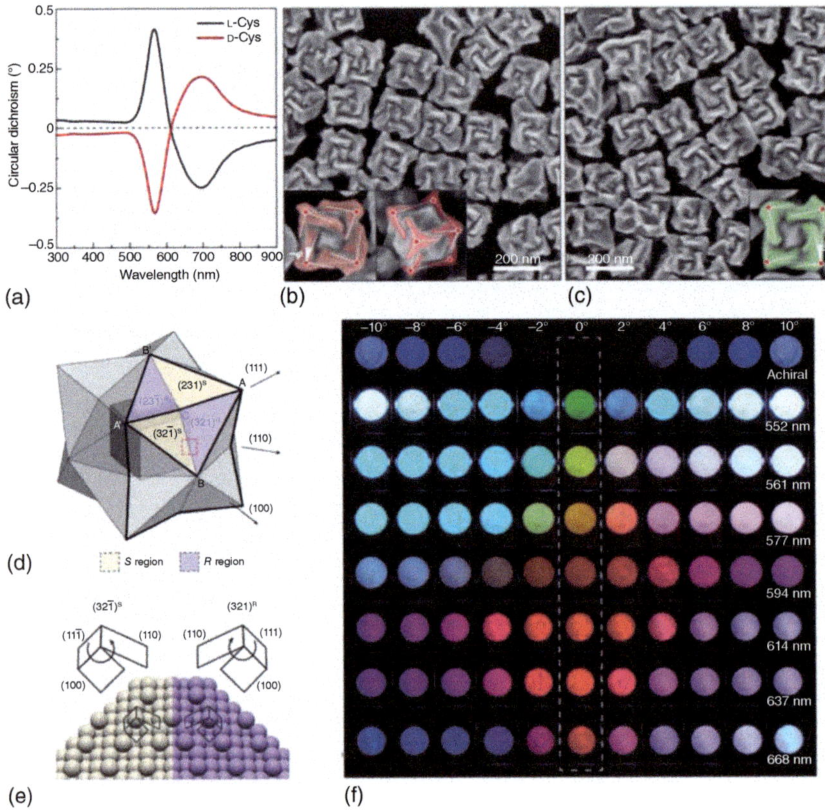


**Figure 2.23** Opposite handedness of 3D plasmonic helicoids through cysteine chirality transfer. (a) CD spectra of chiral nanoparticles prepared based on the L-Cys and D-Cys. (b) SEM image of L-Cys nanoparticles. (c) SEM image of D-Cys nanoparticles. (d) Scheme of a stellated octahedron. (e) Comparison of the atomic arrangement of the *R* and *S* Au surfaces in the region. (f) Polarization-resolved colors of light transmitted through different helicoid solutions with different max values and an achiral nanoparticle solution. Source: Lee et al. [92]/with permission of Nature Publishing group.

catalysis. The magnetic field may interact with magnetic moments of NPs [5]. Following affection of the molecules' chiral state by magnetic state, interesting phenomena may occur. By decomposition of carbonyl iron [$Fe(CO)_5$], $Fe_xO_y$-rich core and Pd-rich shell can be obtained by the subsequent reduction of palladium acetylacetonate [$Pd(acac)_2$] in oleic acid and oleylamine. Followed by the ligand-exchange reaction, chiral magnetic FePd NPs can be prepared. (Figure 2.24A,B) [93]. The obtained NPs were functionalized with chiral 2,20-bis(diphenylphosphino)-1,10-binaphthene (BINAP), and they exhibited typical CD Cotton effect (Figure 2.24C).

There were many reports on the bottom-up preparation of chiral NPs. Many advanced technologies were used to adjust the chiroptical activity of NPs based on the mild reaction environment and simple synthetic procedures. Recently, Kotov's group reported a novel paramagnetic chiromagnetic NP, cobalt oxide ($Co_3O_4$) [94].

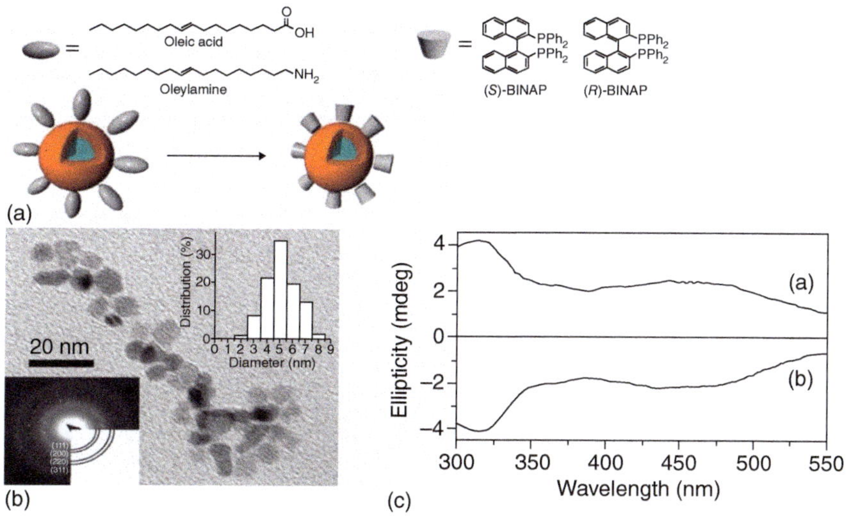


**Figure 2.24** (A) Ligand exchange of FePd NPs from oleic acid/oleylamine to chiral BINAP. (B) TEM image and electron diffraction pattern of FePd NPs capped by (*S*)-BINAP. (C) CD spectra of FePd NPs capped by (a) (*R*)-BINAP and (b) (*S*)-BINAP in $CHCl_3$. Source: Mori et al. [93]/with permission of Royal Society of Chemistry.

D- and L-enantiomers of cysteine (Cys) were used as surface ligands for the chiral $Co_3O_4$ NP preparation. As shown in the TEM images, the synthesized chiral $Co_3O_4$ NPs were approximately ~5 nm. The CD response exhibited mirror-symmetry for both L-/D-cysteine and L-/D-$Co_3O_4$ NPs. Their maximum *g*-factor was 0.02 at 556 nm (Figure 2.25). Moreover, they also elaborately explored the spin and orbital magnetic states for the distorted crystal lattices, which were resulted from the L- or D-enantiomers of cysteine molecules due to their intense chiroptical activity. Importantly, they also encapsulated the $Co_3O_4$ NP into soft gels, which exhibited the same optical characteristics as colloidal solutions. This study has paved a new pathway for the preparation of chiral devices and many other optoelectronic devices.

Another representative study was on $Cu_xO$ nanoparticles. Hao et al. reported chiral molecule-mediated porous $Cu_xO$ nanoparticles [95]. To prepare the nanostructure, Cu(II) ions were selected as the inorganic precursor, L-phenylalanine (L-Phe) as the structure-directing agent, and PVP used as the stabilizer. Firstly, Cu(II)–Phe complex was prepared by coordination into Cu(II) ions. $Cu(OH)_2$–Phe was synthesized by adding $OH^-$ into the copper salt solution. A portion of the CuO was reduced to $Cu_2O$ by the D-(+)-glucose; thus a complex of CuO and $Cu_2O$ NCs was prepared, which was designated "$Cu_xO$ NCs" (Figure 2.26). As a convenient surface ligand for CuO, phenylalanine can bind either by the carboxyl group or the amino group. The coordination bond between the amino group and $Cu^{2+}$ was intense in CuO. And some of the oxygen atoms on the CuO surface were replaced by the covalent bonds with $Cu^{2+}$ of the oxygen atoms at the –COO– end. Importantly, the aromatic phenylalanine group produced an intense hydrophobic attraction between the NPs,

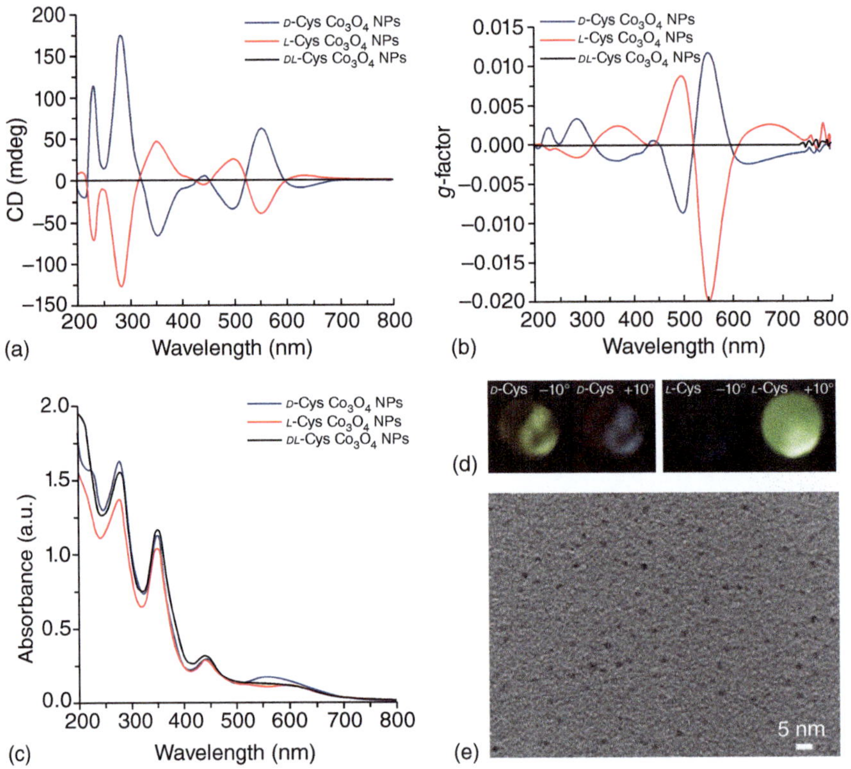


**Figure 2.25** Chiral $Co_3O_4$ NPs characterization. (a) CD, (b) *g*-factor, (c) UV–vis absorption spectra. (d) Photographs of light transmitting through the NPs; (e) TEM image. Source: Yeom et al. [94]/with permission of American Association for the Advancement of Science.

forming the superparticle assemblies. The repulsive electrostatic interactions can be counterbalanced by those attractive hydrophobic interactions, and finally gave rise to NPs.

Similar to NCs with L-Phe, $Cu_xO$ NCs with D-Phe exhibited a similar nanostructure. A sheet-like structure was achieved in the absence of any Phe ligands. Furthermore, more $Cu_xO$ NCs with varied morphologies and sizes were also synthesized based on other amino acids, including L-glutamic acid (L-Glu), L-lysine (L-Lys), L-tyrosine (L-Tyr), L-arginine (L-Arg), and L-aspartic acid (L-Asp). In the case of L-Tyr or L-Asp, ellipsoid-shaped NPs ($Cu_xO$–Tyr, $Cu_xO$–Asp) of about 100 nm were observed. In comparison, L-Glu reduced rod-like NPs ($Cu_xO$–Glu). Interestingly, in the case of either ligand L-Lys or L-Arg, no uniform diameter and morphology ($Cu_xO$–Lys, $Cu_xO$–Arg) could be produced. It was believed that different affinity between copper ions and corresponding amino acids caused the various sizes and morphological $Cu_xO$ NCs. Thus, $Cu_xO$–Phe displayed a distinctive morphology among all these $Cu_xO$ structures made from those amino-acid ligands. It demonstrated the critical role of Phe ligands in final NC structures. Interestingly, these prepared $Cu_xO$ nanostructures displayed the mimic activities of bioenzymes,

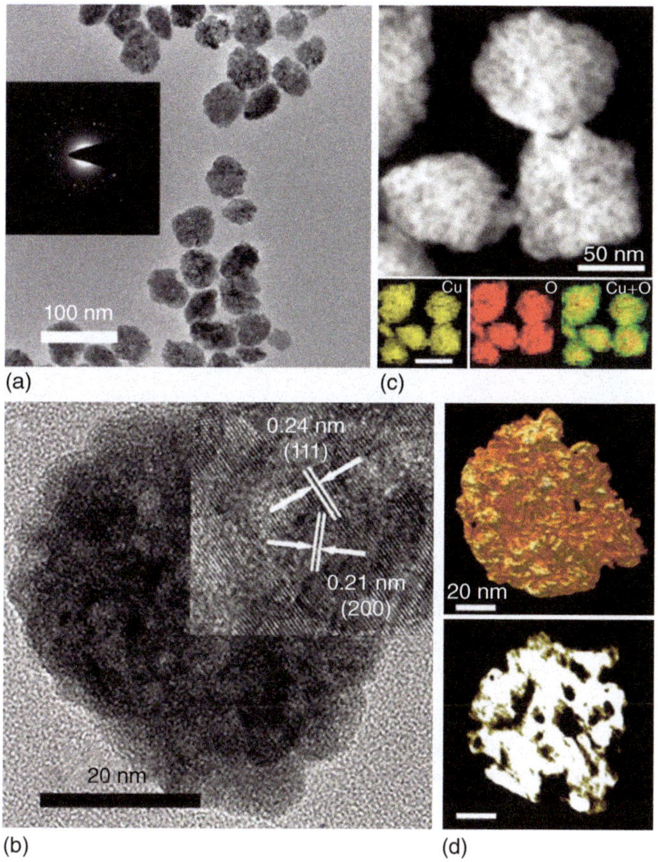


**Figure 2.26** (a) TEM image, (b) HRTEM image and SAED pattern, (c) STEM and EDS mapping, and (d) 3D electron tomography reconstruction images of $Cu_xO$ NCs with L-Phe. Source: Hao et al. [95]/with permission of American Chemical Society.

such as superoxide dismutase, catalase, peroxidase, and glutathione peroxidase. They also demonstrated their biological activity exampled by the neurotoxicity inhibition in a cellular model of Parkinson's disease and can even rescued the memory loss of mice with Parkinson's disease. Notably, multiple enzyme-mimicking activities of $Cu_xO$ NCs provided promising potential in biosensing, biomedicine, and biocatalysis.

More recently, Ye and colleagues reported another kind of chiral oxide nanoparticles, gadolinium oxide ($Gd_2O_3$) nanohelices [96]. $Gd_2O_3$ nanohelices were prepared based on the bottom-up synthetic route by the thermal decomposition of gadolinium acetylacetonate precursors. Unlike the reported flexible bending inorganic nanocoils, 3D rigid nanohelices were prepared from the continuous helical growth by the built-in lattice mismatch of two adjacent crystal planes. As shown in the TEM images in Figure 2.27, the diameter $d$ and width $b$ of the helices were measured to be $10.4 \pm 0.4$ and $3.0 \pm 0.3$ nm, respectively. The helical pitch $p$ was about $4.8 \pm 0.4$ nm, and their helical angle $\alpha$ was about ~8.4°. Each nanohelix

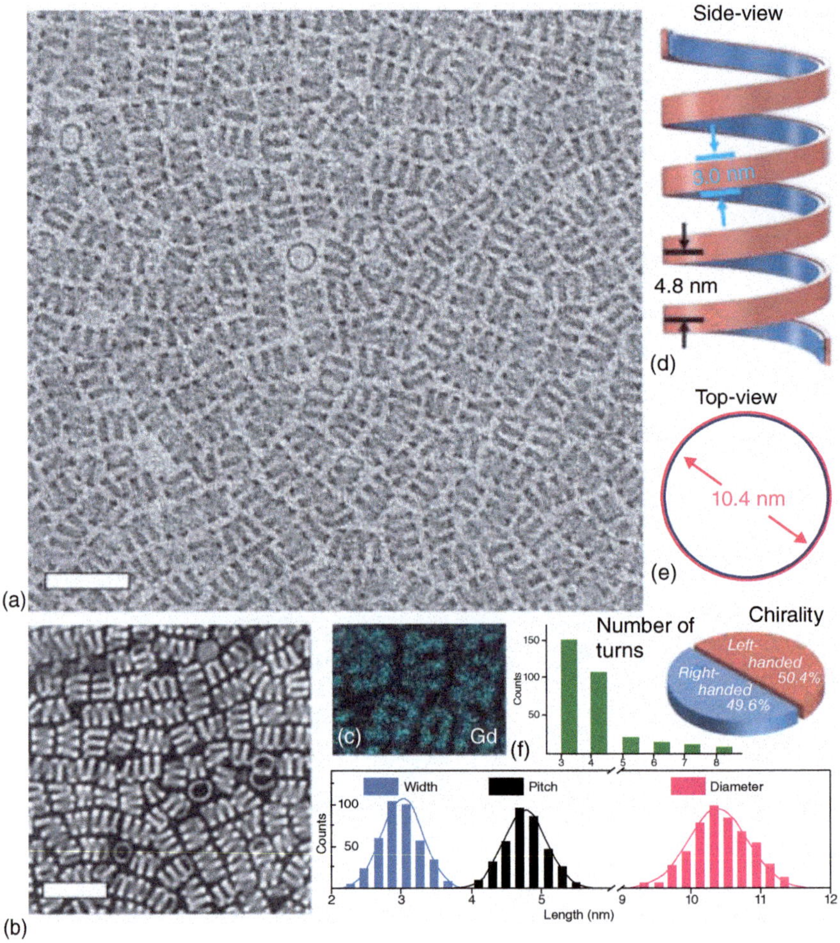


**Figure 2.27** (a) Representative TEM image, (b) HAADF-STEM image, and (c) Gd elemental map of $Gd_2O_3$ helices. (d, e) Scheme of individual $Gd_2O_3$ helices: (d) side view; (e) top view. (f) Statistical analysis of morphology of helices. Source: Liu et al. [96]/with permission of American Chemical Society.

displayed a bilayer structure. Their outer and inner layers resulted from the {111} and {100} planes of bulk $Gd_2O_3$.

### 2.4.2 Chiral Sulfide/Selenide Nanoparticles

As a well-known general nanoparticle preparation method, seed growth technique has generally been used to construct inherent chiral-shaped nanomaterials. Ouyang and colleagues recently fabricated a chiral cinnabar mercury sulfide (HgS) nanocrystal through cooperative chirality interaction [97]. They prepared chiral nanostructures with twisted triangular bipyramids by the collective interaction of crystallographic and geometric chirality. In addition, the surface of the nanocrystal lattice was reconstructed; therefore, achiral structures were transformed into a chiral nanostructure along the epitaxial growth process. By changing the stoichiometry of Hg and S precursors, accurate CD signal of HgS were achieved. When the nanostructure twist angle $\theta$ was 0°, their CD response corresponded to α-HgS NPs, which were confirmed by the electromagnetic simulation results. The periodic chiral lattices determined their chirality. However, chirality of NPs depended on their geometry, when the chiral parameter ($\xi$) was zero. Such a flexible chirality generation mechanism provided a novel method for the synthesis of chiral-shaped NPs with strong CD activity [98].

Another chiral sulfide NP is nanocrystalline copper sulfide, which is also a p-type semiconductor. The copper sulfide NPs show a wide range of potential applications in biology, chemistry, and physics, especially in bioapplications, such as biomedicine, and optoelectronic conversion (photovoltaics, photocatalysis, and thermoelectrics). Guo et al. fabricated chiral ultrasmall copper (I) sulfide nanoparticles [99]. As shown in Figure 2.28, the monodispersed $Cu_2S$ NPs were observed with an average diameter of $4 \pm 0.5$ nm, using L-or D-peptide penicillamine-phenylalanine-tryptophan (Pen-Phe-Trp) as the ligands. The NPs showed weak signals at 530 and 880 nm. Similar and mirror-asymmetrical CD bands were achieved for the L-and D-$Cu_2S$ NPs. The CD intensity reached 46 mdeg at 530 nm.

Based on the above research, Xu's group reported chiral $Cu_{2-x}S$ NPs in the presence of D/-L-Pen [100]. The prepared NPs were uniform, and the average diameter was $4 \pm 0.5$ nm. As shown in Figure 2.29, two binding energy (BE) values of Cu 2p at 932.3 and 952.1 eV were observed, which were assigned to the BE of $Cu^+$ $2p_{3/2}$ and $Cu^+$ $2p_{1/2}$, respectively. Moreover, they observed two $Cu^{2+}$ oxidation states (935.1 and 954.8 eV). There were two doublets from the deconvolution of the S 2p spectrum, which were assigned to the BE values for sulfides ($S^{2-}$) and disulfides ($(S_2)^{2-}$).

After being attached to the NPs, the typical L- and D-Pen CD bands at 225 nm disappeared in the ultraviolet range. Meanwhile, the CD band at 205 nm was shifted to 201 nm, with greatly enhanced intensity (Figure 2.30). Resulting from the interaction between L-/D-Pen and the NP surface, a new chiral band 270 nm appeared. Moreover, a bisignated Cotton effect with peaks at 608 and 863 nm was detected for D/L-NPs. It was noteworthy that the bisignated line shape displayed a zero-crossing point located at ~745 nm, which was assigned to the localized surface plasmon resonance peak of $Cu_{2-x}S$.

With the increasing concentration of chiral L/D-Pen, their plasmonic CD intensity gradually increased. Notably, the largest g-factor value even reached 0.01, much higher than any other reported value of chiral-ligand-capped QDs or even higher than that of high-order chiral assemblies. The large g-factor values may be attributed to their marked coordinate bond between Cu and S atoms.

They proposed two hypotheses for the intense CD activity. The first hypothesis was the inherent chiral NP core. The individual NPs can produce chiroptical activity in the presence of chiral shape. However, this hypothesis was ruled out by examining more than 100 TEM images. Most of the $Cu_{2-x}S$ NPs showed centrosymmetric shapes; therefore, it was not possible to display chirality from the geometry of NPs. The second hypothesis ascribed the chiroptical activity to the chirality transfer from

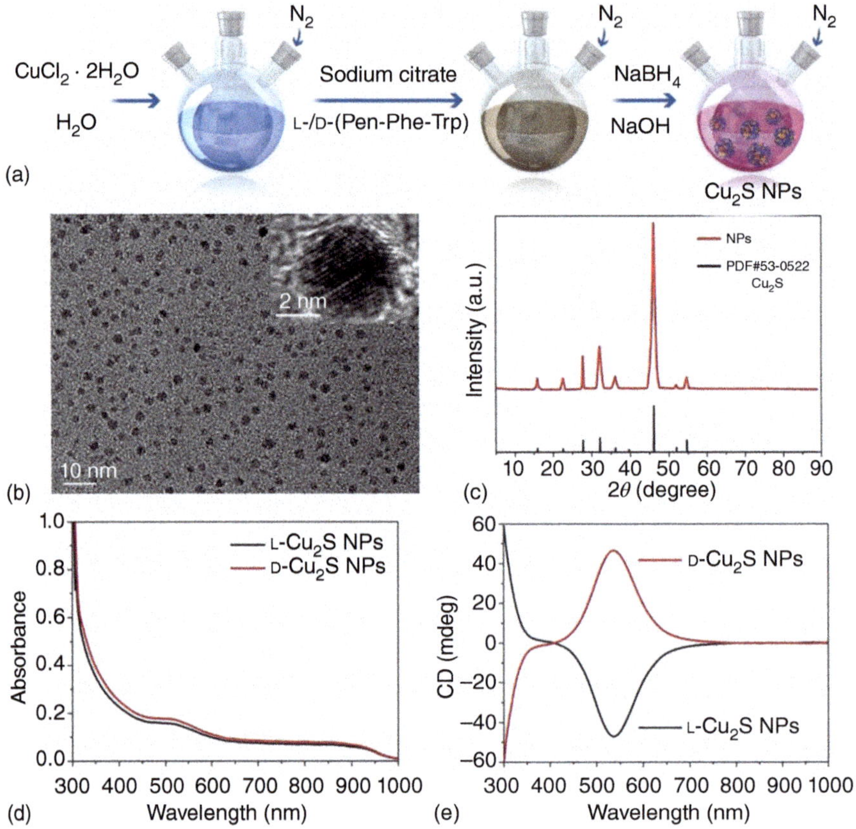


**Figure 2.28** (a) Chiral $Cu_2S$ NPs preparation, (b) TEM and HRTEM image, (c) XRD pattern of $Cu_2S$ NPs, (d) UV-vis, and (e) CD spectra of L/D-$Cu_2S$ NPs. Source: Guo et al. [99]/with permission of Wiley.

the adsorbed chiral Pen shell to the achiral core based on the surface distortion in a chiral pattern. However, due to a much smaller dielectric constant of a semiconductor nanocrystal, the dipolar interaction was a relatively weak effect. Thus, the chirality of the $Cu_{2-x}S$ NPs was most likely derived from the Coulomb interactions of electrons and/or holes.

Besides sulfide, selenide element has also been used for the preparation of chiral NPs. A representative study was reported by Zhang et al., where chiral iron copper selenide NPs ($Fe_xCu_ySe$ NPs) were fabricated [101]. Based on the interaction between the transition metal ions and the thiol groups, D- or L-Pen was conjugated to the surfaces of the $Fe_xCu_ySe$ NP. The special morphology and NIR absorbance of the $Fe_xCu_ySe$ NPs were closely corresponding to the $Fe^{3+}$ concentration. After adding $Fe^{3+}$, a porous nanostructure was produced. No significant variation in surface structures was observed, when $Fe_xCu_ySe$ NPs was modified with chiral Pen. For the absorbance spectrum, two characteristic peaks were observed at 550 and 850 nm

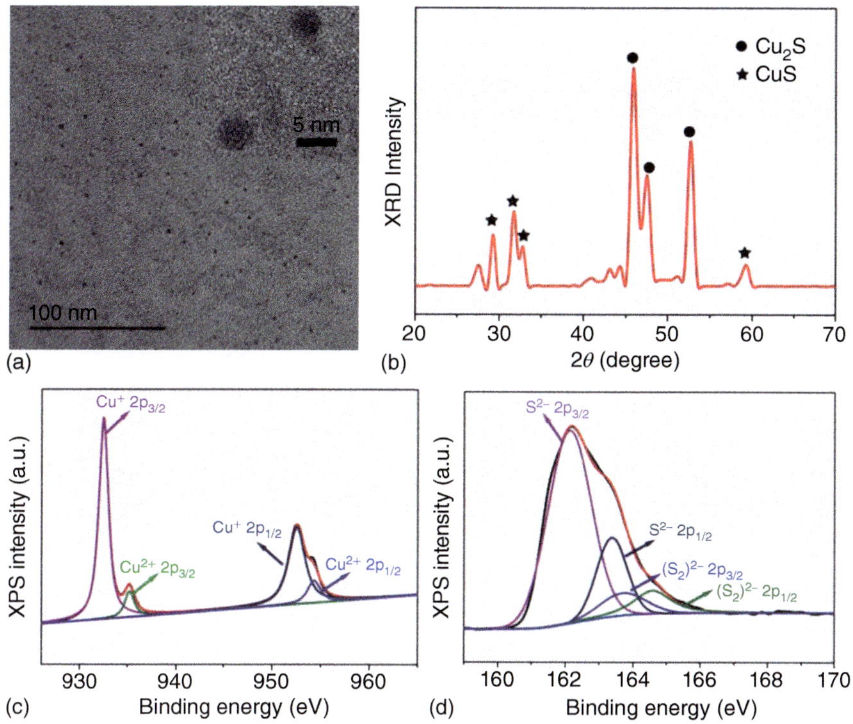


**Figure 2.29** (a) TEM image of L-$Cu_{2-x}S$. (b) XRD spectrum of L-$Cu_{2-x}S$. XPS spectra of L-$Cu_{2-x}S$: (c) Cu 2p, (d) S 2p. Source: Hao et al. [100]/with permission of Wiley.

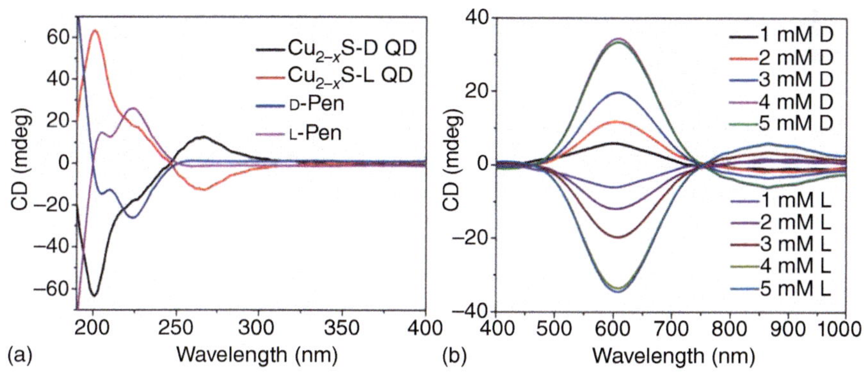


**Figure 2.30** (a) CD spectra of L-/D-Pen and the $Cu_{2-x}S$ NPs. (b) CD spectra of the $Cu_{2-x}S$ NPs under different concentrations of L-/D-Pen. Source: Hao et al. [100]. Reproduced with permission of Wiley.

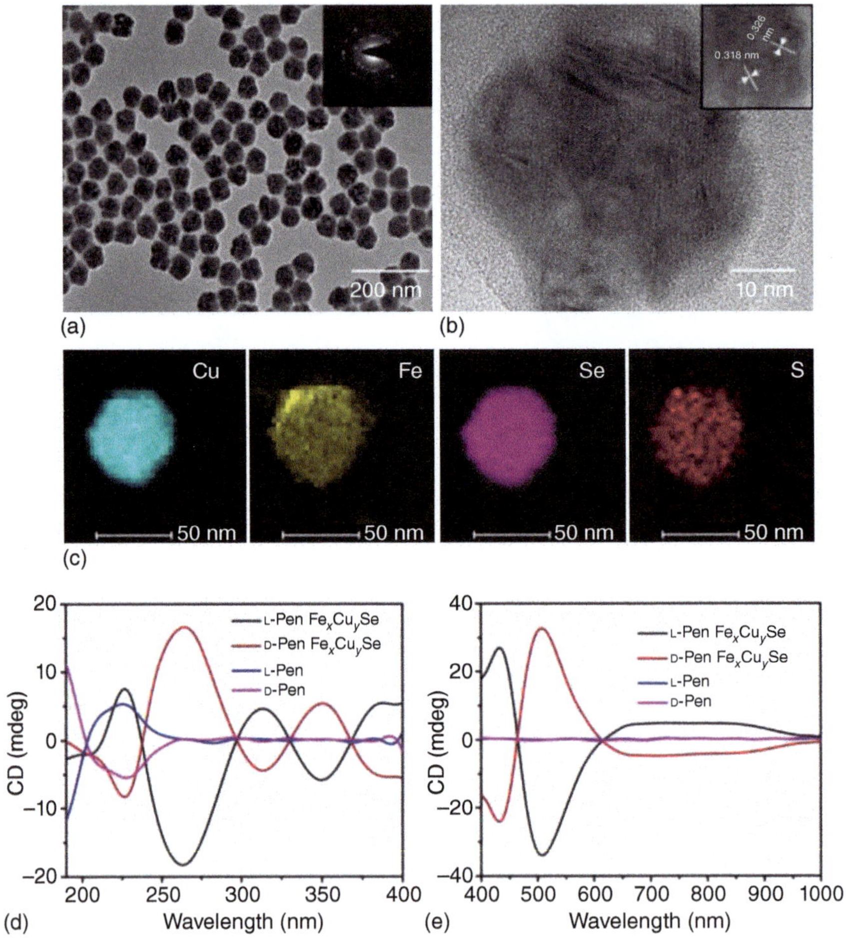


**Figure 2.31** (a) TEM image and SAED pattern. (b) HR-TEM image. (c) STEM and EDS mapping of $Fe_xCu_ySe$ prepared using penicillamine. (d, e) CD properties of L/D-Pen and L/D-NPs at 190–400 nm (d) and 400–1000 nm (e). Source: Zhang et al. [101]/with permission of Wiley.

for $Fe_xCu_ySe$ NPs; however, one peak at 850 nm was obtained for $Cu_{2-x}Se$ NPs without $Fe^{3+}$ addition. As for the CD spectra, opposite in sign and equal in intensity of the CD bands were measured at 190–1000 nm (Figure 2.31). Especially, a series of CD bands of the L/D-NPs were mainly assigned to L/D-Pen interacting with the metal core of the $Fe_xCu_ySe$ NPs, such as typical bands at 262, 312, 350, 440, 510, and 750 $cm^{-1}$. Interestingly, more reactive oxygen species (ROS) were produced by chiral $Fe_xCu_ySe$ NPs than that of $Cu_{2-x}Se$ NPs with irradiation of NIR-light. It is interesting to find that D-NPs attenuate the Aβ42 adhesion to membranes in the living MN9D cells. The D-$Fe_xCu_ySe$ NPs showed effective protection against neuronal damage by Aβ42 deposition and alleviated symptoms in a mouse model of AD, resulting in cognitive competence recovery.

### 2.4.3 Ceramic Nanohelices

Taking advantage of helical geometry systems, ceramic nanohelices can be formed. Single nanowires may be chiral; however, the preparation of ceramic helices was not an enantioselective process. Helical NWs could be achieved in confined environments with various morphologies [102]. The mesoscopic structure of silica through confined growth within cylindrical nanochannels produced chirality; however, the fabrication process was not enantioselective. The double-helical silica was spontaneously produced from the inside individual alumina nanochannels. Geometry transition was observed from a coiled cylindrical to a spherical cage-like geometry in the mesopore morphology as confinement tightened. The mesostructures can serve as a template for the manufacture of highly ordered NWs.

Besides, some studies reported enantioselective processes for the preparation of helical NWs. Through crystallization, double-stranded and cylindrical helical $BaCO_3$ nanofibers were spontaneously generated on the organosilane-coated substrate and controlled by a phosphonated block copolymer [103]. The alternative method was firstly reported by Antonietti and colleagues. By employing the double hydrophilic block copolymers (DHBCs), achiral $BaCO_3$ nanocrystals formed helices, indicating helical alignment of racemic polymers via selective adsorption on the (110) face of nanocrystals [104]. For a long time, helical nanostructure preparations were limited to ZnO, Si, and SiC materials. Ahn and coworkers used ZnSe NWs to synthesize $ZnGa_2O_4$/ZnSe nanovines and $ZnGa2O_4$ nanosprings through thermal evaporation. A common structure was found in the nanovines and nanosprings, with the same zigzag angle for the helical $ZnGa_2O_4$NW growth direction [105]. Based on the vapor transport method, a single crystalline $CrSi_2$ hexagonal nanoweb was prepared on silicon substrates in a horizontal tube furnace [106]. By studying distinct helical NWs, different preparation steps enabled the formation mechanism of inorganic helices and illustrated the chiral law of natural construction.

More and more studies were reported on porous silica nanostructures, especially with helical shape or helical pore channels. In the presence of achiral surfactants, interfacial interactions can cause the formation of chiral channels and the surface free energy reduction can drive the spontaneous formation of the helical morphology (Figure 2.32a,b) [107]. However, the ee value was very small, with the ratio of 1.1 for right- to left-handed helices. Chiral nanostructures were made up of chiral anionic surfactants and silicates, with chiral surfactants as templates (Figure 2.32c,d) [108]. However, the chiral template molecule was not the sole driving force, which was exemplified by the formation of helical mesostructured silica by achiral cationic surfactants. Generally speaking, distinct design criteria, synthesis technology, and stabilizing mechanisms were required for the preparation of helical nanostructures with relatively rigid inorganic materials. Inorganic helices showed importance and high expectations in many aspects. Firstly, not all of the formation mechanisms for various inorganic helices were clear, and the exploration of formation mechanisms paved a novel way to unravel the rule of natural

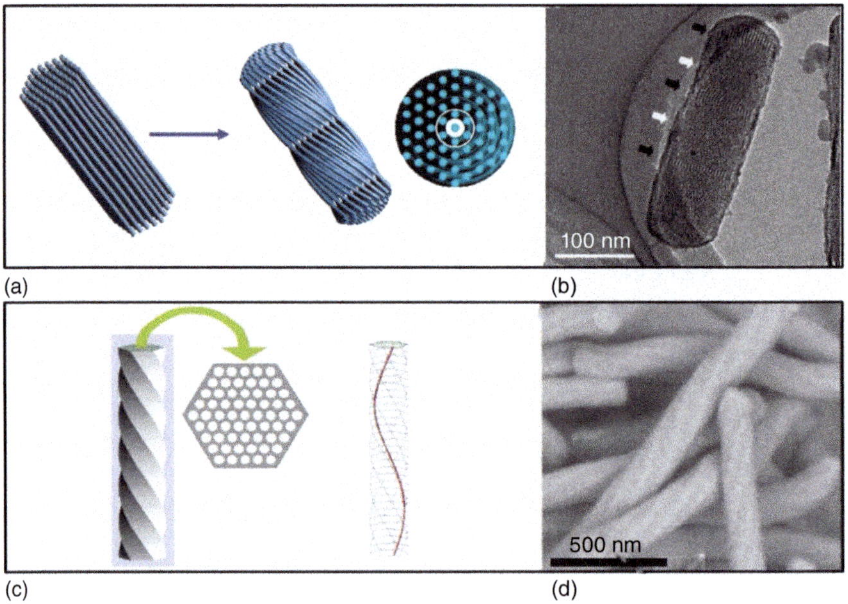


**Figure 2.32** (a) Schematic illustration of helical mesostructured rod formation. (b) TEM image of helical silicon [107]. (c) Schematic illustration of a structural model and cross section for the chiral mesoporous material. (d) SEM image and scheme illustration of a structural model of chiral mesoporous silica [108]. Source: (a, b) Yang et al. [107]/with permission of American Chemical Society; (c, d) Che et al. [108]/with permission of Macmillan Publishers Ltd.

construction. Secondly, micro-scale of inorganic helices were expected to display promising and fascinating physical and optical properties

Although slightly beyond the boundaries of inorganic nanostructures, it is also useful to briefly introduce polymer nanostructures with close relation to NPs and NRs. Most of the reported researches of chiral organic NPs were focused on organic polymers and biological macromolecules. A type of organic NP was prepared based on the complexation of divalent metals (i.e., $Ca^{2+}$, $Ba^{2+}$) and poly(phenylacetylene) polymers with MPA pendants with cis-isomer [109]. The prepared helical polymer–metal complex NPs exhibited two interesting properties. One was the tunable diameter via the adjusted polymer/metal ion ratio. The other one was the switchable handedness of the helix both in the interior and on the surface. The optical activity response can be tailored by the cofactors such as solvent, the metal ion, and helicity of the polymer in the aggregation. The chiral NPs capability of encapsulation various materials provided a way for the interesting hybrid supramolecular assemblies, especially in the case of composites prepared by the chiral NPs inclusion. Even if semiempirical, parameterized models were used, the diameter of the polymeric NPs was larger than 100 nm, and this size was not amenable to quantum chemical modeling. Nevertheless, the classical coarse-grained models may be able to simulate helical nanostructures and provide insights into the solvent and metal ion effects.

## References


1 Ma, W., Mao, J.T., Hao, C.L. et al. (2019). Chiral semiconductor nanorod heterostructures with high photocatalysis activity. *Applied Catalysis B: Environmental* 245: 691–697.

2 Ben-Moshe, A., Wolf, S.G., Sadan, M.B. et al. (2014). Enantioselective control of lattice and shape chirality in inorganic nanostructures using chiral biomolecules. *Nature Communications* 5 (1): 4302.

3 Huo, S., Duan, P., Jiao, T. et al. (2017). Self-assembled luminescent quantum dots to generate full-color and white circularly polarized light. *Angewandte Chemie International Edition* 56 (40): 12174–12178.

4 Noguez, C. and Garzón, I.L. (2009). Optically active metal nanoparticles. *Chemical Society Reviews* 38 (3): 757–771.

5 Ma, W., Xu, L.G., de Moura, A.F. et al. (2017). Chiral inorganic nanostructures. *Chemical Reviews* 117 (12): 8041–8093.

6 Elliott, S.D., Moloney, M.P., and Gun'ko, Y.K. (2008). Chiral shells and achiral cores in CdS quantum dots. *Nano Letters* 8 (8): 2452–2457.

7 Farrag, M., Tschurl, M., and Heiz, U. (2013). Chiral gold and silver nanoclusters: preparation, size selection, and chiroptical properties. *Chemistry of Materials* 25 (6): 862–870.

8 Harada, T., Simeon, F., Vander Sande, J.B., and Hatton, T.A. (2010). Formation of magnetic nanotubes by the cooperative self-assembly of chiral amphiphilic molecules and $Fe_3O_4$ nanoparticles. *Physical Chemistry Chemical Physics* 12 (38): 11938–11943.

9 Gao, P.X., Ding, Y., Mai, W. et al. (2005). Conversion of zinc oxide nanobelts into superlattice-structured nanohelices. *Science* 309 (5741): 1700–1704.

10 Ostrikov, K. and Mehdipour, H. (2011). Thin single-walled carbon nanotubes with narrow chirality distribution: constructive interplay of plasma and Gibbs–Thomson effects. *ACS Nano* 5 (10): 8372–8382.

11 Jung, J.H., Ono, Y., Hanabusa, K., and Shinkai, S. (2000). Creation of both right-handed and left-handed silica structures by sol–gel transcription of organogel fibers comprised of chiral diaminocyclohexane derivatives. *Journal of the American Chemical Society* 122 (20): 5008–5009.

12 Hou, K., Zhao, J., Wang, H. et al. (2020). Chiral gold nanoparticles enantioselectively rescue memory deficits in a mouse model of Alzheimer's disease. *Nature Communications* 11 (1): 4790.

13 Fan, Z. and Govorov, A.O. (2012). Chiral nanocrystals: plasmonic spectra and circular dichroism. *Nano Letters* 12 (6): 3283–3289.

14 Silvera Batista, C.A., Larson, R.G., and Kotov, N.A. (2015). Nonadditivity of nanoparticle interactions. *Science* 350 (6257): 1242477.

15 Schaaff, T.G. and Whetten, R.L. (2000). Giant gold–glutathione cluster compounds: intense optical activity in metal-based transitions. *The Journal of Physical Chemistry B* 104 (12): 2630–2641.

**16** Losada, M. and Xu, Y. (2007). Chirality transfer through hydrogen-bonding: experimental and ab initio analyses of vibrational circular dichroism spectra of methyl lactate in water. *Physical Chemistry Chemical Physics* 9 (24): 3127–3135.

**17** Suzuki, N., Wang, Y., Elvati, P. et al. (2016). Chiral graphene quantum dots. *ACS Nano* 10 (2): 1744–1755.

**18** Feng, W., Kim, J.-Y., Wang, X. et al. (2017). Assembly of mesoscale helices with near-unity enantiomeric excess and light-matter interactions for chiral semiconductors. *Science Advances* 3 (3), e1601159.

**19** Zhao, X.L., Xu, L.G., Sun, M.Z. et al. (2016). Gold-quantum dot core-satellite assemblies for lighting up microRNA in vitro and in vivo. *Small* 12 (34): 4662–4668.

**20** Purcell-Milton, F., Visheratina, A.K., Kuznetsova, V.A. et al. (2017). Impact of shell thickness on photoluminescence and optical activity in chiral CdSe/CdS core/shell quantum dots. *ACS Nano* 11 (9): 9207–9214.

**21** Ben-Moshe, A., Teitelboim, A., Oron, D., and Markovich, G. (2016). Probing the interaction of quantum dots with chiral capping molecules using circular dichroism spectroscopy. *Nano Letters* 16 (12): 7467–7473.

**22** Chen, G.C., Tian, F., Zhang, Y. et al. (2014). Tracking of transplanted human mesenchymal stem cells in living mice using near-infrared Ag-2 S quantum dots. *Advanced Functional Materials* 24 (17): 2481–2488.

**23** Moloney, M.P., Gun'ko, Y.K., and Kelly, J.M. (2007). Chiral highly luminescent CdS quantum dots. *Chemical Communications* 38: 3900–3902.

**24** Li, Y., Zhou, Y., Wang, H.-Y. et al. (2011). Chirality of glutathione surface coating affects the cytotoxicity of quantum dots. *Angewandte Chemie International Edition* 50 (26): 5860–5864.

**25** Zhou, Y., Zhu, Z., Huang, W. et al. (2011). Optical coupling between chiral biomolecules and semiconductor nanoparticles: size-dependent circular dichroism absorption. *Angewandte Chemie International Edition* 50 (48): 11456–11459.

**26** Zhou, Y., Yang, M., Sun, K. et al. (2010). Similar topological origin of chiral centers in organic and nanoscale inorganic structures: effect of stabilizer chirality on optical isomerism and growth of CdTe nanocrystals. *Journal of the American Chemical Society* 132 (17): 6006–6013.

**27** Ben-Moshe, A., Govorov, A.O., and Markovich, G. (2013). Enantioselective synthesis of intrinsically chiral mercury sulfide nanocrystals. *Angewandte Chemie International Edition* 52 (4): 1275–1279.

**28** Govan, J.E., Jan, E., Querejeta, A. et al. (2010). Chiral luminescent CdS nano-tetrapods. *Chemical Communications* 46 (33): 6072–6074.

**29** Gallagher, S.A., Moloney, M.P., Wojdyla, M. et al. (2010). Synthesis and spectroscopic studies of chiral CdSe quantum dots. *Journal of Materials Chemistry* 20 (38): 8350–8355.

**30** Ben Moshe, A., Szwarcman, D., and Markovich, G. (2011). Size dependence of chiroptical activity in colloidal quantum dots. *ACS Nano* 5 (11): 9034–9043.

**31** Govorov, A.O., Fan, Z., Hernandez, P. et al. (2010). Theory of circular dichroism of nanomaterials comprising chiral molecules and nanocrystals: plasmon enhancement, dipole interactions, and dielectric effects. *Nano Letters* 10 (4): 1374–1382.

**32** Gao, X., Zhang, X., Deng, K. et al. (2017). Excitonic circular dichroism of chiral quantum rods. *Journal of the American Chemical Society* 139 (25): 8734–8739.

**33** Tohgha, U., Varga, K., and Balaz, M. (2013). Achiral CdSe quantum dots exhibit optical activity in the visible region upon post-synthetic ligand exchange with D- or L-cysteine. *Chemical Communications* 49 (18): 1844–1846.

**34** Qiu, F., Han, Z., Peterson, J.J. et al. (2016). Photocatalytic hydrogen generation by CdSe/CdS nanoparticles. *Nano Letters* 16 (9): 5347–5352.

**35** Gao, X., Han, B., Yang, X., and Tang, Z. (2019). Perspective of chiral colloidal semiconductor nanocrystals: opportunity and challenge. *Journal of the American Chemical Society* 141 (35): 13700–13707.

**36** Gao, X., Zhang, X., Zhao, L. et al. (2018). Distinct excitonic circular dichroism between wurtzite and zincblende CdSe nanoplatelets. *Nano Lett* 18 (11): 6665–6671.

**37** Sun, M., Xu, L., Qu, A. et al. (2018). Site-selective photoinduced cleavage and profiling of DNA by chiral semiconductor nanoparticles. *Nature Chemistry* 10 (8): 821–830.

**38** Boersma, A.J., Coquière, D., Geerdink, D. et al. (2010). Catalytic enantioselective syn hydration of enones in water using a DNA-based catalyst. *Nature Chemistry* 2 (11): 991–995.

**39** Wang, W., Hao, C., Sun, M. et al. (2018). Peptide mediated chiral inorganic nanomaterials for combating gram-negative bacteria. *Advanced Functional Materials* 28 (44): 1805112.

**40** Ben-Shahar, Y., Scotognella, F., Kriegel, I. et al. (2016). Optimal metal domain size for photocatalysis with hybrid semiconductor-metal nanorods. *Nature Communications* 7 (1): 10413.

**41** Niu, Z., Becknell, N., Yu, Y. et al. (2016). Anisotropic phase segregation and migration of Pt in nanocrystals en route to nanoframe catalysts. *Nature Materials* 15 (11): 1188–1194.

**42** Simon, T., Carlson, M.T., Stolarczyk, J.K., and Feldmann, J. (2016). Electron transfer rate vs recombination losses in photocatalytic $H_2$ generation on Pt-decorated CdS nanorods. *ACS Energy Letters* 1 (6): 1137–1142.

**43** Zhuang, T.-T., Liu, Y., Li, Y. et al. (2016). Integration of semiconducting sulfides for full-spectrum solar energy absorption and efficient charge separation. *Angewandte Chemie International Edition* 55 (22): 6396–6400.

**44** Du, L., Shi, G., Zhao, Y. et al. (2019). Plasmon-promoted electrocatalytic water splitting on metal–semiconductor nanocomposites: the interfacial charge transfer and the real catalytic sites. *Chemical Science* 10 (41): 9605–9612.

**45** Bierman, M.J., Lau, Y.K.A., Kvit, A.V. et al. (2008). Dislocation-driven nanowire growth and Eshelby twist. *Science* 320 (5879): 1060–1063.

**46** Sone, E.D., Zubarev, E.R., and Stupp, S.I. (2005). Supramolecular templating of single and double nanohelices of cadmium sulfide. *Small* 1 (7): 694–697.

**47** Qiao, Y., Lin, Y., Wang, Y. et al. (2009). Metal-driven hierarchical self-assembled one-dimensional nanohelices. *Nano Letters* 9 (12): 4500–4504.

**48** Sone, E.D., Zubarev, E.R., and Stupp, S.I. (2002). Semiconductor nanohelices templated by supramolecular ribbons. *Angewandte Chemie International Edition* 41 (10): 1705–1709.

**49** Yeom, B., Zhang, H., Zhang, H. et al. (2013). Chiral plasmonic nanostructures on achiral nanopillars. *Nano Letters* 13 (11): 5277–5283.

**50** Mark, A.G., Gibbs, J.G., Lee, T.-C., and Fischer, P. (2013). Hybrid nanocolloids with programmed three-dimensional shape and material composition. *Nature Materials* 12 (9): 802–807.

**51** Gautier, C. and Bürgi, T. (2009). Chiral gold nanoparticles. *ChemPhysChem* 10 (3): 483–492.

**52** Grzelczak, M., Pérez-Juste, J., Mulvaney, P., and Liz-Marzán, L.M. (2008). Shape control in gold nanoparticle synthesis. *Chemical Society Reviews* 37 (9): 1783–1791.

**53** Boisselier, E. and Astruc, D. (2009). Gold nanoparticles in nanomedicine: preparations, imaging, diagnostics, therapies and toxicity. *Chemical Society Reviews* 38 (6): 1759–1782.

**54** Knoppe, S., Azoulay, R., Dass, A., and Bürgi, T. (2012). In situ reaction monitoring reveals a diastereoselective ligand exchange reaction between the intrinsically chiral $Au_{38}(SR)_{24}$ and chiral thiols. *Journal of the American Chemical Society* 134 (50): 20302–20305.

**55** Schaaff, T.G., Knight, G., Shafigullin, M.N. et al. (1998). Isolation and selected properties of a 10.4 kDa gold:glutathione cluster compound. *The Journal of Physical Chemistry B* 102 (52): 10643–10646.

**56** Gautier, C. and Bürgi, T. (2006). Chiral N-isobutyryl-cysteine protected gold nanoparticles: preparation, size selection, and optical activity in the UV–vis and infrared. *Journal of the American Chemical Society* 128 (34): 11079–11087.

**57** Cathcart, N. and Kitaev, V. (2010). Silver nanoclusters: single-stage scaleable synthesis of monodisperse species and their chirooptical properties. *The Journal of Physical Chemistry C* 114 (38): 16010–16017.

**58** Gautier, C. and Bürgi, T. (2010). Vibrational circular dichroism of adsorbed molecules: BINAS on gold nanoparticles. *The Journal of Physical Chemistry C* 114 (38): 15897–15902.

**59** Petty, J.T., Zheng, J., Hud, N.V., and Dickson, R.M. (2004). DNA-templated Ag nanocluster formation. *Journal of the American Chemical Society* 126 (16): 5207–5212.

**60** Yao, H. (2012). On the electronic structures of $Au_{25}(SR)_{18}$ clusters studied by magnetic circular dichroism spectroscopy. *The Journal of Physical Chemistry Letters* 3 (12): 1701–1706.

**61** Jadzinsky, P.D., Calero, G., Ackerson, C.J. et al. (2007). Structure of a thiol monolayer-protected gold nanoparticle at 1.1 A resolution. *Science* 318 (5849): 430–433.

**62** Tlahuice, A. and Garzón, I.L. (2012). On the structure of the $Au_{18}(SR)_{14}$ cluster. *Physical Chemistry Chemical Physics* 14 (11): 3737–3740.

**63** Hauser, A.W., Mardirossian, N., Panetier, J.A. et al. (2014). Functionalized graphene as a gatekeeper for chiral molecules: an alternative concept for chiral separation. *Angewandte Chemie International Edition* 53 (37): 9957–9960.

**64** Negishi, Y., Chaki, N.K., Shichibu, Y. et al. (2007). Origin of magic stability of thiolated gold clusters: a case study on $Au_{25}(SC_6H_{13})_{18}$. *Journal of the American Chemical Society* 129 (37): 11322–11323.

**65** George, J. and Thomas, K.G. (2010). Surface plasmon coupled circular dichroism of Au nanoparticles on peptide nanotubes. *Journal of the American Chemical Society* 132 (8): 2502–2503.

**66** Kelly, S.M., Jess, T.J., and Price, N.C. (2005). How to study proteins by circular dichroism. *Biochimica et Biophysica Acta* 1751 (2): 119–139.

**67** Wu, Z., Gayathri, C., Gil, R.R., and Jin, R. (2009). Probing the structure and charge state of glutathione-capped $Au_{25}(SG)_{18}$ clusters by NMR and mass spectrometry. *Journal of the American Chemical Society* 131 (18): 6535–6542.

**68** Maoz, B.M., van der Weegen, R., Fan, Z. et al. (2012). Plasmonic chiroptical response of silver nanoparticles interacting with chiral supramolecular assemblies. *Journal of the American Chemical Society* 134 (42): 17807–17813.

**69** Liu, W., Zhu, Z., Deng, K. et al. (2013). Gold nanorod@chiral mesoporous silica core–shell nanoparticles with unique optical properties. *Journal of the American Chemical Society* 135 (26): 9659–9664.

**70** Cathcart, N., Mistry, P., Makra, C. et al. (2009). Chiral thiol-stabilized silver nanoclusters with well-resolved optical transitions synthesized by a facile etching procedure in aqueous solutions. *Langmuir* 25 (10): 5840–5846.

**71** Molotsky, T., Tamarin, T., Moshe, A.B. et al. (2010). Synthesis of chiral silver clusters on a DNA template. *The Journal of Physical Chemistry C* 114 (38): 15951–15954.

**72** Nishida, N., Yao, H., Ueda, T. et al. (2007). Synthesis and chiroptical study of D/L-penicillamine-capped silver nanoclusters. *Chemistry of Materials* 19 (11): 2831–2841.

**73** Si, S., Gautier, C., Boudon, J. et al. (2009). Ligand exchange on $Au_{25}$ cluster with chiral thiols. *The Journal of Physical Chemistry C* 113 (30): 12966–12969.

**74** Hidalgo, F. and Noguez, C. (2013). Optical activity of achiral ligand $SCH_3$ adsorbed on achiral $Ag_{55}$ clusters: relationship between adsorption site and circular dichroism. *ACS Nano* 7 (1): 513–521.

**75** Lechtken, A., Schooss, D., Stairs, J.R. et al. (2007). $Au_{34}^-$: a chiral gold cluster? *Angewandte Chemie International Edition* 46 (16): 2944–2948.

**76** Lieberman, I., Shemer, G., Fried, T. et al. (2008). Plasmon-resonance-enhanced absorption and circular dichroism. *Angewandte Chemie International Edition* 47 (26): 4855–4857.

**77** Graf, P., Mantion, A., Haase, A. et al. (2011). Silicification of peptide-coated silver nanoparticles – a biomimetic soft chemistry approach toward chiral hybrid core–shell materials. *ACS Nano* 5 (2): 820–833.

**78** Řezanka, P., Záruba, K., and Král, V. (2011). Supramolecular chirality of cysteine modified silver nanoparticles. *Colloids and Surfaces A: Physicochemical and Engineering Aspects* 374 (1): 77–83.

**79** Shemer, G., Krichevski, O., Markovich, G. et al. (2006). Chirality of silver nanoparticles synthesized on DNA. *Journal of the American Chemical Society* 128 (34): 11006–11007.

**80** Leroux, F., Gysemans, M., Bals, S. et al. (2010). Three-dimensional characterization of helical silver nanochains mediated by protein assemblies. *Advanced Materials* 22 (19): 2193–2197.

81 Li, Y. and Liu, M. (2008). Fabrication of chiral silver nanoparticles and chiral nanoparticulate film via organogel. *Chemical Communications* 43: 5571–5573.

82 Wu, X., Xu, L., Ma, W. et al. (2015). Gold core-DNA-silver shell nanoparticles with intense plasmonic chiroptical activities. *Advanced Functional Materials* 25 (6): 850–854.

83 Hao, C., Xu, L., Ma, W. et al. (2015). Unusual circularly polarized photocatalytic activity in nanogapped gold–silver chiroplasmonic nanostructures. *Advanced Functional Materials* 25 (36): 5816–5822.

84 Wang, J.C., Wu, X.L., Ma, W., and Xu, C.L. (2020). Chiral AuCuAu heterogeneous nanorods with tailored optical activity. *Advanced Functional Materials* 30 (17): 2000670.

85 Wang, Y., Wang, Q., Sun, H. et al. (2011). Chiral transformation: from single nanowire to double helix. *Journal of the American Chemical Society* 133 (50): 20060–20063.

86 Kondo, Y. and Takayanagi, K. (2000). Synthesis and characterization of helical multi-shell gold nanowires. *Science* 289 (5479): 606–608.

87 Velázquez-Salazar, J.J., Esparza, R., Mejía-Rosales, S.J. et al. (2011). Experimental evidence of icosahedral and decahedral packing in one-dimensional nanostructures. *ACS Nano* 5 (8): 6272–6278.

88 Nakagawa, M. and Kawai, T. (2018). Chirality-controlled syntheses of double-helical Au nanowires. *Journal of the American Chemical Society* 140 (15): 4991–4994.

89 Xie, J.J. and Che, S.A. (2012). Chirality of anisotropic metal nanowires with a distinct multihelix. *Chemistry-A European Journal* 18 (50): 15954–15959.

90 Wang, G.Y., Hao, C.L., Ma, W. et al. (2021). Chiral plasmonic triangular nanorings with SERS Activity for ultrasensitive detection of amyloid proteins in Alzheimer's disease. *Advanced Materials* 33 (38): 2102337.

91 Golze, S.D., Porcu, S., Zhu, C. et al. (2021). Sequential symmetry-breaking events as a synthetic pathway for chiral gold nanostructures with spiral geometries. *Nano Letters* 21 (7): 2919–2925.

92 Lee, H.-E., Ahn, H.-Y., Mun, J. et al. (2018). Amino-acid- and peptide-directed synthesis of chiral plasmonic gold nanoparticles. *Nature* 556 (7701): 360–365.

93 Mori, K., Kondo, Y., and Yamashita, H. (2009). Synthesis and characterization of FePd magnetic nanoparticles modified with chiral BINAP ligand as a recoverable catalyst vehicle for the asymmetric coupling reaction. *Physical Chemistry Chemical Physics* 11 (39): 8949–8954.

94 Yeom, J., Santos, U., Chekini, M. et al. (2018). Chiromagnetic nanoparticles and gels. *Science* 359: 309–314.

95 Hao, C.L., Qu, A.H., Xu, L.G. et al. (2019). Chiral molecule-mediated porous $Cu_xO$ nanoparticle clusters with antioxidation activity for ameliorating Parkinson's disease. *Journal of the American Chemical Society* 141 (2): 1091–1099.

96 Liu, Y., Li, Y.D., Jeong, S. et al. (2020). Colloidal synthesis of nanohelices via bilayer lattice misfit. *Journal of the American Chemical Society* 142 (29): 12777–12783.

**97** Wang, P.-p., Yu, S.-J., Govorov, A.O., and Ouyang, M. (2017). Cooperative expression of atomic chirality in inorganic nanostructures. *Nature Communications* 8 (1): 14312.

**98** Zhang, H.Y., Li, S., Qu, A.H. et al. (2020). Engineering of chiral nanomaterials for biomimetic catalysis. *Chemical Science* 11 (48): 12937–12954.

**99** Guo, X., Sun, M.Z., Gao, R. et al. (2021). Ultrasmall copper (I) sulfide nanoparticles prevent Hepatitis B virus infection. *Angewandte Chemie International Edition* 60 (23): 13073–13080.

**100** Hao, C., Gao, R., Li, Y. et al. (2019). Chiral semiconductor nanoparticles for protein catalysis and profiling. *Angewandte Chemie International Edition* 58 (22): 7371–7374.

**101** Zhang, H.Y., Hao, C.L., Qu, A.H. et al. (2020). Light-induced chiral iron copper selenide nanoparticles prevent β-amyloidopathy in vivo. *Angewandte Chemie International Edition* 59 (18): 7131–7138.

**102** Wu, Y., Cheng, G., Katsov, K. et al. (2004). Composite mesostructures by nano-confinement. *Nature Materials* 3: 816–822.

**103** Zhu, J.-H., Yu, S.-H., Xu, A.-W., and Cölfen, H. (2009). The biomimeticmineralization of double-stranded and cylindrical helical $BaCO_3$ nanofibres. *Chemical Communications* 9: 1106–1108.

**104** Yu, S.-H., Cölfen, H., Tauer, K., and Antonietti, M.J.N.M. (2005). Tectonic arrangement of $BaCO_3$ nanocrystals into helices induced by a racemic block copolymer. *Nature Materials* 4: 51–55.

**105** Bae, S.Y., Lee, J., Jung, H. et al. (2005). Helical structure of single-crystalline $ZnGa_2O_4$ nanowires. *Journal of the American Chemical Society* 127 (31): 10802–10803.

**106** Zhang, J., Hao, J., Wei, Y. et al. (2010). Nanoscale chiral rod-like molecular triads assembled from achiral polyoxometalates. *Journal of the American Chemical Society* 132 (1): 14–15.

**107** Yang, S., Zhao, L., Yu, C. et al. (2006). On the origin of helical mesostructures. *Journal of the American Chemical Society* 128 (32): 10460–10466.

**108** Che, S., Liu, Z., Ohsuna, T. et al. (2004). Synthesis and characterization of chiral mesoporous silica. *Nature* 429 (6989): 281–284.

**109** Freire, F., Seco, J.M., Quiñoá, E., and Riguera, R. (2012). Nanospheres with tunable size and chirality from helical polymer–metal complexes. *Journal of the American Chemical Society* 134 (47): 19374–19383.

# 3

# Chiral Nanoassemblies

*Chuanlai Xu, Xiaoling Wu, and Changlong Hao*

*Jiangnan University, International Joint Research Laboratory for Biointerface and Biodetection, State Key Lab of Food Science & Technology, and School of Food Science and Technology, Wuxi, Jiangsu Province, 214122 China*

Self-assembly is a scalable and diverse method to arrange individual NPs into sophisticated nanostructures. Benefiting from particular structural designing, a variety of nanostructures with emerging collective characteristics and functions can be fabricated, which is determined by the size, geometry, and spatial configuration of building blocks [1–3]. In recent decades, distinct chiroplasmonic, chiroexcitonic, and other "chiral" technologies are rapidly developed based on the nanoscale assemblies by taking advantage of this technology [4–6]. A variety of methods have been developed to construct chiral nanostructures via the self-assembly from different kinds of building blocks [7]. The self-assembly process may take place in colloidal dispersions or on interfaces [8, 9]. Importantly, higher order superstructures can display chiroptical activity, while individual building blocks show no chirality [10, 11]. The chirality of these assemblies originates from particular chiral geometry patterns of the NP nanoassemblies. By connecting with distinct bridges, individual NPs orderly assemble into different geometries. Chiral nanoassemblies are composed of several thousands of nanoparticles (NPs) serving as building blocks [12]. Therefore, the routine nanoscale of the nanoassemblies may range from 100 to 1000 nm, or even larger. By regulating a multitude of noncovalent bonds between different nanoparticles, nanoassemblies with a variety of geometric shapes, sizes, and forms can be established. Moreover, the symmetry properties and optical characteristics of nanostructures are quite different.

DNA functionalization has been a popular choice for stabilization of metal NPs in bioimaging and biosensing applications in the past decade. Similar to stereochemical control in organic chemistry, highly controllable modification of the DNA sequence on the NP surface enables metal nanocrystal assemblies with different geometries, which include dimer, trimer, pyramidal, and octahedral configurations [13]. Such engineered nanostructures constitute the basis for the preparation of metal NP chiral assemblies. In the past decades, many asymmetric configurations of chiral assemblies were reported based on DNA techniques, such as dimers, tetramer assemblies, and helical structures [14–17]. The obtained chiroptical activity is

*Chiral Nanoprobes for Biological Applications*, First Edition. Edited by Chuanlai Xu.

mainly derived from dissymmetry and free rotation of the nanoassemblies in their spatial arrangement. Therefore, the types of chirality for nanoscale assemblies differ from those for individual NPs. The geometry of these patterns provides a convenient classification protocol and serves as a method to organize the chiral systems into subgroups. Herein, according to the different geometries and constituents, the chiral nanoassemblies can be divided into several categories. Thus, in this book, we simply classify the chiral nanoassemblies into seven subclasses: twisted nanorod (NR) pairs, nanoparticle dimers, tetramer assemblies, nanoparticle helices, chiral nanochains, core–satellite nanoassembly, and nanoplatelets and chiral films.

## 3.1 Twisted Nanorod Pairs

Due to their strong and tunable plasmon resonances, called biological transparency windows in the nanomedicine field, gold nanorods (Au NRs) have been widely used [18, 19]. The Au NRs are commonly used in the chiral system. For better understanding of chiral plasmonic effects, the simplest chiral nanoassemblies are considered as the scissor-like dimer of NRs separated by a certain distance. A dihedral angle $\varphi$ always fluctuates from 0° and 90°, which indicates the absence of a symmetry plane [20]. In recent decades, many studies and reports about the fabrication and optical properties of the twisted metal NRs with such geometries have emerged . The research has focused not only on the establishment of these chiral structures but also on the practical applications of chiroplasmonics [21–23]. Theoretical description/simulation of the nanoscale assemblies is also involved.

The Au NR dimers are very essential for the manipulation of tailorable chiroptical activity, as well as in theoretical studies. One of the representative studies was reported by Ma et al. in 2013 [24]. They assembled twisted pairs and "ladders" with respect to each other by side-by-side (SBS) and end-to-end patterns. Due to the balance of attractive and repulsive forces, the twisting of NRs is thermodynamically favorable. Thus, by the minimization of electrostatic repulsion, the dihedral angle of twisted NRs increases from 0° to 90°, similar to charged rod-like units of many biological assemblies. Strong chiroptical activity was observed even with a relatively small dihedral angle between the NRs, i.e. between 9° and 13° (Figure 3.1a–d). Based on different assembly triggers (sodium citrate (SCI), or sodium carbonate (SCA), and DNA), two enantiomeric conformations, positive and negative dihedral angles, were obtained. Especially, SCI and SCA induced (+) enantiomers, whereas DNA gave rise to (−) enantiomers. It showed a strong preference for one specific enantiomer. The chirality of Au NR dimers was further verified by the circular dichroism (CD) spectra of the single NR pairs. From the single-particle spectroscopy and transmission electron microscopy (TEM) tomography (Figure 3.1e–h), scissor-like geometry of NR pairs was ensured. There were rich spectral transitions in the CD spectra of dimers. Significantly, reversal of the polarity of the CD peaks can be observed along with the sign switch of their dihedral angle from positive to negative (Figure 3.1i–k). During the dynamic process, there was a strong

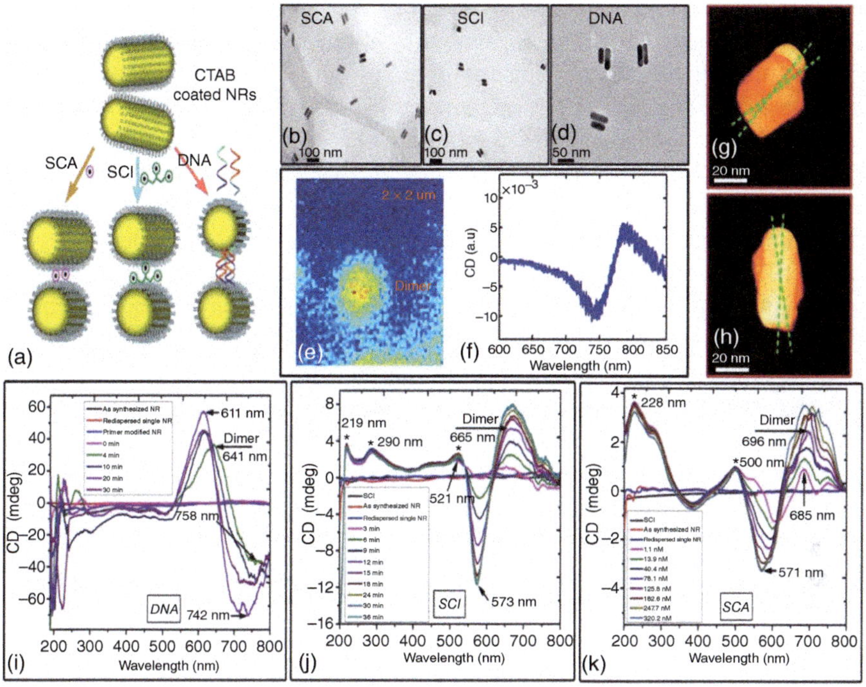


**Figure 3.1** (a) Graphical illustration of Au NR dimers triggered through SCI, SCA, and DNA. (b–d) Representative TEM images for Au NR dimers assembled based on SCA (b), SCI (c), and DNA (d). (e, f) Representative dark-field scattering image of single NR dimers. (e) The CD spectra for single dimers (f). (g, h) Representative TEM tomography images of NR dimers by SCA. (i–k) The CD spectra of NR assemblies and their dynamics/concentration profiles triggered by DNA, SCI, and SCA, respectively. Source: Ma et al. [24]/with permission of Springer Nature.

correlation between spectral shifts of both negative and positive CD bands and growing intensities of the maxima absorption at the transverse and longitudinal NR plasmons. Therefore, the chirality of Au NRs is ascribed to the small dihedral angle between the Au NRs. Mimicking the organic chemistry, the twisted NR dimer is similar to the conformation of a chiral molecule of two isolated chromophores such as BINAP. Furthermore, electronic transitions of charge transfer between different NRs are possible by relative rotation among the NRs, which is analogous to those transitions detected from tilted aromatic chromophores with a certain angle [25]. Therefore, combining with plasmon hybridization terminology, the chiral origin explanation of the plasmonic structure makes sense [20]. This study demonstrates the dynamic chiroplasmonic assemblies with chemical modulation and elucidates the origin of the chiroptical activity in 3D geometry of the nanoscale assemblies [20].

Apart from triggering by the DNA strand, DNA origami also received great attention as assembly templates. Taking advantage of precision engineering of the superstructures, DNA origami templates are beneficial for the construction of chiral assemblies [26, 27]. Recently, Professor Liu's group executed reconfigurable 3D plasmonic metamolecules by the assembly of the Au NR dimers via the special DNA origami [28]. By connecting two bundles with a tunable angle, two NRs were hosted on a tailorable DNA origami template. Furthermore, the handedness of the 3D chiral nanoassemblies was tailorable by the switchable DNA locks. Thus, distinct 3D conformations can be obtained by altering the relative angle of the DNA bundles. Especially, the NRs exhibited tunable left-handed, relaxed, and right-handed states by the DNA-regulated conformational changes of NRs. Bisignate CD spectral bands of the NRs were observed with opposite chirality. In the relaxed state, the CD spectrum possessed right-handed preference. Furthermore, it displayed stronger CD response for the right-handed state than that for the left-handed state (Figure 3.2).

Another representative study of the NRs triggered by DNA origami was the research by Liu and coworkers [29]. Different from the above switchable states of the NR dimers, the authors demonstrated the directional, reverse, and stepwise walking on nanoscale DNA origami of the Au NRs. Here, the plasmonic walking device was composed of a "body" of Au NRs and stepping "feet" of conjugated DNA strands (Figure 3.3). Two Au NRs were assembled together on a double-layer DNA origami template; one of them was as a walker and the other one as a stator. Along with the stepwise movements of the walker on a walking track, a variety of conformational changes took place and it gave rise to immediate chiral spectral response changes, which can be read out optically, through activating the stator with subsequent near-field interaction changes. These studies demonstrated a powerful approach of origami for the fabrication of 3D nanostructures with tailored optical chirality. Moreover, specific reconfiguration of the chiral nanostructures can be tunable by external stimuli (pH, light, DNA, magnetic) [28, 30]. For example, light drives the switchable chiral configurations of Au NR dimers through the molecular motion of azobenzene-hosted nanostructures [31]. Another research demonstrated that pH triggered the structural regulation of the chiral Au NR dimers in a programmable manner over a range of pH values [30].

The plasmonic NR dimers can not only assembly in vitro but can also be assembled in living cells. Professor Xu's group demonstrated an intracellular chiral assembly of Au NRs for the spatially resolved detection of intracellular biomarkers [19]. After the tip-blocking process, the researchers modified specific single-stranded DNA sequence onto Au NR side facets, which triggered the formation of twisted SBS dimers (Figure 3.4). Simultaneously, a methoxy polyethylene glycol sulfhydryl group (mPEG-SH) and DL-dithiothreitol (DTT) was functionalized with the above phosphorothioate DNA, forming a dense packing at highly curved areas. To enhance the intracellular stability and biocompatibility of NR-DNA, additional surface capping with poly(styrene-*b*-acrylic acid) (PS-PAA) was added into the system. The modification of cell-penetrating peptides (TAT) on the Au NR surfaces was beneficial to their transport through the cellular membrane. Therefore, after incubation with living cells, the Au NR SBS dimers can be observed to initiate assembly by the intracellular target miRNA.

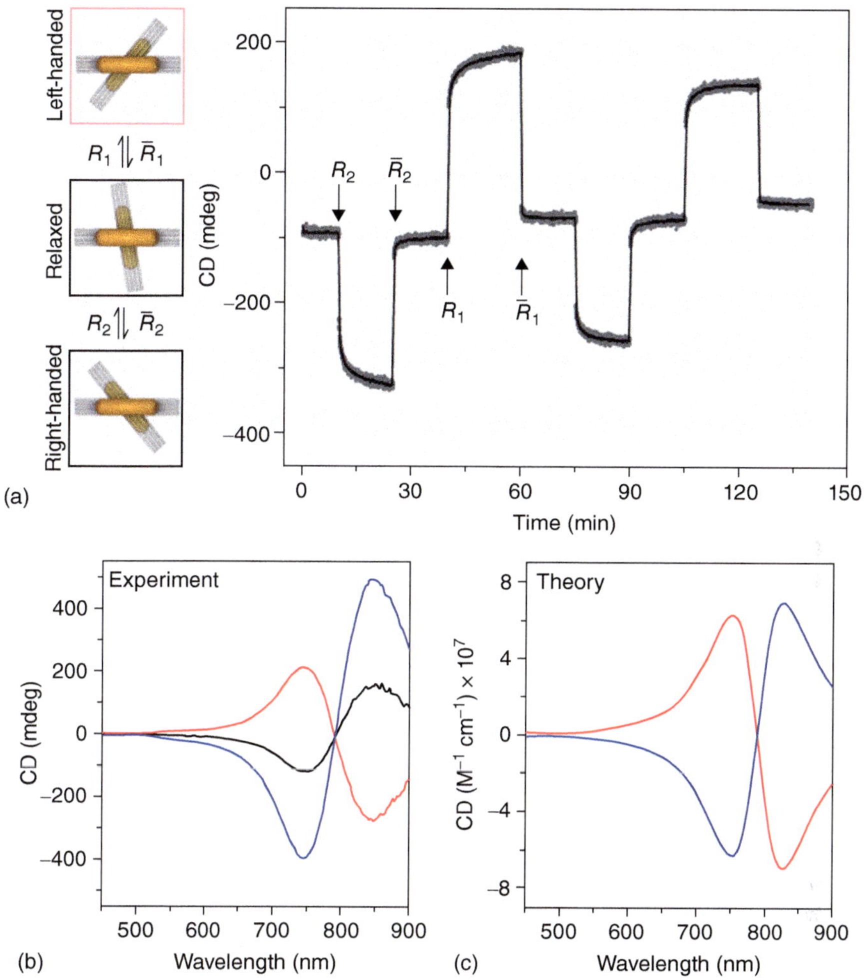


**Figure 3.2 The 3D plasmonic NR dimers between three states.** (a) The plasmonic NR dimers can be triggered to either left- or right-handed state by adding $R_1$ or $R_2$ and back to the relaxed state by adding $\bar{R}_1$ or $\bar{R}_2$. (b) Measured CD spectra in the first cycle of (a). (c) The theoretically simulated CD spectra of the NR dimers. The red, black, and blue curves correspond to the CD spectra of the left-handed, relaxed, and right-handed states, respectively. Source: Kuzyk et al. [28], figure 3 (p. 2)/with permission of Springer Nature.

Importantly, the authors modified specific fluorescent dyes on the surface of NRs, by labeling cyanine 3 (Cy3) and cyanine 5 (Cy5) phosphoramidite to DNA1 and DNA2, respectively. These fluorescent dyes helped to confirm the intracellular assembly. Briefly, green fluorescence in the confocal images was from Cy3, which represented the internalization of probe-1 by the MCF-7 cells. On the other hand, upon dimer formation, with a shortened distance between NRs-DNA@PS-PAA, red fluorescence originated from Cy5. The brightest fluorescence in the cells was obtained at 12 hours (Figure 3.5), consistent with the CD spectra recorded from cells. In CD measurements, Au NR SBS dimers assembled in the living cells displayed a

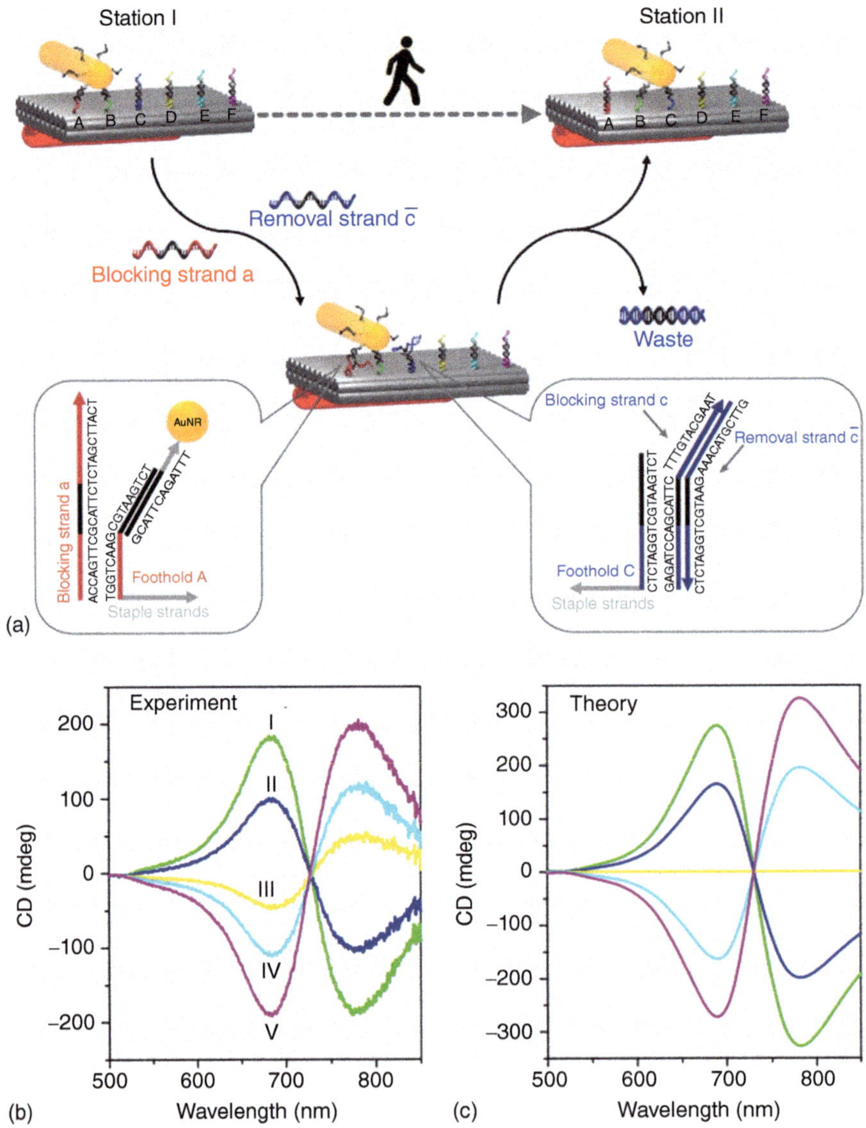


**Figure 3.3** Walking mechanism and CD spectra at different stations. (a) Walking mechanism. Firstly, the walker hosts at station I. Two toehold-mediated strand-displacement reactions happen concurrently after the addition of the blocking strand and removal strand. Blocking strand *a* induces the disconnection of the walker's feet from footholds. (b) Experimental CD spectra at different stations. (c) Calculated CD spectra at different stations. Source: [29], figure 2 (p. 3)/Springer Nature/CC BY-4.0.

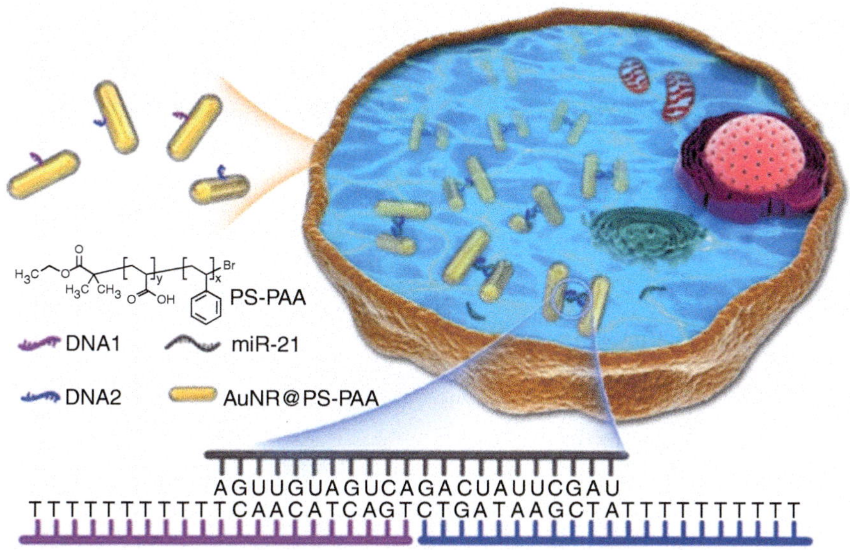


**Figure 3.4** Schematic illustration of special miRNA-21 sequence triggered the formation of chiral Au NR side-by-side (SBS) dimers. Source: Xu et al. [19]. Reproduced with permission of Wiley.

positive CD peak at 815 nm and a negative CD dip at 657 nm. As a comparison, the CD spectra of Au NR SBS dimers assembled in vitro, also with PS-PAA coating shell, presented a dip (740 nm, negative) and a peak (665 nm, positive). Thus, when Au NR SBS dimers were formed in an environment with a high content of proteins and salts, Au NR SBS dimers showed a reverse CD effect, indicating a reconfiguration of NR dimers. This study was essential as it demonstrated the successful assembly of chiral NR dimers in living cells and their reverse chiroptical activity in a different bio-microenvironment.

## 3.2 Chiral Nanoparticle Dimers

The electromagnetic theory and simulations facilitate the precise prediction of their optical activity, such as CD spectra. These calculations require the NP shapes and the 3D geometries of the assembled nanostructures. Importantly, reported simulation studied the chiroptical activities of the NP dimers affected by the slight deviations of the NP shapes from idealized spherical configurations. Dimers assembled from NPs can be selectively prepared based on the DNA hybridization [32]. As shown in Figure 3.6a, theoretical calculations displayed the CD enhancement of a molecule by plasmonic hot spots between symmetric configurations from idealized NPs dimers. The simulated CD signal enhancement were 150-fold and 12-fold for the Ag NPs and Au NPs, respectively (Figure 3.6b). As a comparison, the enhancement factors were 4 and 12 as for randomly oriented chiral molecules (Figure 3.6c) [33, 35]. Au

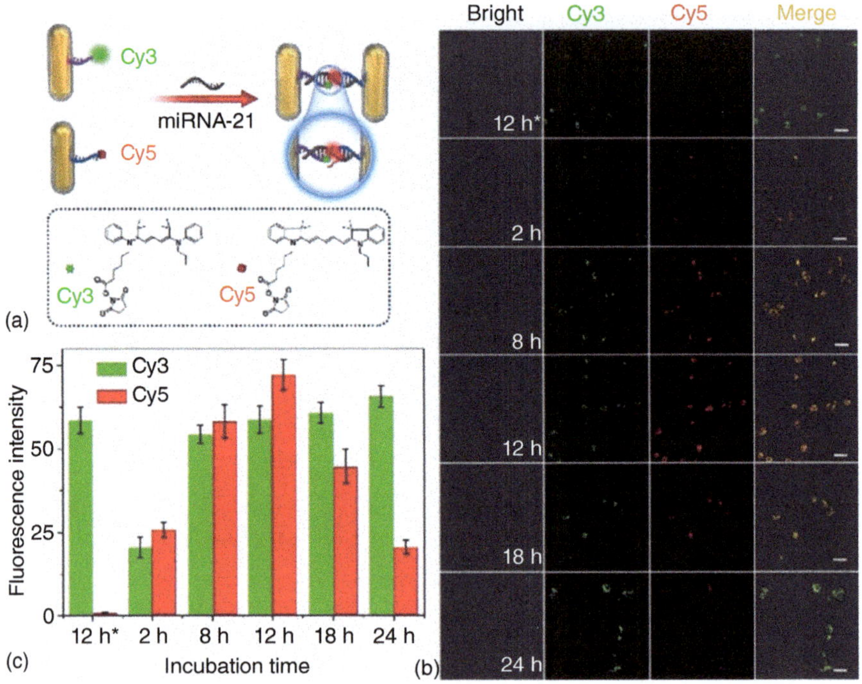


**Figure 3.5** Confirmation of the chiral NR dimers formation through fluorescence resonance energy transfer (FRET). (a) Schematic illustration of FRET in SBS NR@PS-PAA dimers. (b) Confocal images and (c) statistical analysis of the fluorescence intensity, which were incubated with NR@PS-PAA probes for 2, 8, 12, 18, and 24 hours. Source: Xu et al. [19]/with permission of Wiley.

NP–protein composite nanoassemblies exhibited molecular chirality transfer to the plasmon resonances [36]. Typically, the hot-spot CD amplification of the organic molecules was usually considered a local effect. The whole structure chirality should be considered for the assembly. Importantly, polarization rotation induced by the configuration of the nanoassemblies as a whole may exhibit much greater chiral activity due to the comparable assembly dimensions to the wavelengths of visible photons. Intense CD enhancement (about tenfold) from the hot-spot enhancement of the dimer asymmetry geometry as a whole was demonstrated based on electrodynamic simulations. Huge electrical fields can be produced in the single nanometer scale, which breaks the limits of the small gap demand [37]. Importantly, the property of being non-superimposable with their mirror image was required for the NPs in the dimers. None of the prepared NPs displayed perfect metallic spheres; they showed a slight asymmetry between the long and short axes with an aspect ratio of about 1 : 1.1 [38, 39]. Even very small nonspherical NPs generated a strong CD signal for highly polarizable assemblies, which resembled the geometry of twisted rod assemblies [24]. As shown in Figure 3.6d–f, small dihedral angles (10°) of dimers from slightly oblong NPs were constructed through DNA bridges and ionic assemblies. In DNA- and NaCl-assembled heterodimers,

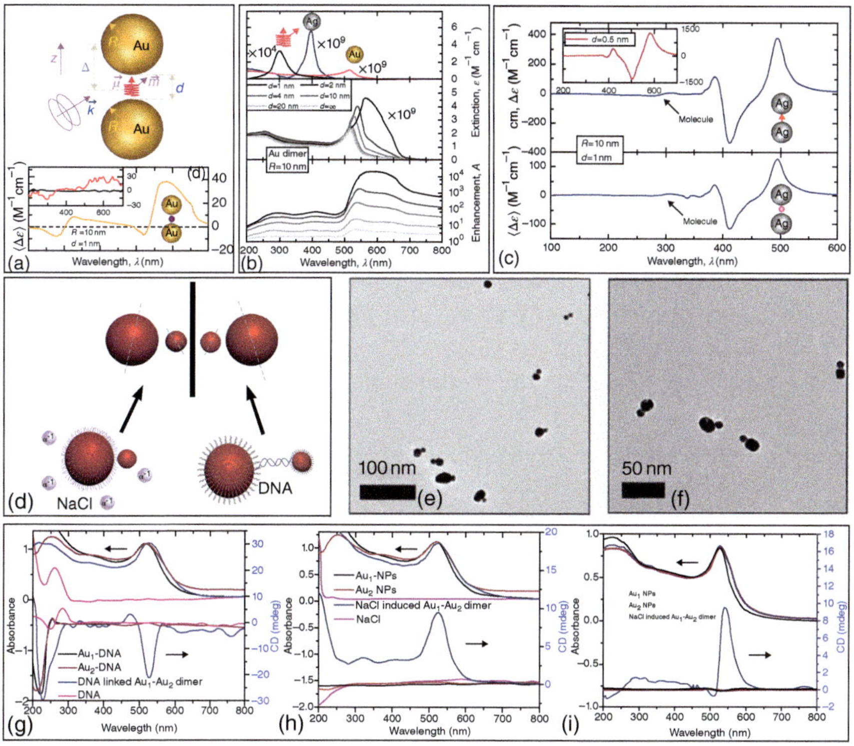


**Figure 3.6** (a–c), Chiral dimers from Au NPs [33]. (a) Schematics of an Incorporating Au NP dimer and a chiral molecule. (b) Extinction spectrum of single molecule and single noble metal NPs (top); extinction spectrum of Au NP dimers for different gaps (middle). (c) CD curve of Ag NP dimers for $d = 1$ nm. (d) Schematics of preparation of chiral NP heterodimers. (e, f) The typical TEM images of heterodimers triggered DNA hybridization (e) and screening of ionic repulsion (f); (g, h) The UV-vis and CD curve of NP heterodimers assembled with DNA (e) and NaCl (h). The calculated spectra of NP heterodimers triggered by NaCl (I) [34]. Source: (a–c) Reproduced with permission from Zhang and Govorov [33]. Copyright 2013 American Physical Society; (d–i) Yan et al. [34]/with permission of American Chemical Society.

a magnitude of (−) 20 or (+) 10 mdeg of CD signal was observed at the plasmon frequency of Au NPs (Figure 3.6g–i) [39]. A certain ee of inorganic nanostructures was required for secondary effects of chiral optical activity in CD or optical rotatory dispersion (ORD). As for DNA-bridged NP dimers, helicity of the oligonucleotides was one of ee origins, which caused positive dihedral angles rather than the negative ones. Note that other ee sources are also certainly possible.

Importantly, these experiments offered new insights into the chiral origins in plasmonic assemblies. For instance, constituent dimers reports were of one the first publications on NP assembly with chiral assembly and strong chiral activity, which partially contributed to strong polarization rotation [32, 40]. And importantly, the scissor-like orientation of the long axes on NP ellipses breaks the symmetric nature of two uniform spheres, and the chiral geometries of NP assemblies are confirmed by simulations [24, 39].

Dynamic NP dimer assemblies from varied NP components or NP shapes can give rise to tunable complex chiral optical activity [41]. For instance, dynamic CD transition was detected for heterodimers consisting of various sized Au NPs, Au-Ag NP dimers [42]. Commonly, CD signal was partly dependent on the production yield, which is lower for heterodimers than that of homodimers. In recent years, a series of studies established various chiroptical dimers using distinct NP components, including Au-Ag NP dimers [38, 43], Au-Au NP dimers [44, 45], and Au-quantum dot (QD) NP dimers [46]. Importantly, different NP building blocks in the assembled dimers resulted in distinct characteristic CD bands. Taking Au-Ag dimers as an example, the CD activity for Au-Ag dimers displayed a special absorption spectrum at the position of both the Au and Ag NP regions (Figure 3.7a) [38]. Wu and coworkers assembled the Au-Ag dimers based on the immune reaction of an antibody and antigen

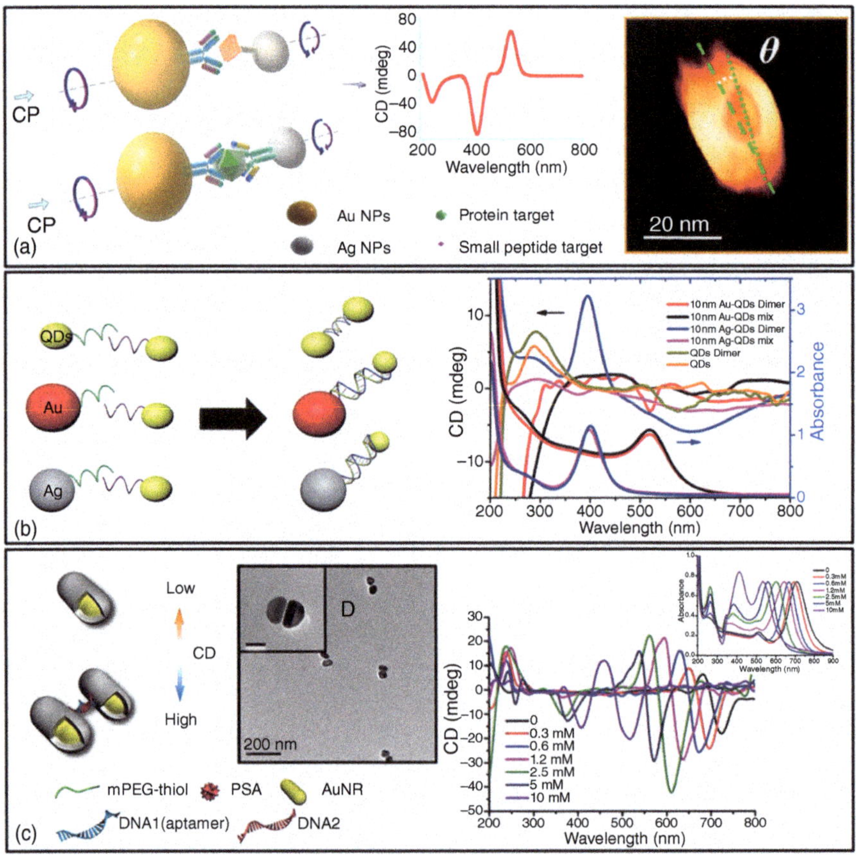


**Figure 3.7** (a) Chiral Au-Ag NP heterodimers and biological analysis [38]. (b) Chiral Au-QD, QD-QD dimers, and relative CD spectra [46]. (c) Au@Ag core–shell NR dimers and corresponding chiroptical activity. Source: (a) Wu et al. [38]/with permission of American Chemical Society; (b) Reproduced from the permission of Copyright 2014, The Royal Society of Chemistry; (c) Tang et al. [47]/with permission of American Chemical Society.

immunocomplex. The prepared NPs dimer showed scissor-like conformation. The long axes of NPs displayed a dihedral angle of ~9°. Moreover, the specific conformation of the chiral force, bridging biomacromolecules, disturbed the enantiomeric equivalence of the NP dimers. Thus, the small dihedral angle resulted in the obvious CD activity of the NP dimers.

For other plasmons, such as metal–semiconductor hybrid nanoassemblies constituting Au, Ag NPs, and QDs, their corresponding CD activities were different from the above studies. Because of the dielectric function of metals and QDs, these NP (Au-QD, Ag-AQ, and QD-QD) dimers produced the CD signal relative to the variation in material compositions (Figure 3.7b) [46]. Interestingly, the assemblies of different material composites enhanced chirality. Some studies demonstrated that chiral activity can be significantly improved after the growth of the silver shell on Au NR dimers (Figure 3.7c) [47]. The enhancement of the CD signal was in favor of the subsequent biological and other emerging applications, such as the analysis of certain targets. This strategy can also be applied to the construction of more complicated superstructures in the future.

Different surrounding conditions may significantly affect or even alter the conformation and morphology of nanomaterials. Recently, Sun et al. demonstrated the chirality reversal of NP dimers by intracellular localization [48]. By real-time spectroscopy, one can use the CD activity to distinguish extracellular or intracellular localization of DNA-bridged NP dimers in ensemble during the internalization of the NPs by the mammalian cells. As shown from the CD peaks in the visible range, chiral signal changed from negative to positive during transmembrane transport (Figure 3.8). Notably, the reversed chirality was accompanied by a spontaneous twisting motion along the DNA bridge. This phenomenon was possibly induced by a large change in electrostatic repulsion between NPs as the movement of the dimers from interstitial fluid to cytosol.

More recently, Professor Xu's group fabricated heterodimers of upconversion nanoparticles (UCNPs) and gold yolk–shell nanoparticles for the quantification of polymyxin-B-resistant *Escherichia coli* (Figure 3.9a–c), based on the dual-model pattern (CD and upconversion luminescence, UCL) [15]. As shown in Figure 3.9d, the Au yolk-shell (YS) nanostructures displayed a weak peak (~17 mdeg) at the plasmonic absorption region (447–596 nm), which was possible due to the amplified plasmon-induced chiral activity of Au YS by coupling to the chiral biomolecules [49]. The Au YS dimers exhibited obvious chirality intensity (~35 mdeg) at a similar plasmonic band position; however, the UV–vis absorption spectrum almost remained unchanged. The chirality of the heterodimers may be attributed to two factors: (i) chiral structure of the scissor-like spatial configuration assembled from elliptically shaped NPs [38, 50, 51] and (ii) intensive plasmonic enhancement by the shell structure [52, 53]. The UCL signal was also observed with extensive quenching under 980 nm excitation, where there were three distinct UCL peaks at 528, 541, and 655 nm, respectively [54]. The extensively quenched luminescence after assembly with Au YS may be attributed to the small distance between UCNPs and plasmonic Au YS [55]. Interestingly, a smaller CD signal was observed for the resistant bacterial strains than outside polymyxin-B-sensitive bacteria, which was

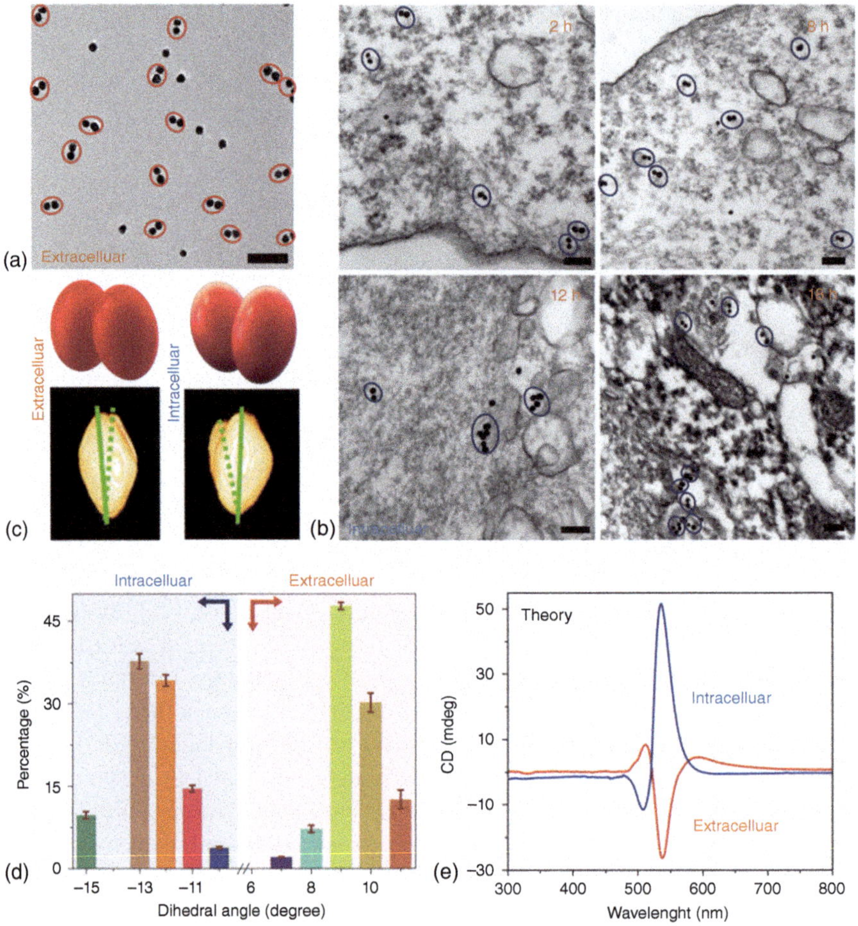


**Figure 3.8 Chiral geometry of NP dimers.** (a) Representative TEM image of NP dimers in cell culture media. (b) Biological TEM images of NP dimers in the living cells. (c) TEM tomography images of NP dimers both outside and inside cells. (d) Statistical analysis of the dihedral angles for NP dimers inside and outside the HeLa cell as observed from cryo-TEM tomography images. (e) Calculated chiral spectra of NP dimers, extra- and intracellular of NP dimers [48].

possibly ascribed to the lower extracellular UCNP concentration. In this study, the dual CD and UCL signal variation of the dimers enabled the authors to achieve the polymyxin-B-resistant level of bacteria, as well as the polymyxin-B-sensitive level of bacteria.

## 3.3 Tetramer Assemblies

Pyramid is one of the special tetramers. First, Mastroianni et al. prepared discrete pyramids with gold NPs at the pyramid tips in solution. Unfortunately, there was no

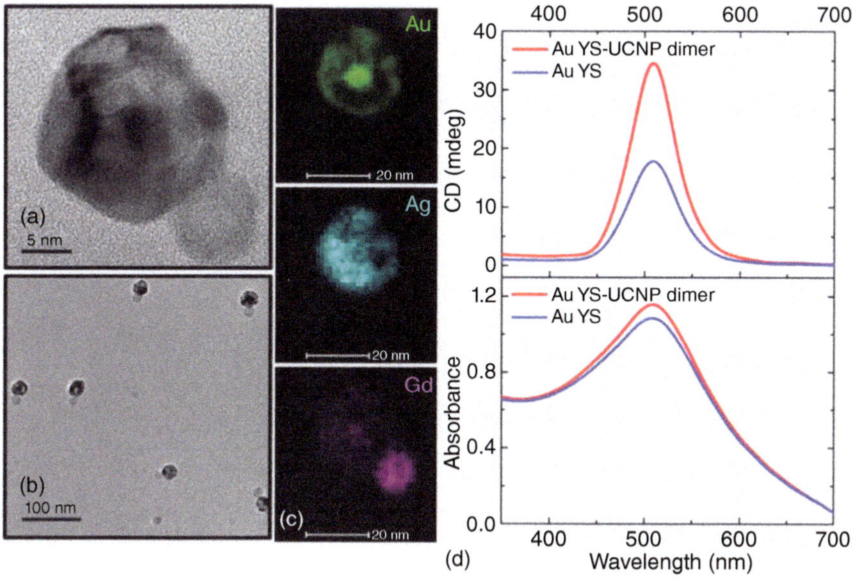


**Figure 3.9** The characterization of the Au YS-UCNP heterodimer. (a) The HRTEM image, (b) TEM images, (c) the EDX mapping, and (d) the CD and UV–vis absorption spectra. Source: Sun et al. [15], figure 2 (p. 2)/Springer Nature/CC BY-4.0.

obvious CD signal for the same Au NP building blocks. They also proposed a reasonable strategy for the construction of chiral pyramids with four different sized NPs, which were confirmed by TEM images (Figure 3.10) [56]. As inspired by the above work, Yan et al. fabricated different types of pyramids with strong chiroptical activity and further explored the structure–activity relationship by systematic investigations of the structures and chiroptical activity via the TEM images and CD spectra [32]. By replacing the NPs with systematic variation of composition and size, they produced six different types of tetramers with pyramidal geometry [54]. Each NP at the apexes was bound to the strand with a unique sequence; it was favorable for the precise positioning of NPs [56]. By exchanging the position of two of these NPs via coupling to the opposite strand, the enantiomeric structures were prepared. They were composed of small Au NPs (10 nm, $Au_1$), medium Au NPs (15 nm, $Au_2$), large Au NPs (25 nm, $Au_3$), CdSe@ZnS NPs (5 nm), and silver NPs (10 nm) (Figure 3.11a–c) [32]. The assembly yield of these pyramids reached 80%, and the good assembly yield allowed for further research and their possible application in optical devices.

Generally, the Xu group constructed a combination of metal and semiconductor NPs and systematically explored the optical activity in this study [32]. Importantly, this research demonstrated the tunable optical activity and capability of synthesis by self-organization, especially compared to electron beam lithography. Moreover, they built the connection with the reported researches of chiral compounds at the molecular-scale assemblies. Notably, it revealed incomplete understanding of chiral geometries of the NP assemblies, and it also demonstrated

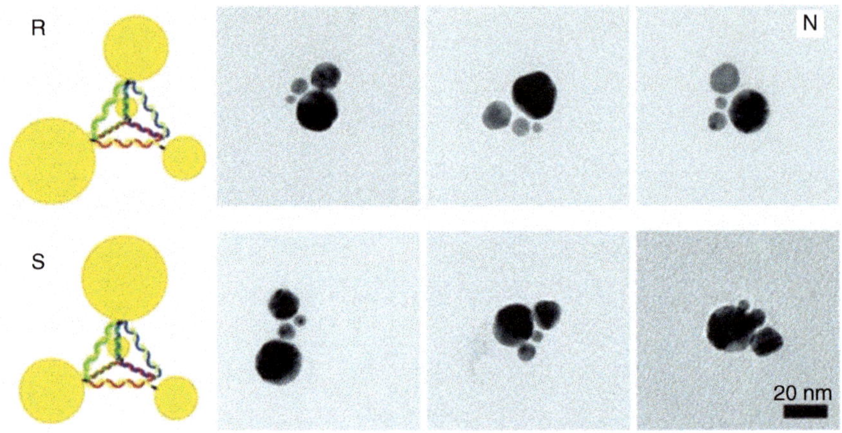


**Figure 3.10** Au NP chiral pyramids assembled on the DNA frame. Source: Mastroianni et al. [56]/with permission of American Chemical Society.

that components of the pyramids, i.e. DNA-bridged NPs, also possess intense chiral activity.

Based on the above strategy of pyramid assembly, Kuang and coworkers constructed gold-upconversion nanoparticle (Au-UCNP) pyramids via the nucleic acid skeleton of pyramids [54]. The pyramids were composed of two gold nanoparticles (Au NPs, $20 \pm 3$ nm, aspect ratio, 1 : 1.2) and upconversion nanoparticles (UCNPs: $NaGdF_4$:$Yb^{3+}$, $Er^{3+}$, $19 \pm 3$ nm). Concurrently, the self-assembled pyramids exhibited an intense CD signal. The luminescence of the UCNPs in the pyramids was significantly quenched, which was due to the luminescence resonance energy transfer (LRET) between the Au NPs and UCNPs (Figure 3.12). The mixture of Au NPs and UCNPs (molar ratio: 1 : 1), similar to the mixture of Au NPs and UCNPs which were dissociated from the pyramids, displayed no obvious CD signal but showed strong luminescence intensity. Taking advantage of the dual signals (CD signal and luminescence signal) of the Au-UCNP pyramids and the ingenious design of the DNA scaffold, the authors achieved ultrasensitive and highly selective detection of miRNA in live cells.

Apart from the above Au-UCNP pyramids, Kuang and coworkers reported a novel hybrid NP pyramid by involving other luminescent NPs [57]. The nanopyramids, composed of Au-$Cu_9S_5$ nanoparticles (NPs), upconversion NPs (UCNPs), and silver sulfide ($Ag_2S$) NPs, exhibited two luminescence signals simultaneously under excitation at 808 nm. One of the luminescence signals was from the UCNPs at 541 nm in the visible band, and the other one was from the $Ag_2S$ NPs at 1227 nm in the second near-infrared (NIR-II) region window. Compared with the original luminescence intensities of the $Ag_2S$ NPs and UCNPs, the luminescence intensities of the pyramids were largely quenched both in the visible and NIR-II regions (Figure 3.13). Based on the luminescence change of the pyramids, the authors developed an ultrasensitive and selective dual miRs detection method in the living cells with ultrasensitive limit of detections. The nanopyramids were also used for achieving precise

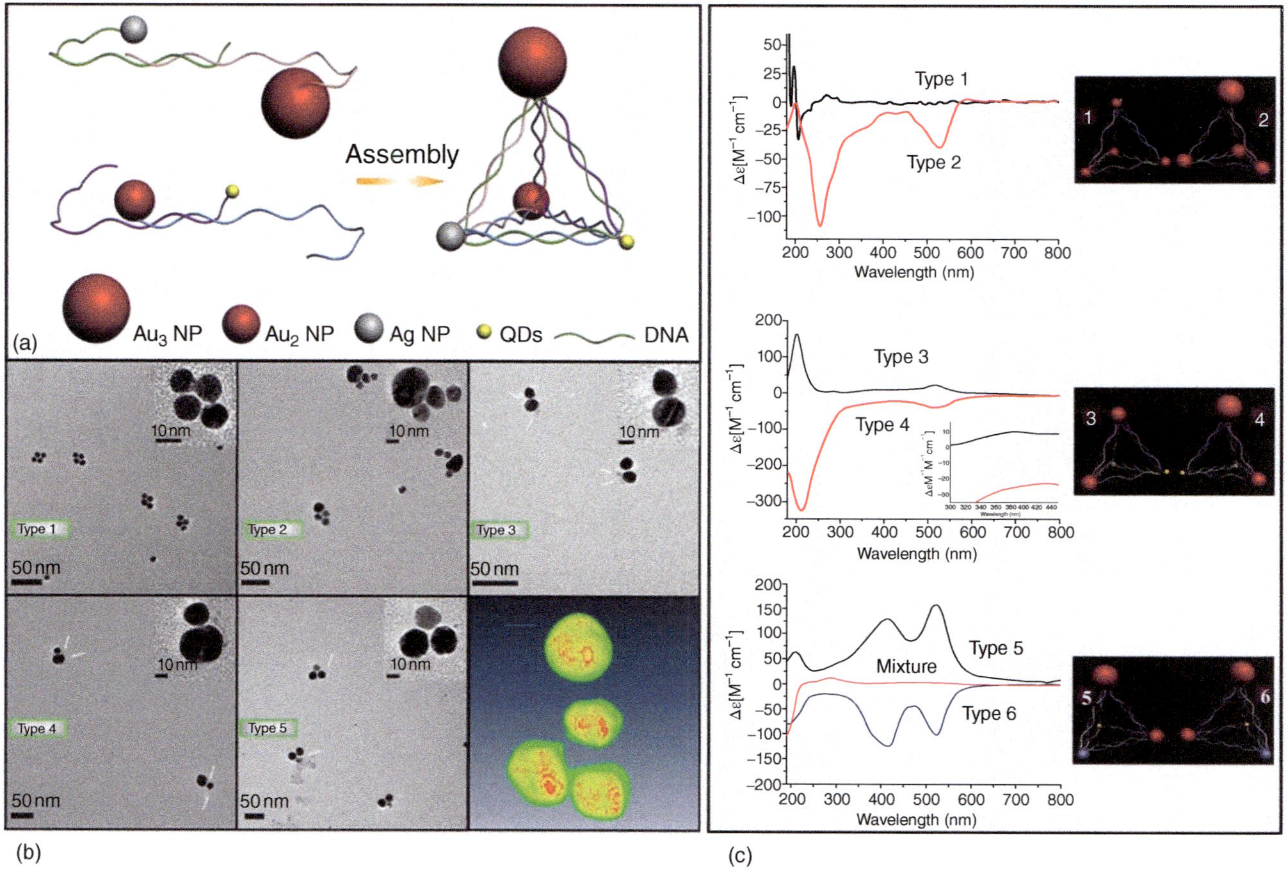


**Figure 3.11** (a) Schematic illustration of chiral pyramid; (b) Representative TEM images of different types of pyramids; (c) CD spectra of assembled pyramid. Type 1 pyramid was composed of four $Au_1$. Type 2 pyramid was made up of three $Au_2 + Au_3$. Type 3 was made up of two $Au_2$ + two QD NPs, Type 4 pyramid was made up of $Au_2 + Au_3$ + two QD NPs, and Type 5 pyramid was made up of $Au_2 + Au_3 + Ag + QD$ as *S*-enantiomers and Type 6 was *R*-enantiomers [32]. Source: Yan et al. [32]/with permission of American Chemical Society.

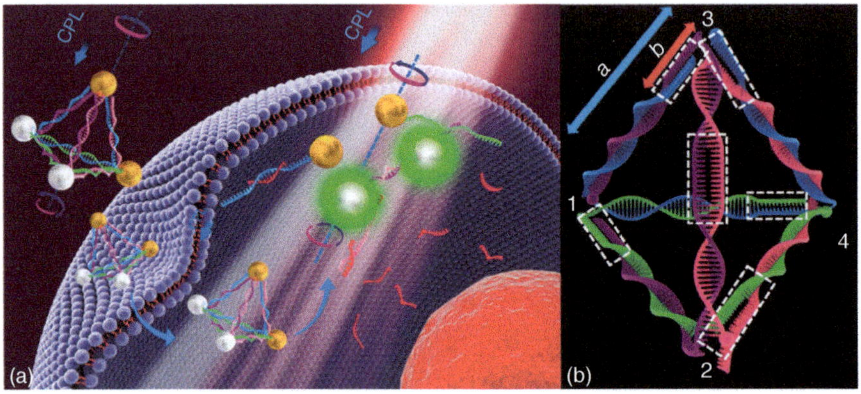


**Figure 3.12** (a) Schemes of Au-UCNP pyramids for miRNA bioanalysis and (b) the nucleic acid frame of pyramid [54]. Source: Reproduced with permission from Li et al. [54]. Copyright 2016 American Chemical Society.

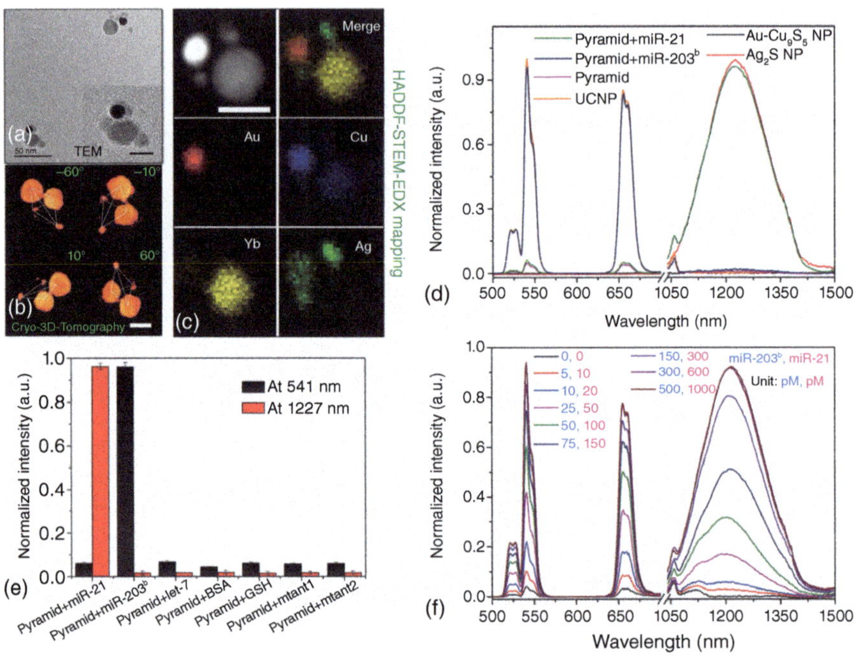


**Figure 3.13** (a) Representative TEM images of nanopyramids consisting of Au-$Cu_9S_5$ NPs, $Ag_2S$ NPs, and UCNPs. (b) Three-dimensional (3D) reconstruction cryo-TEM tomography images of nanopyramids. (c) Scanning TEM elemental mapping images of nanopyramids. (d) Luminescence spectra of nanopyramids with miR-21 and nanopyramids with miR-$203^b$. (e) Luminescence spectra of nanopyramids with miR-21, miR-$203^b$, let-7, BSA, GSH, mutant 1, and mutant 2. (f) Luminescence spectra of nanopyramids with miR-$203^b$ and miR-21 in vitro [57]. Source: Li et al. [57]/with permission of Wiley.

tumor bioimaging in vivo, which included miR-responsive upconversion luminescence and NIR-II luminescence imaging, X-ray computed tomography (CT) imaging, and photoacoustic (PA) imaging. Moreover, the pyramids were demonstrated to possess excellent efficiency for tumor treatment in vivo.

More recently, Professor Kuang group further fabricated a chiral assembly consisting of yolk–shell NPs tetrahedron (UYTe). Interestingly, it can be clearly shown that one UCNP was centralized by four Au NPs through DNA hybridization, observed from the structure of the tetrahedron [58]. Briefly, to achieve the expected tetrahedron, the dimers of yolk–shell NPs (YSNPs) were first assembled by DNA sequences [59]. Then, ATP aptamer sequence-modified UCNP was incubated with the other YSNP dimers. Finally, the dimers and trimers were assembled into a UYTe structure by DNA complementary (Figure 3.14).

Along with the UYTe assembly, the chiroptical activity varied. First, the YSNP dimer exhibited weak signals of $4 \pm 1.5$ mdeg, and the CD intensity

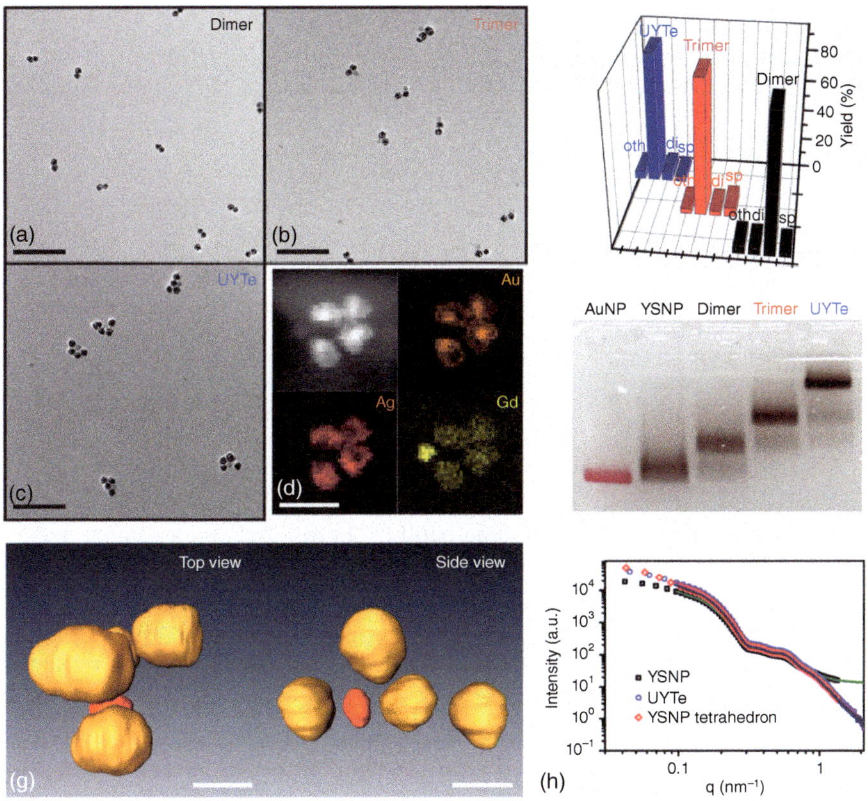


**Figure 3.14** Construction of self-assembled yolk–shell NPs tetrahedron. The TEM images of (a) YSNP dimer, (b) YSNP-YSNP-UCNP trimer, (c) UYTe. (d) The relevant EDX mapping images of UYTe. (e) The yield statistic during the self-assembly process. (f) The electrophoresis image. (g) The corresponding 3D reconstruction cryo-TEM tomography image of UYTe. (h) The SAXS spectrum. Source: Sun et al. [58]/with permission of Springer Nature.

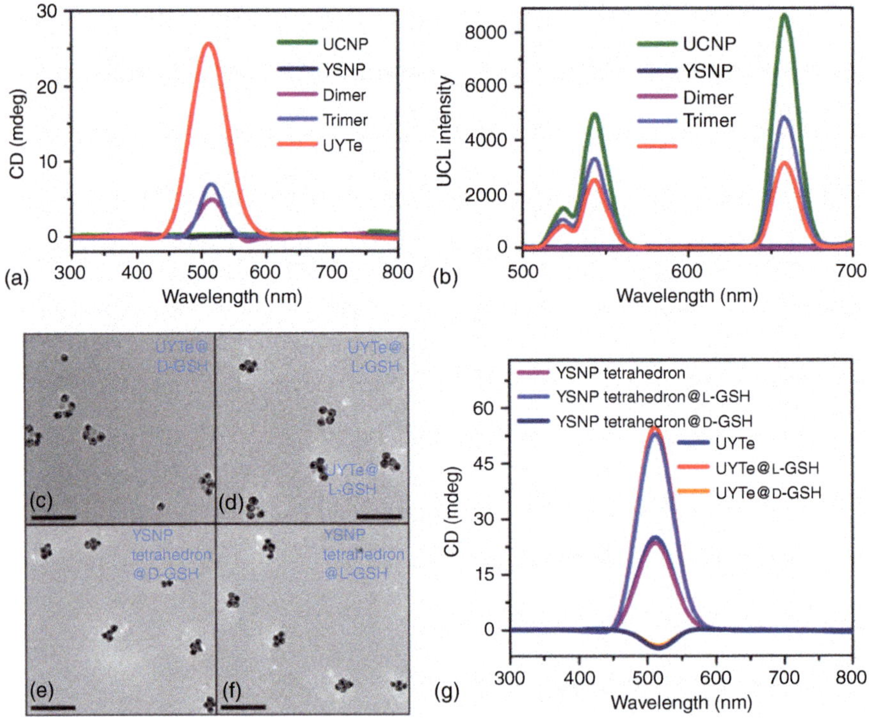


**Figure 3.15** Structure and optical characterization of chiral UYTe assemblies. The (a) CD and (b) UCL spectrum of UCNP, YSNPs, YSNP dimer, YSNP-YSNP-UCNP trimer, and UYTe assemblies. The TEM images of (c) D-GSH- and (d) L-GSH-modified UYTe. The TEM images of (E) D-GSH- and (F) L-GSH-modified YSNP tetrahedrons. (G) The CD spectrum of D-/L-GSH-modified UYTe and YSNP tetrahedrons. Source: Sun et al. [58]/with permission of Springer Nature.

of YSNP-YSNP-UCNP trimer was 7 ± 1.2 mdeg in the region of 470–596 nm (Figure 3.15a). However, the UYTe nanostructure displayed an intense CD signal (about 26 ± 1.3 mdeg) at the plasmonic band (511 nm) after the assembly. Meanwhile, under excitation at 980 nm, the variation of UCL during the assembly process was observed. Three distinct UCL peaks at 528, 541, and 655 nm for UCNP were found (Figure 3.15b) [60]. The YSNP tetrahedron structure was also determined after modification with D- or L-GSH with a high yield of about 85.7% ± 1.2% (Figure 3.15e,f). As indicated by the TEM images, both the D- and L-GSH-modified UYTe nanostructures showed well discrete properties (Figure 3.15c,d). As expected, the CD signals of both kinds of assemblies exhibited nearly the same intensity (Figure 3.15g).

The chirality of the UYTe was mainly attributable to two factors – its chiral configuration and plasmon-induced chirality. Because the CD signal of L-GSH-modified YSNP was positive, the CD intensity of UYTe increased after L-GSH coating [61]. On the contrary, the CD signal of the YSNP modified with D-GSH was negative, and the UYTe showed a weak negative signal because of the counteracted signal between the chiral structure and induced chirality [49]. Notably, the induced chirality was much stronger than the structural chirality, which caused chirality reversal. Third, without UCNP embedded in the nanostructure, the DNA backbone was unchanged. The YSNP tetrahedron nanostructure had nearly the same CD signal as the UYTe structure, indicating that the CD signal mainly originated from the tetrahedron [62].

Besides the pyramid and tetrahedron, Wu and coauthors fabricated another kind of tetramer, a propeller-like nanorod-UCNP tetramer [63]. The tetramers were assembled from NRs and UCNPs based on complementary DNA. First, thiol-terminated DNA was bound to the end facets of the NRs. Then, a single-stranded DNA (DNA1) was coated on the side facets of the NRs. Simultaneously, maleimide-polyethylene glycol (PEGylated) UCNPs were functionalized with the complementary thiolated DNA sequence (DNA2) via the thiol-maleimide reaction. The molar ratio of the DNA to NRs and UCNPs was essential for the tetramer formation. The authors observed that those tetramers possessed chiral geometry, propeller-like conformation similar to chiral molecules, which was well known in coordination chemistry (Figure 3.16). The authors confirmed the propeller-like conformations by the 3D reconstruction of TEM tomography, where three constitutive NRs in the tetramer were not parallel to one another. The special conformation was probably due to the balancing of electrostatic attraction and repulsion forces, similar to many other nanoscale assemblies with biopolymers [64–67].

By switching the building blocks, such as length of DNA sequences, sizes of UCNPs, and aspect ratios of NRs, similar propeller-like tetramers could be constructed. Various and tailorable chiral activities can also be detected in the visible region (Figure 3.17). For example, a strong chiroptical activity in the 300–800 nm region of the tetramers was observed, if the tetramers were constructed from

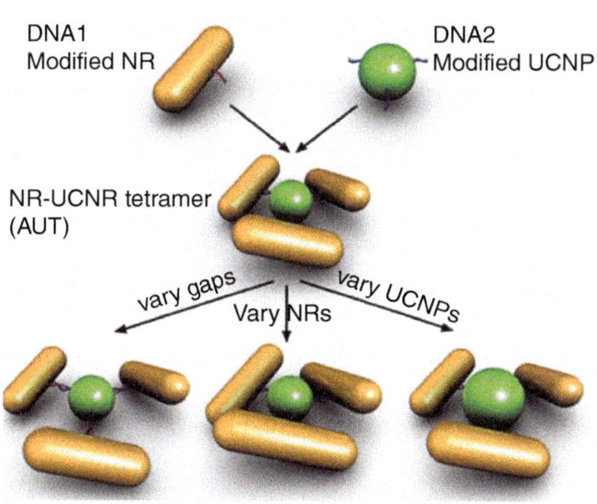


**Figure 3.16** Schematic illustration of the propeller-like NR-UCNP tetramer assembly. Source: Wu et al. [63], figure 1a (p. 2)/with permission of John Wiley & Sons.

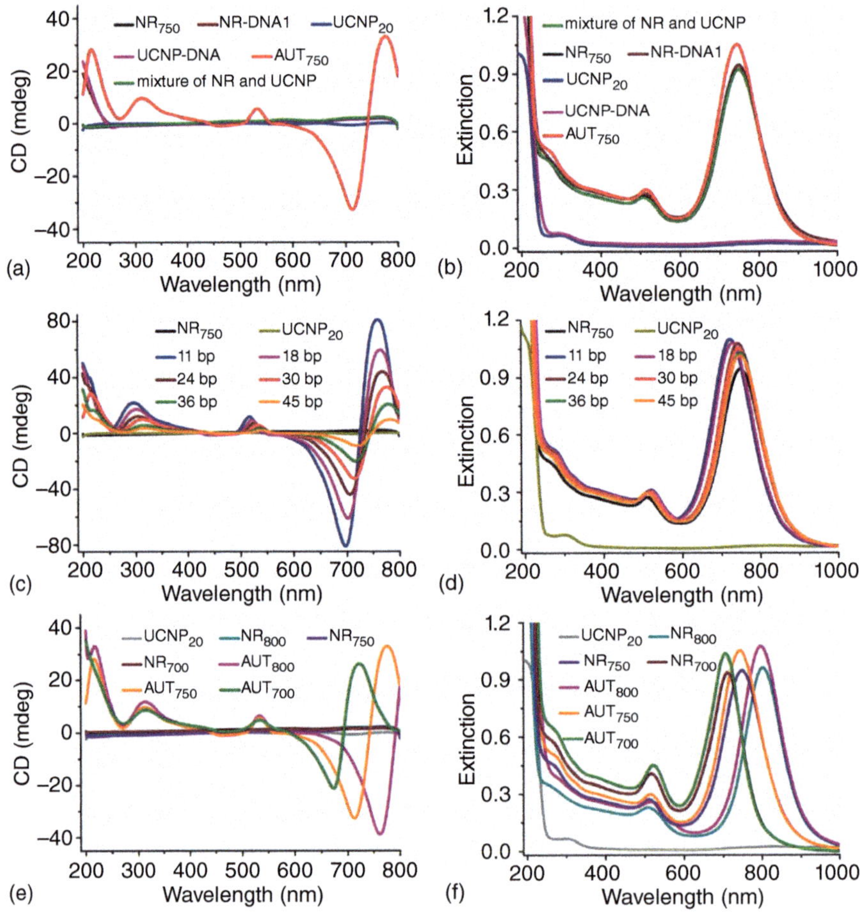


**Figure 3.17** CD spectra of the NR-UCNP tetramer assemblies [63]. (a, c, e) The CD and (b, d, f) UV–vis absorption spectra for NR-UCNP tetramer assemblies with varied lengths of (c, d) DNA sequences, and (e, f) longitudinal absorption. Source: Wu et al. [63], figure 2 (p. 3)/with permission of John Wiley & Sons.

the building blocks of 20 nm UCNPs, NRs with an aspect ratio of 3.3 (average length 50.2), and a DNA sequence of 30 bp (Figure 3.17a). The CD curves were negative for 600–750 nm (−32.7 mdeg at 713 nm) and positive for 750–800 nm (32.9 mdeg at 776 nm). The chiral anisotropy factor, $g$, of the assembled tetramers was $1.06 \times 10^{-2}$. If the DNA sequence from the assembled tetramers decreased from 45 to 11 bp, the CD curves remarkably moved to blue by 22 nm. The CD spectra exhibited a similar bisignate signal, and their intensity gradually increased from 9.3 to 80.9 mdeg (Figure 3.17c,e). Notably, the g-factor reached the largest value of $2.1 \times 10^{-2}$, which was much higher than that of the molecular system. Based on the frequency-domain finite integral method, the authors also carried out the computational simulation of chirality to reveal the origin of chirality of the tetramer. Herein, the researchers found out that the induced plasmonic CD activity was originated from plasmon–plasmon coupling of the NRs. Then the NR–NR interaction led to an intense plasmonic field. The electronic "imprint" of the DNA helix on electronic oscillations also contributed to the strong CD response.

Apart from the chiral activity, the authors observed luminescence enhancement after the tetramer assembly. As shown in Figure 3.18, upconversion luminescence emission of Yb/Er-doped UCNPs was detected at the typical peaks of 529, 546, and 662 nm. The UCL intensity gradually rose during the first 75 minutes of the tetramer formation process and then the luminescence signal remained stable (Figure 3.18a,b). The UCL intensity of emission bands (from 500 to 700 nm) of the

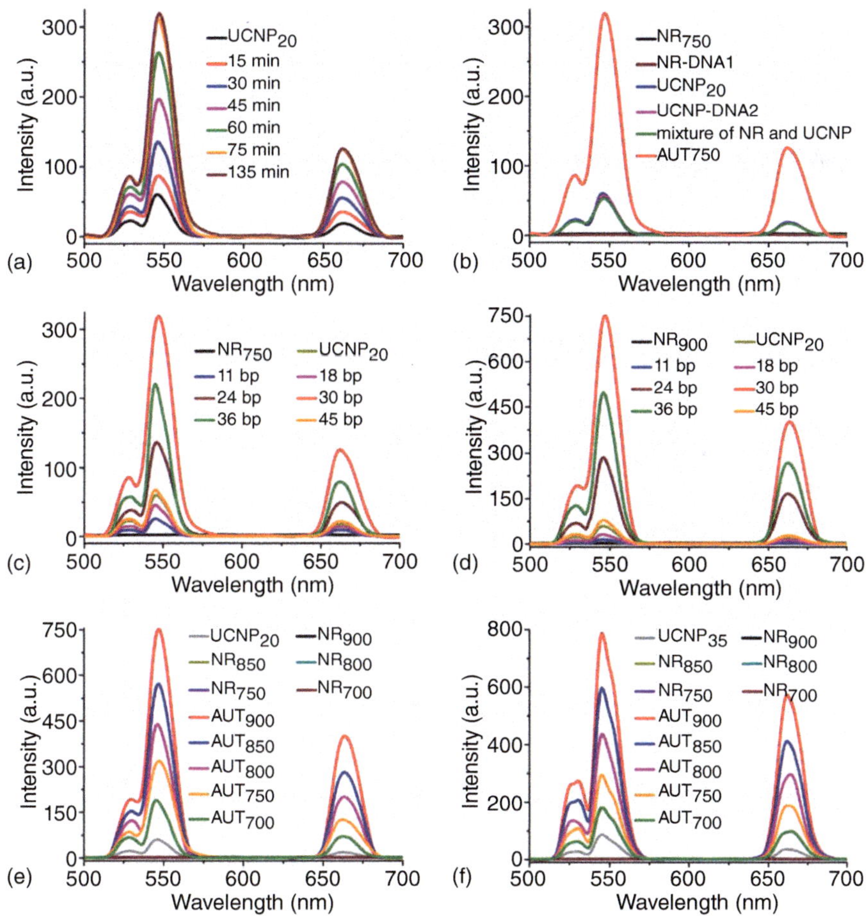


**Figure 3.18** Upconversion luminescence properties of the NR-UCNP tetramer assemblies with 980 nm excitation [63]. (a) Dynamic luminescence spectrum. (b) Luminescence spectrum. Tetramer assembled from the longitudinal absorption peak of NRs of (c) 750 nm and (d) 900 nm. Tetramers assembled from different NRs, where UCNPs were (e) 20 nm and (f) 35 nm. Source: Wu et al. [63], figure 3 (p. 5)/with permission of John Wiley & Sons.

tetramers was enhanced by 3.8–6.6 times. Different from the situation of the CD signal, as the DNA sequence shortened to lower than 18 bp, especially with a short DNA sequence of 11 bp, the UCL intensity quenched intensely (Figure 3.18c,d). It was probably due to the nonradiative energy transfer to the plasmonic states in gold NRs. The maximum enhancement of UCL was observed at 30 bp of DNA sequence. Notably, the enhancement factor reached 21.3 times corresponding to the naked UCNPs (Figure 3.18e,f).

The propeller-like nanoscale tetramers provided intense CD response and enhanced UCL signals. More importantly, the CD signal and luminescence can be tailored by changing the building blocks. Furthermore, the dual-signal response of the tetramers allowed for ultrasensitive biosensors, and DNA detection with an unusually low limit of detection (LOD) was achieved in this study.

In addition to the tetramers, other chiral oligomers were prepared. As reported by Xu and coworkers, they prepared various NP assemblies (Au NRs [37], Au NPs [2, 32, 40], semiconductor NPs [68], and magnetic NPs [68]) based on the polymerase chain reaction (PCR) approach [69]. By regulating the temperature regimes and PCR cycles (in the process of initialization, denaturation, annealing, extension, and final elongation), the researchers could control the programming assemblies and produce the predetermined complexity of NPs. Importantly, they obtained self-assemblies with strong chiroptical response via ssDNA modified on the surface of plasmonic gold NPs [40]. A series of products with varied geometries were observed, including dimers, trimers, tetramers, and even very complex agglomerates based on the distinct reaction cycle numbers and density of DNA primers bound to the NP surface [40]. Chiral optical active nanoassemblies with a little weak peak at the plasmonic band of 650 nm from the CD spectra were clearly observed followed by the purification via gel electrophoresis. Certain enantiomers were formed, which was ascribed to the specific chirality of DNA [70]. Au NP oligomers were also prepared by the immuno-driven assembly of two different types of Au NPs. The yield of heterodimers, such as heterotrimers and heterotetramers, can be as high as 96.6%. Compared to absorption intensity, the self-organized chiral oligomers showed strong and reliable CD signals. The CD bands (at 520 nm) responded to varied molecules specifically, and it resulted in a decrease in CD intensity. The chiral responsive oligomers enabled the establishment of a novel chiral biosensing platform for the detection of various targets.

## 3.4 Nanoparticle Helices

NP helical assemblies are another example of nanoscale structures that exhibit asymmetry from the assembly of individual NPs. They can show the cooperative effects that can be combined with the field effects specific to inorganic materials and 3D organization of nanoscale structural units, thereby giving rise to a set of attractive chemical and physical phenomena. Similar to the tetrahedrons discussed above, it was important to make fundamental parallels between the helical configurations and supramolecular assemblies from organic building blocks showing similar geometries, and the CD activity can be produced from organic and inorganic helicoids.

The 3D chiral templates played an important role in assemblies of NPs into chiral geometries [71]. Due to the chiral dimensions in the nanoscale range, it is possible that the DNA and original molecular templates may limit the assemblies to small NPs (few nanometers). Because of the controlled diameter, length, and pitch from several nanometers to microns, DNA origami [72], peptides [73], and fibers [41] were attractive and promising platforms for NP assemblies in 3D space. In the past decades, helical geometries such as single spirals, stacked rings, double spirals, and nested spirals were successfully achieved based on DNA nanotubes (Figure 3.19A,B) [74]. As revealed from electron tomography, left-handed chirality was observed in the spiral tubes and double-helical wall tube features. Moreover, the double helix displayed a better ordered pattern of assemblies by peptides (Figure 3.19C) [73]. Later, a left-handed helix of DNA-origami-organized NPs at the plasmonic band was obtained (Figure 3.19D) [72]. More complex geometries were achieved by using the toolbox with 9 helically arranged attachment sites from the 24-helix bundles for plasmonic particles (Figure 3.19E,F) [26]. The CD activity of both NP helices showed the anticipated spectral properties in the 500–600 nm region (Figure 3.19G). The helical NP assemblies produced a strong chiroptical activity with a bisignate signature at the plasmonic resonance frequency, and they were corresponding to the handedness of the helices (Figure 3.19H) [75]. As indicated from the polarization-resolved transmission images, the Au helical droplets rotated the polarization angle of the linearly polarized light (Figure 3.19I).

The NP helical nanostructures assembled by DNA and similar templates, such as DNA origami [72] and peptides, usually displayed a few nanometer pitch length (<100 nm) [76]. Other templates based on fibers or lithographic techniques exhibited micrometer dimensions of the helices [77]. Unfortunately, helices assembled from metal NPs with a micrometer (or submicrometer) inorganic template are very rare. Recently, researchers reported an Au NPs/silica helix in the aqueous phase. Based on the sol–gel polycondensation on organic templates of amphiphilic molecules, silica helices were first achieved. They were further functionalized with (3-aminopropyl)triethoxysilane (APTES) [78]. These helices showed intense interaction with Au NPs. This method offered inorganic helix-templated NP assemblies; however, defects were found in the helical patterns.

The assembly of NPs mediated by interparticle interactions offers an alternative approach to chiral morphologies [9]. The "native" interparticle forces incorporate a series of interactions, which included van der Waals force, electrostatic repulsion/attraction, hydrophobic effects, hydrogen bonding, electrostatic forces, dipolar interactions, and entropic-driven effects. Moreover, in external fields, such as light, electrical, magnetic, or gravitational field, an external force can also be introduced. Due to the larger electrical polarizability and magnetic susceptibility, the assemblies formed by external forces can exhibit more inorganic NP characteristics compared to organic building blocks of the same size. Toroidal and normal helical NP assemblies with high yields of both left- and right-handed enantiomers exhibited the typical positive or negative Cotton effects at the plasmonic regions of the metal NPs.

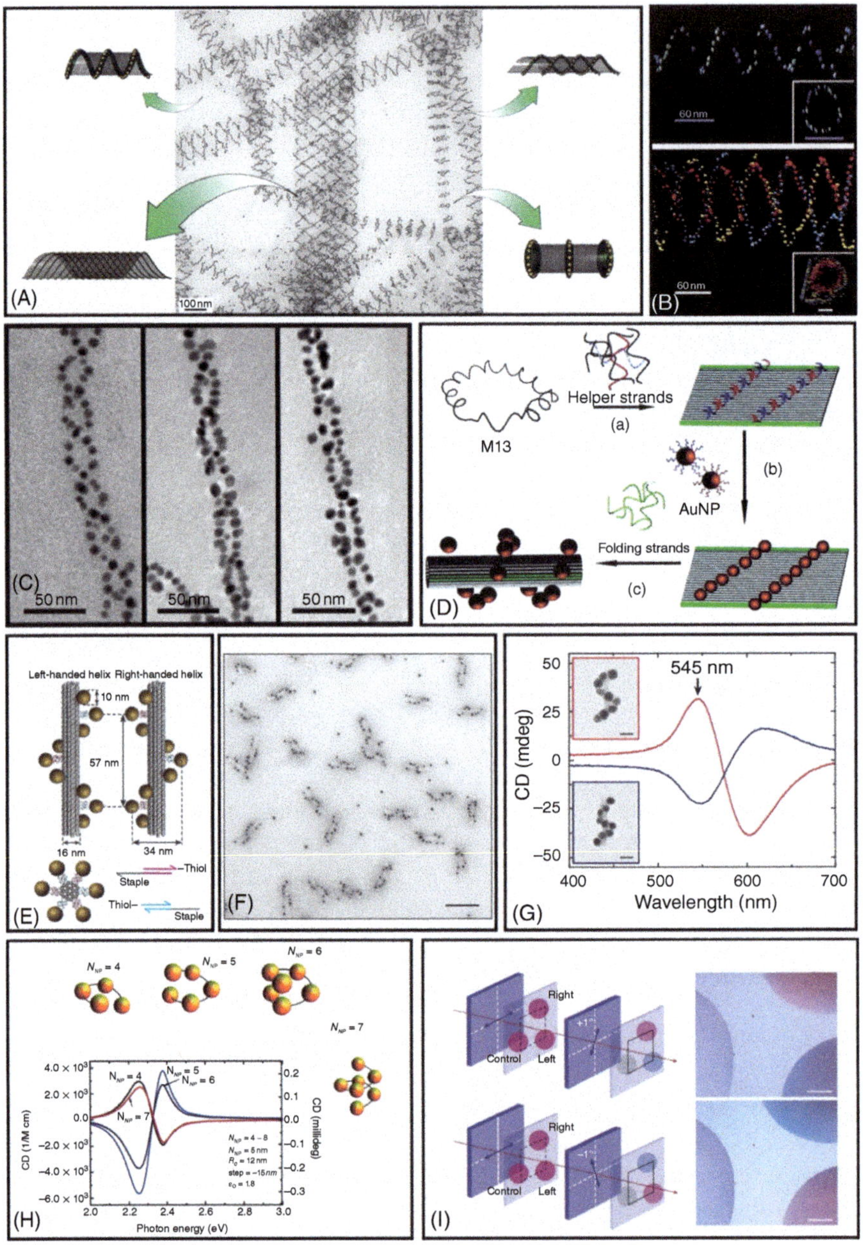


**Figure 3.19** Chiral NPs helices. (A) A single TEM image of different DNA tube assembled NP conformations. (B) Tomogram images of a single-spiral Au NP tube (up) and a double-spiral AuNP tube with a single NP spiral inside each (bottom) [74]. (C) TEM images of Au NPs double nanohelices based on peptide-conjugated molecules [73]. (D) Schemes of Au NPs arranged into a 3D nanohelix [72]. (E, F) Au NPs nanohelices assembled with DNA origami (E) and relevant TEM image (F) [26]. (G) Experimental CD spectra of NP nanohelices [26]. (H) Simulated CD spectra of nanohelical complexes [75]. (I) Optical rotatory dispersion of Au NP assembled helices [26]. Source: (A, B) Sharma et al. [74]; (C, D, H) Reproduced with permission from Refs. [26, 72, 73], respectively. Copyright 2010, 2012, and 2010 American Chemical Society; (E–G, I) Reproduced with permission from Fan and Govorov [75]. Copyright 2012 Macmillan Publishers Ltd.

Similar to Au NP helices, Au NR helices and twisted superstructures exhibited CD bands at the plasmonic band of the longitudinal region. Based on the PCR method, a series of twisted Au NR superstructures can be constructed by altering length of DNA, cycles, and primer densities bound to the NR surfaces. Taking advantage of precisely regulated properties of DNA origami, controllable enantiomer configurations and chiral nanostructures can be achieved, which enables promising bioapplications in the future.

One representative study of Au NR helices was reported by Wang and coworkers [79]. Based on a programmable DNA origami pattern, they constructed anisotropic Au NR helical nanostructures with tunable CD activity (Figure 3.20). First, DNA capturing strands were specially organized, with an "X" pattern of the arrangement on both sides of a two-dimensional DNA origami template. The Au NRs were modified with the complementary DNA sequences. The modified Au NRs were placed

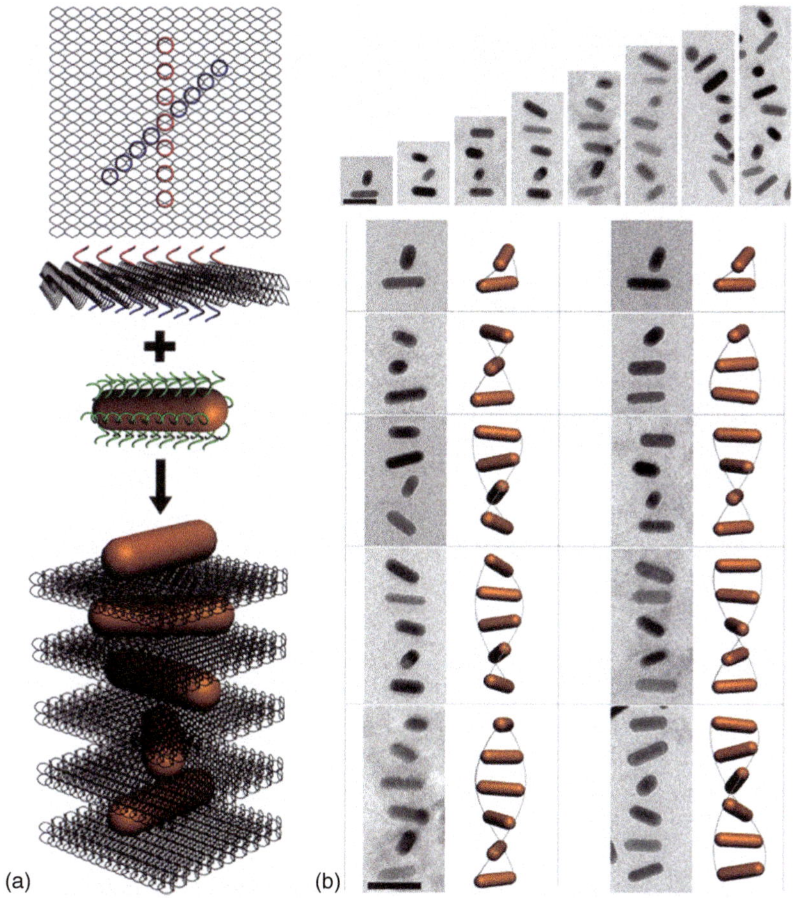


**Figure 3.20** (a) Schematic illustration of the self-assembly of RH-Au NR helices. (b) Cryo-TEM images of the assembled Au NR helices with varying numbers of Au NRs. Source: Lan et al. [79]/with permission of American Chemical Society.

on the origami. Thus, Au NR helices were assembled with the origami intercalated between neighboring Au NRs. By simply tuning the mirrored-symmetric "X" patterns of the capturing strands on the origami, left-handed (LH) and right-handed (RH) Au NR helices were conveniently achieved. The average number of Au NRs in the helices can be tunable from 2 to 4 and 9 by altering the molar ratio of Au NRs to the origami during the assembly system. Intense CD activities were obtained from longest Au NR helices. The maximum anisotropy factor reached ~0.02, which was highly comparable to the reported macroscopic Au NR assemblies.

Noncovalent forces drive the assembly, including van der Waals forces, hydrogen bond formation, electrostatic interactions, hydrophobic interactions, and coulombic forces. Those nanoassemblies can be regulated by the polarity of the solvent, pH changes, and ionic strength. As reported, by regulating the solvent from water to methanol, CdTe NPs (with chiral cysteine ligand coating layer) assembled into same-handedness semiconductor helices [80]. In this research, specific chiral dissymmetric collections were formed by an orderly arrangement of semiconductor subunits through the electrostatic attraction forces between chiral CdTe NPs (Figure 3.21). Importantly, their optical properties can be effectively regulated by the critical performance indicators (length, diameter, pitch, and thickness) of the chiral nanohelices. The solvent-driven nanoassemblies improved the topography-dependent chiroptical activity.

In another important representative study, Kotov and coworkers elaborately explained the cooperative effects for controlling the formation of chiral assemblies, which included different solvent proportions, pH changes, precursor addition, and coordination bonding interactions [81]. The final output and conformation of the nanostructures can be directly affected by the mixture solvent, water, methanol, and different molar ratios (Figure 3.22). Moreover, the acid environment allowed for increased surface chiral cysteine ionization and resulted in uniformly twisted fibrous bundles. Furthermore, the precise regulation of the Cd and Cys amount can lead to distinct stacked morphologies and obtain highly purified enantiopure isomers. Notably, hierarchical chiral mesoscale assemblies with good CD activity were prepared through the coordination effect driven by the interstitial lattice. Therefore, it is beneficial to the structural tunability and specific space, which offers more possibilities for high-performance biocatalysis.

Because it is relatively difficult to manipulate the geometries, very few studies have been reported on chiral NP assemblies driven by proteins and polypeptides. Recently, Au NR assemblies were synthesized by the assembly on anionic rod-like viruses. Right-handed helical twisting morphology was observed, and it exhibited special CD activity (Figure 3.23a,b) [82]. In this study, chiral Au NP double-helical superstructures were prepared by the peptide [$C_{11}H_{23}CO$]-AYSS-GAPPMPPF with both left- and right-handed structures. The absorption and CD spectra exhibited time-dependent transition characteristics. The CD intensity gradually increased with decreasing absorption (Figure 3.23b). Such signal response was conducive to protein biosensing. Similar to peptides, Au NP helices were reported by controlled growth on supramolecular fibers by UV-light mediated reduction of Au(I) ions.

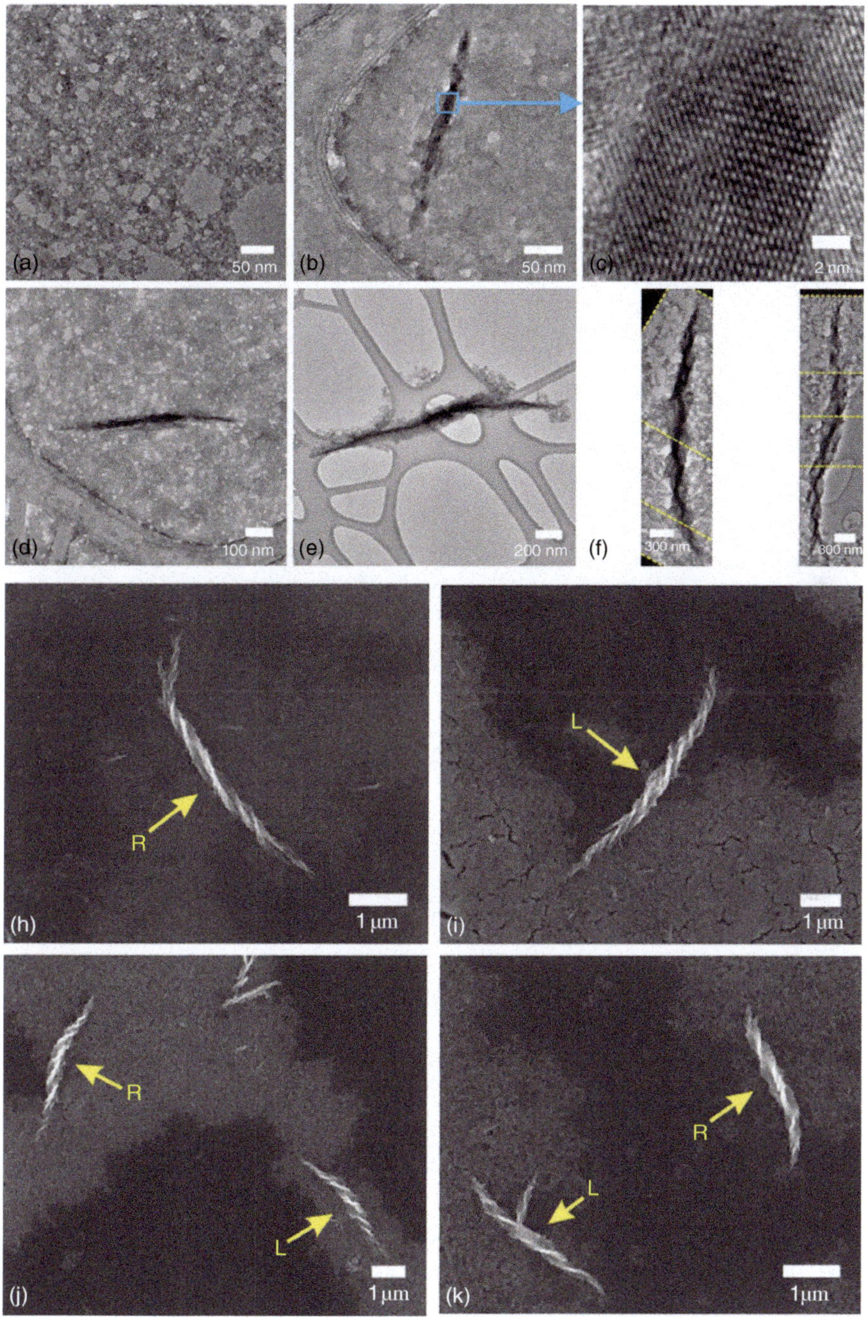


**Figure 3.21** Formation illustrations. (a–g) TEM images of self-assembly of D-Cys CdTe NPs at different times: (a) 0 hour, (b) 1 hour, (c) 1 hour, (d) 3 hours, (e) 5 hours, (f) 6 hours, and (g) 8 hours. (h–k) scanning electron microscope (SEM) images of different assemblies, (h) D-Cys CdTe NP assembly, (i) L-Cys CdTe NP assembly, (j) DL-Cys CdTe NP assembly, and (k) D-Cys CdTe/L-Cys CdTe NP assembly. Source: Yan et al. [80].

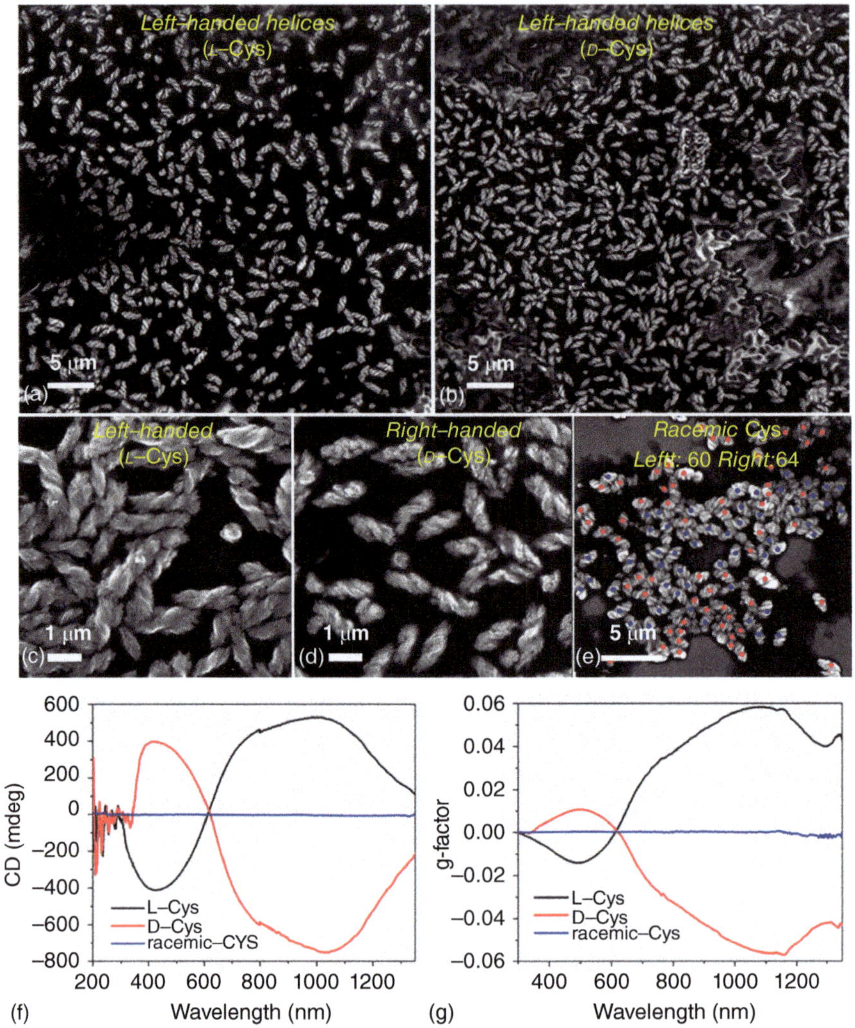


**Figure 3.22** SEM images of left-handed helices (a) and right-handed helices (b). SEM images of nanohelices were prepared from (c) L-, (d) D-, and (e) D/L-Cys ligand. (f) CD and (g) *g*-factor results for nanohelices. Source: Yan et al. [81]/with permission of American Chemical Society.

Typical chiroptical bands of Au NPs were observed. Furthermore, an anodic aluminum oxide template with cylindrical confinement can drive the double-helical Au NP assembly, which showed a defined helical morphology. In recent years, a template-free technology has also been developed for chiral assemblies. For instance, evaporation-driven Ag nanoprism helical columnar assemblies were obtained with a scale of 2–3 μm length based on the strong binding thiol chiral ligand (Figure 3.23c) [83]. In addition, chiral ribbons with a pitch of 400 nm were prepared by twisting CdSe nanoscale platelets driven by chiral molecules. Interestingly, right- or left-handed twisted ribbons have been reported by illumination of

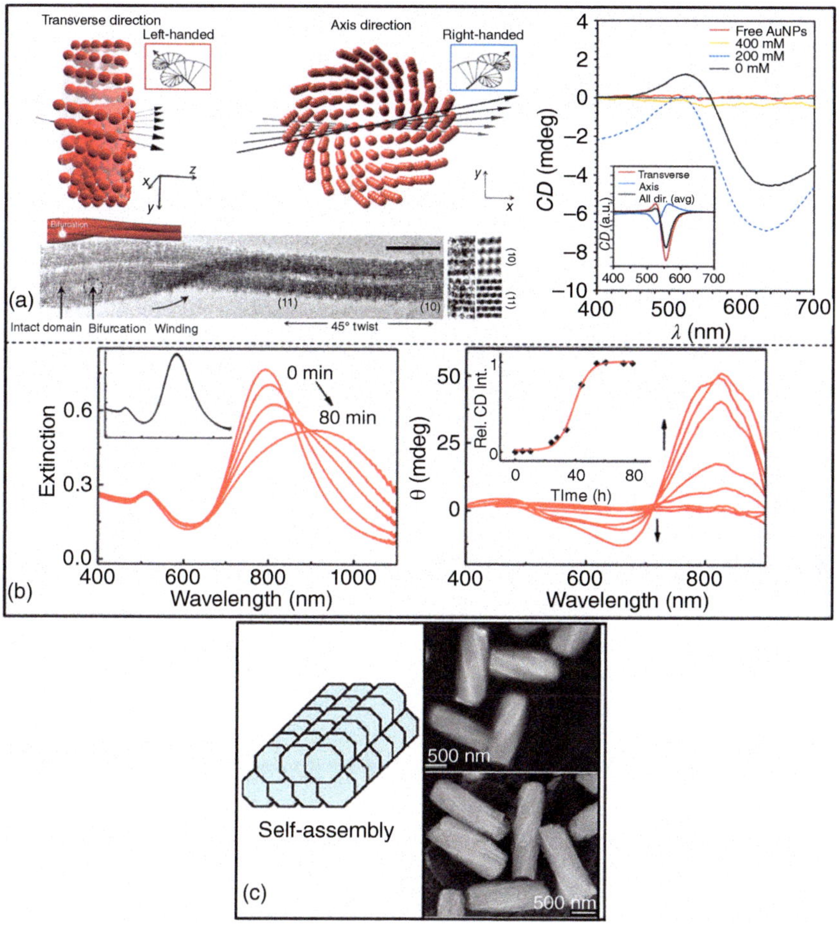


**Figure 3.23** (a, b) The Au NP helical assembly on anionic rod-like virus (a) and CD spectral changes of Au NRs during the 80 minutes followed by the α-synuclein fibrils (80 nM) addition (b). Source: (a) Liljestrom et al. [82]/with permission of Springer Nature, (b) Liljestrom et al. [82], figure 4 (p. 5)/Springer Nature/CC BY-4.0; (c) Ag prisms helical assemblies. Source: Cathcart and Kitaev [83]/American Chemical Society.

racemic CdTe NPs with right- or left-handed circularly polarized light (CPL). The total length of the prepared twisted ribbons was $3 \pm 0.5$ μm and their pitch length was $800 \pm 20$ nm. Notably, whether a template (DNA origami, or inorganic template) or template-free form, it is still challenging to scale up the chiral assemblies for real applications.

One of the typical soft templates is hydrogels. Hydrogels are usually made up of low-molecular-weight gelators and water and have been widely applied in tissue engineering, microfluidic devices, cell culture, and environmental applications. Recent studies demonstrated them as chiral templates for helical assemblies. For example, Hao et al. reported chiroptical activity of iron disulfide quantum dot hydrogel helices, which can be tailorable under CPL [85]. They demonstrated that chiral

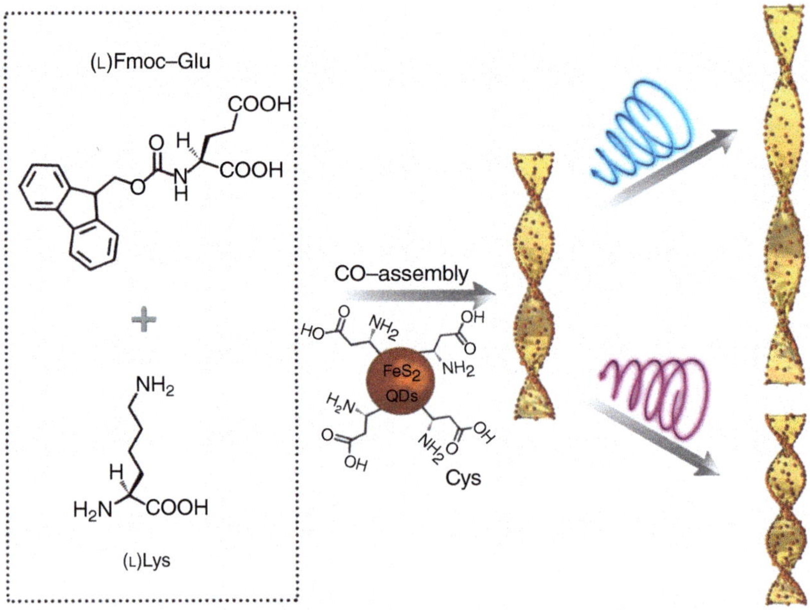


**Figure 3.24** The illustration of the cogel formation and its CPL responsiveness. Source: Hao et al. [85], figure 1 (p. 2)/with permission of John Wiley & Sons.

$FeS_2$ QDs could assemble with chiral hydrogels to form a helical superstructure, which exhibited intense CD and circularly polarized luminescence signal in the visible plasmonic band. Moreover, the CD responses of the assemblies can be regulated with CPL because of alteration of the helical pitch length by CPL irradiation (Figure 3.24).

Briefly, to fabricate hydrogels with helical nanostructures, chiral lysine (Lys) and well-known gelators *N*-(9-fluorenylmethox-ycarbonyl)-protected glutamic acid (Fmoc-Glu) were used (Figure 3.25a,b) [86]. Left-handed helical structures were formed with the L-gelators and D-gelators for right-handed nanohelical structures with the opposite molecular chirality. As shown in Figure 3.25e,f, the helical structures were confirmed through scanning electron microscopy. Importantly, the molecular chirality of the L- and D-gelators was transferred to the self-assemblies at the nanoscale during gelation. The chiral $FeS_2$ QDs with the L/D-cysteine (L/D-Cys) capping ligand were then synthesized. By mixing the gelators with chiral $FeS_2$ QDs, assembled helical fibers were observed, where the electrostatic interactions between the carboxylate groups and the amine groups contributed to the cogels of chiral $FeS_2$ QDs and the gelators. Interestingly, the chiral $FeS_2$ QDs were organized along the helices (Figure 3.25c–h).

As displayed in the figure, the helical pitch of D/L-Gel and D/L-(Gel + $FeS_2$) was in the range of 149–167 nm, and the diameters were about 50 nm. Mirror image spectra of D-Gel and L-Gel were detected, with CD bands of 423 mdeg at 304 nm. However, after assembled with $FeS_2$ QDs, the CD bands of the cogels shifted to

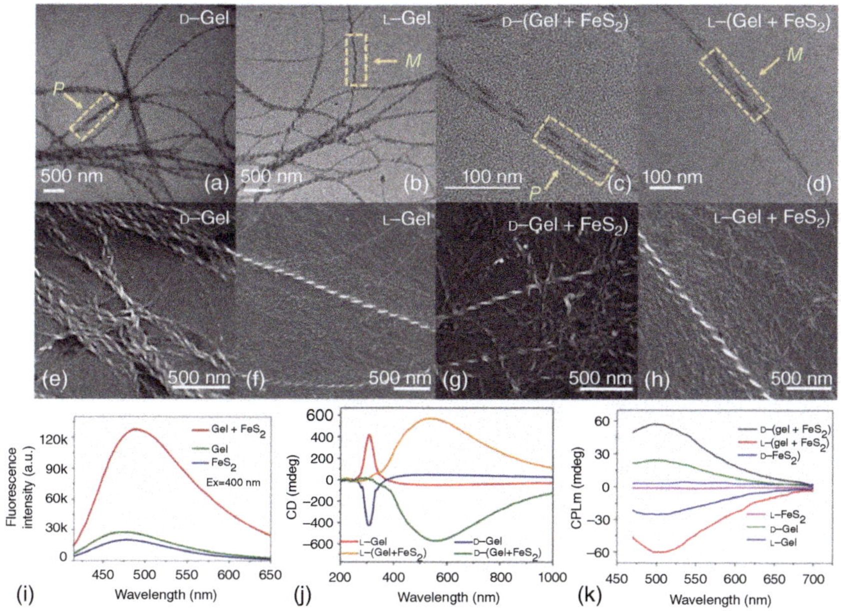


**Figure 3.25** (a–d) TEM images of D-Gel nanofibers, L-Gel nanofibers, D-(Gel + $FeS_2$) nanofibers, and L-(Gel + $FeS_2$) nanofibers. (e–h) SEM images of D-Gel nanofibers, L-Gel nanofibers, D-(Gel + $FeS_2$) nanofibers, and L-(Gel + $FeS_2$) nanofibers. (i) FL curves of L-(Gel + $FeS_2$), L-Gel, and L-$FeS_2$. (j) CD curves of L-Gel, D-Gel, L-(Gel + $FeS_2$), and D-(Gel + $FeS_2$). (k) Circularly polarized luminescence (CPLm) emission curves of cogels. Source: Hao et al. [85]/with permission of Wiley.

537–556 nm with intensity around 566 mdeg. Another CD peak with 21 mdeg at 255 nm was attributed to $FeS_2$. Interestingly, opposite arrangement direction of the hydrogel structure and Cys on the $FeS_2$ surfaces were observed. The strong signal around 447 mdeg from the gel molecules can influence the $FeS_2$ CD signals in the range of 272–391 nm, which may be induced by a scattering effect on the CD measurements [87]. Notably, the cogels exhibited mirror peaks around 566 mdeg at 537 nm; however, no CD signals were observed for gelators or the chiral $FeS_2$ QDs (Figure 3.25j). Besides the CD signal, strong fluorescence intensity was also observed for cogels (Figure 3.25i). Moreover, as shown in Figure 3.25k, a negative circularly polarized luminescence (CPLm) of L-(gel + $FeS_2$) was detected at 496 nm. Meanwhile, mirror-imaged CPLm of D-(gel + $FeS_2$) was also observed. Compared to gels, cogel assemblies displayed much stronger CPLm intensity, which can be ascribed to chirality transfer between the helix and the chiral $FeS_2$ QDs from the cogel assemblies. By the chiral template of the hydrogel, the distribution of the chiral $FeS_2$ QDs was crucial to chirality transfer [88]. Importantly, such intense CPLm response enabled the possibility for potential applications in various fields, including data storage, photoelectric nanodevices, optical nanoprobes, and spin-optoelectronic circuits.

More importantly, cogels showed CPL responsiveness. In the absence of left circular polarized (LCP) light irradiation, L-(Gel + $FeS_2$) exhibited a chiroptical signal

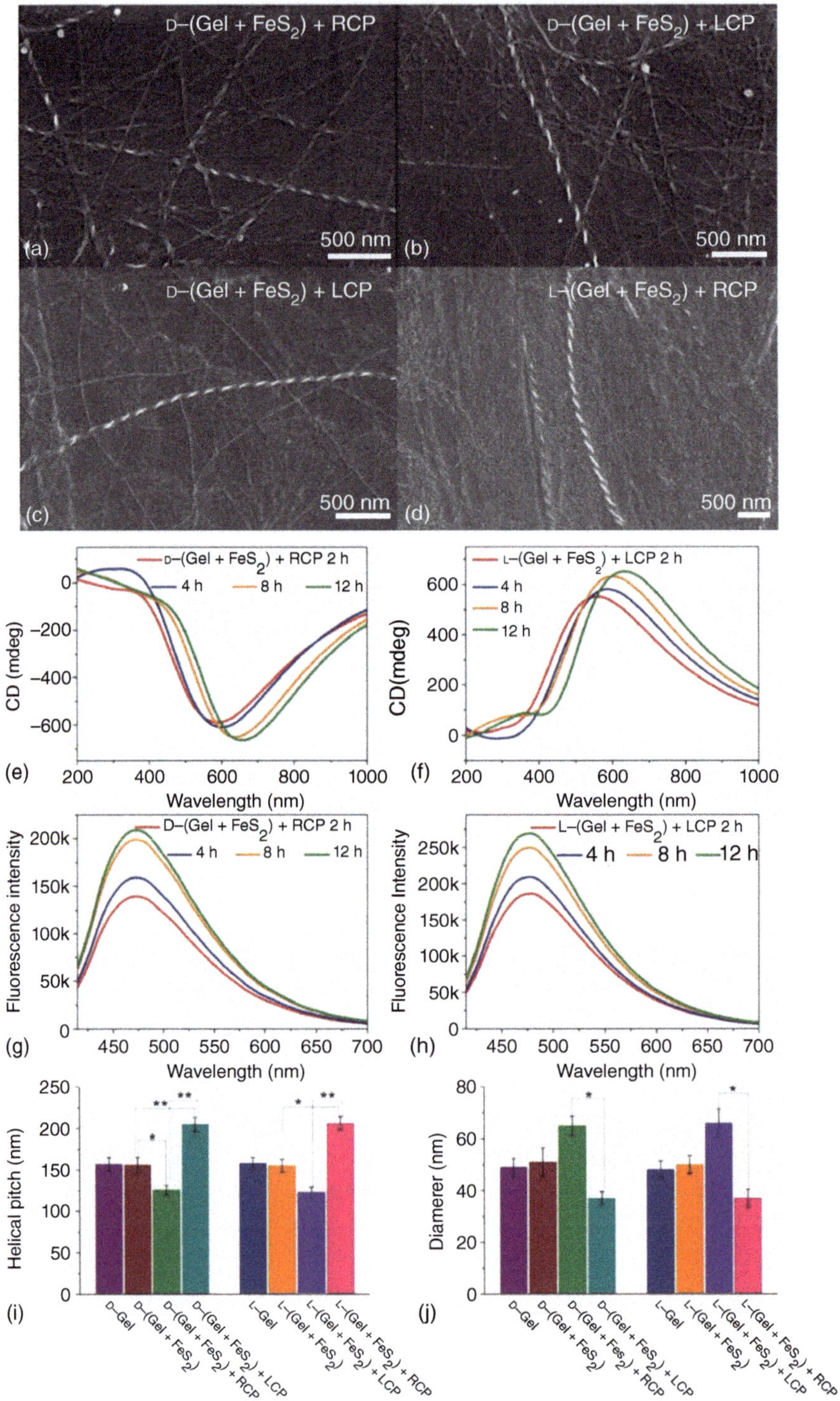


**Figure 3.26** (a–d) SEM images of cogel helical nanoassemblies under different CPL. (e, f) CD spectra of cogel helical nanoassemblies under corresponding CPL. (g, h) FL spectra of cogel helical nanoassemblies under corresponding CPL. (i, j) The corresponding statistics of helical pitches and diameters. Source: Hao et al. [85]/with permission of Wiley.

with a positive peak at 537 nm. After LCP light irradiation for different times, the CD signal shifted to 670 nm. Simultaneously, CD intensity showed a gradual increase from 558 to 664 mdeg. This phenomenon was probably ascribed to the enlarged space between the $FeS_2$ QDs. When LCP light was replaced by linearly polarized (LP) light or LCP light at 532 nm, the intensity of the CD signal of D-(Gel + $FeS_2$) decreased and a significant redshift was observed. Notably, a much smaller increase in CD and fluorescence (FL) signal was detected for pure L-Gel and D-Gel after irradiation with LCP light (Figure 3.26). As known, nanohelical chirality was closely related to the helical pitch and the helical diameter of the coassemblies. Importantly, along with the optical activity variation, the pitch of the helices was shortened under illumination. Since different CPL with various CPL energy was absorbed for different chiral composite nanohelices, the distinct CPL energy gave rise to the helices of the hydrogel to expand and contract. Under irradiation, the pure gel can capture photon energy and offer the driving force that stretches the hydrogel. The chiral $FeS_2$ QDs facilitated the photoinduced electron transfer in the cogel system, making the cogel stretch to a larger degree than the pure gel. Therefore, chiral $FeS_2$ QDs displayed intrinsic characteristics of light response over a broad spectral band and the increased surface amplitude of the photogenerated excitons. This study demonstrated CPL-responsive cogel helical nanostructures and showed promising potential applications of amplification and modulation of chirality in dynamic soft matter and functional nanomaterials.

## 3.5 Chiral Nanochains

In recent years, there have been many researches about the chiral nanochains. Selective DNA hybridization allowed for the geometrically precise assembly of chiral isomers. One of the representative studies is shown in Figure 3.27; nanochains of regulated length were observed using the PCR process. The heterochains were assembled by alternating 25 and 10 nm Au NPs (Figure 3.27a,b) [89]. The number of NPs in each chain increased along with increased PCR cycles from 2 to 20. Because of the specific length of the DNA linkers, the gaps between adjacent NPs from the nanochains can be well controlled. Interestingly, the nanochains exhibited intense CD activity and surface-enhanced Raman scattering activity. Moreover, branched chains consisting of different sized Au NPs by PCR were constructed [2]. By regulating the PCR cycles and primer density, two distinct nanostructures (namely, asymmetric and symmetric) were observed (Figure 3.27c). The size of the structures increased from 35 to 600 nm when the number of PCR cycles increased (Figure 3.27d). The asymmetry of their geometry was confirmed by TEM tomography, which resulted in chiroptical activity [2].

Controllable CD activity in the NIR region is significant in the application fields such as target biosensing, chiral catalysis, and optical nanodevices. Therefore, instead of assemblies from Au NPs, linear chains based on the Au NRs by cysteine (Cys) showed intense CD activity (Figure 3.27e,f). Cys molecules were preferentially bound to the ends of Au NRs. Therefore, end-to-end assemblies of Au NRs

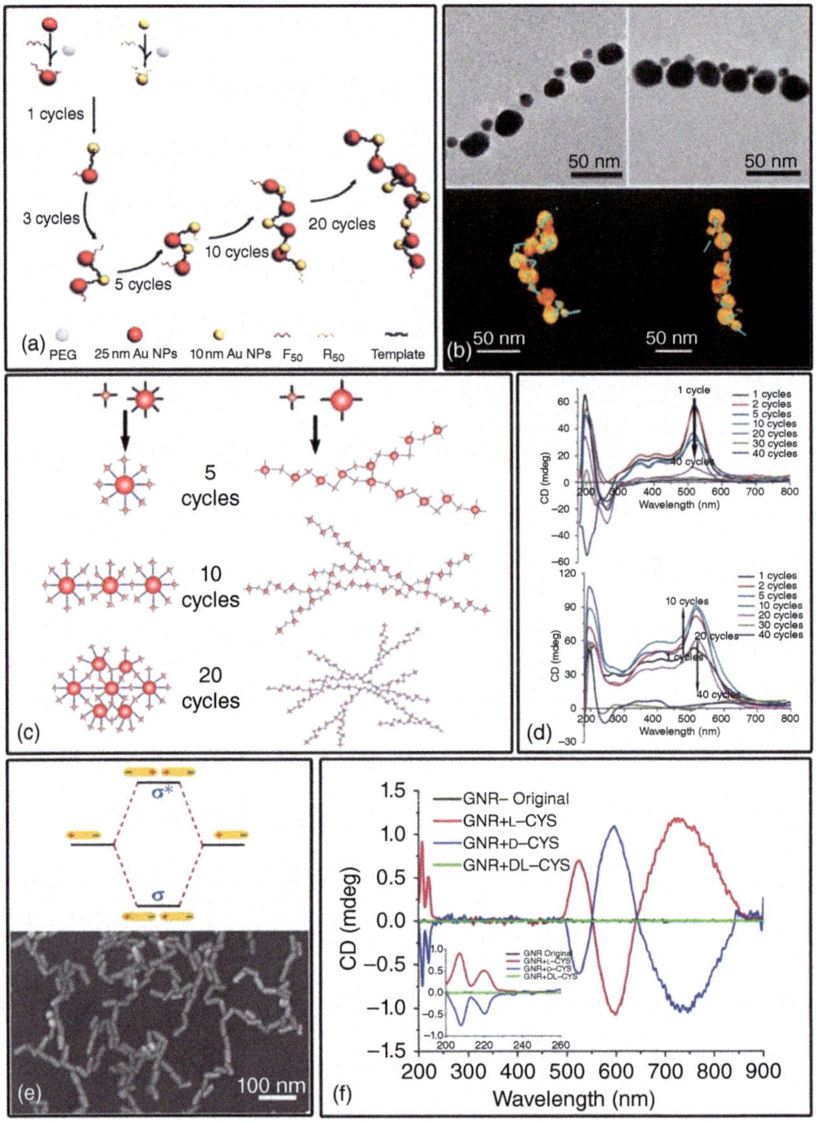


**Figure 3.27** The chiral nanochains and branched heterochains. (a) Scheme of NP chiral chains by PCR. (b) Representative TEM images and the tomography 3D reconstruction of NPs chains with 20 PCR cycles. Source: Zhao et al. [89], figure 1 (p. 642)/with permission of American Chemical Society. (c) Scheme of branched NPs chains via asymmetric and symmetric PCR. (d) The CD spectra for branched chains. Source: [2], scheme 1 (p. 5576)/Royal Society of Chemistry. (e) The nanorod bonding and anti-bonding mode and the SEM image for branched nanorod chains. (f) The CD spectra of original gold nanorods (GNRs) and assembled GNRs by different cysteine. Source: Zhu et al. [90], figure 1 (p. 2327)/with permission of American Chemical Society.

nanochains formed, which was confirmed by the TEM images [90]. CD signal in both the UV and visible–NIR light regions was detected. By changing the aspect ratios of Au NRs in the assembly, the CD signal can be regulated from 550 nm to more than 900 nm (Figure 3.27f) [90]. The bisignate CD wave was attributed to the coupling of chiral molecules with longitudinal surface plasmon resonance of Au NRs at both symmetric (antibonding $\pi^*$-peak) and antisymmetric (bonding $\pi$-peak) hybrid modes (Figure 3.27e). As indicated from the CD spectra of individual Cys-attached Au NRs and Cys-assembled Au NPs, the assembly formation and the shape of building blocks were important for both the intensity and the line shape of the CD band. Notably, the gaps here were shorter than that of many other NR and NP assemblies, leading to potentially comparable plasmon-enhanced CD and polarization rotation from the asymmetric scissor-like geometry of dimers. Therefore, the CD properties of these nanostructures may be considered as complex superposition of chiral geometries at multiple scales. As a whole, one should consider the asymmetries corresponding to the configuration of the assembled oblong NPs, the individual CD activity of NP building blocks, hot-spot enhancement effects, and 3D configurations of the NP chains.

Furthermore, the SBS assemblies of NRs have also been prepared with chiral conformations [1]. The $Hg^+$ ions resulted in the assembly of Au NRs into a ladder nanostructure and produced a varied chiroplasmonic signal in the CD curve (Figure 3.28a–f) [37]. The positive and negative dihedral angles of the adjacent NRs were corresponding to the left rotating enantiomers and right rotating enantiomers, respectively (Figure 3.28g,h). Those dihedral angles for the dimers, trimers, tetramers, and pentamers were about −9.0°, −7.1°, −8.0°, and −7.0°, respectively. A special intense bisignate CD curve of these SBS assemblies, between 500 and 800 nm, was observed (Figure 3.28i,j). The strong intensity of polarization rotation was probably ascribed to the twisted structure, which was confirmed by TEM tomography. The g-factor increased from $1.6\times10^{-3}$ to $2.3\times10^{-3}$ under the cycle numbers of 2–10 (Figure 3.28k,l). The simulated CD wave was in agreement with above experimental data, which further confirmed the results (Figure 3.28m,n).

## 3.6 Core–Satellite Nanoassembly

Cai et al. developed a shell–core gold satellite structure for ochratoxin A (OTA) biosensing [91]. The core–satellite superstructure was composed of a shell NP and gold NPs based on the DNA assembly. The shell NPs were fabricated from an Au–Ag alloy hollow sphere via a galvanic replacement reaction between Ag NPs and $HAuCl_4$. After reduction of $HAuCl_4$, a thin layer of Au shell was coated on the hollow sphere surface. The Au shell NPs were about $24.8\pm4.2$ nm and functionalized with OTA aptamers. Correspondingly, complementary sequences were coupled to the Au NP satellites. The core–satellite structure can be gradually observed after incubation (Figure 3.29). Along with more Au NPs coupled to the shell NPs, increased CD intensity was measured, which was likely due to the long-range coupling between Au NPs and shell NPs. The strong enhanced coupling

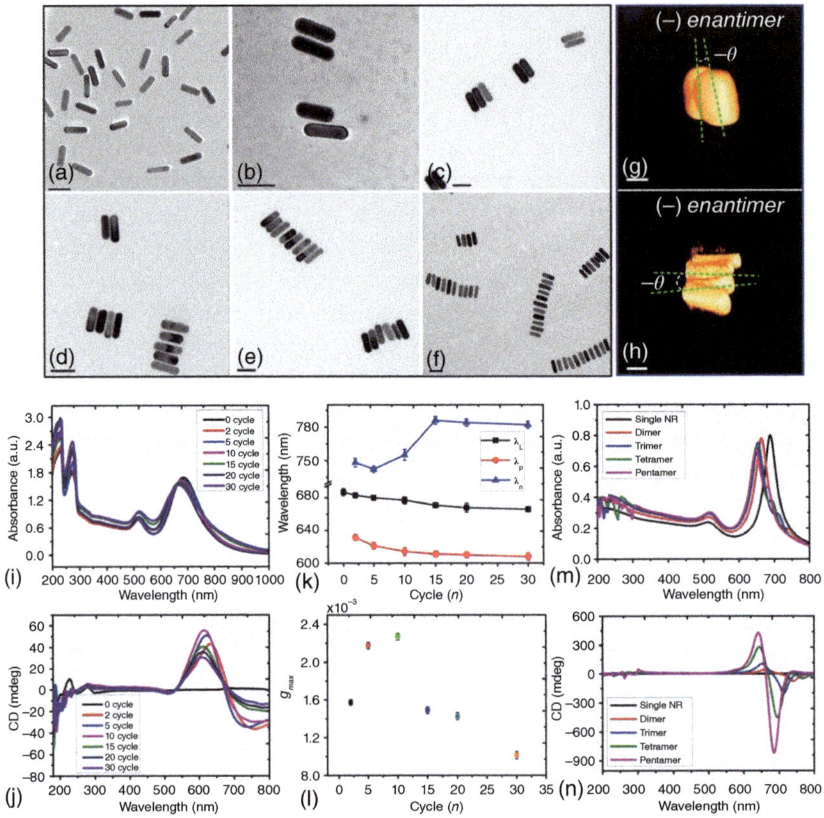


**Figure 3.28** Structure and optical properties of Au NR SBS nanostructures. (a–f) Representative TEM images for SBS nanostructures with different cycles (0, 2, 5, 10, 15 PCR cycles). (g, h) Representative cryo-TEM tomography images for NR SBS assembled trimers and pentamers. (i, j) CD spectra and UV–vis spectra. (k) Profiles of SBS NR nanoassembly spectral features. (l) Maximum chiral anisotropy factor with optimized cycles. (m, n) Simulated CD spectra and absorbance spectra for SBS nanostructures. Source: Ma et al. [37]/with permission of Springer Nature.

was confirmed by the improved electric field of the Au NP–Au NP gap junctions, where the E-field of surrounded Au NPs on shell NPs reached 29 $Vm^{-1}$. Taking the intense CD signal into consideration, such developed assembled nanostructures were applied for mycotoxin detection.

Lanthanide-doped upconversion nanoparticles (UCNPs) can emit higher energy light than the excitation source with longer fluorescence lifetime and narrower band emission [63, 92–99]. The UCNPs are very popular and the typical building blocks for the assemblies because of their good fluorescence activity and good biocompatibility [15, 54, 60, 100–102]. Recently, Hao et al. reported a hybrid nanoassembly made up of an UCNP core and a zeolitic imidazolate framework-8 (ZIF) shell encapsulated with chiral NiSx NPs [92]. For the fabrication of a hydrophilic UCNP@ZIF core–shell structure, the authors first prepared hexagonal plate-like UCNPs. Then UCNPs were deposited with polyvinylpyrrolidone (PVP)

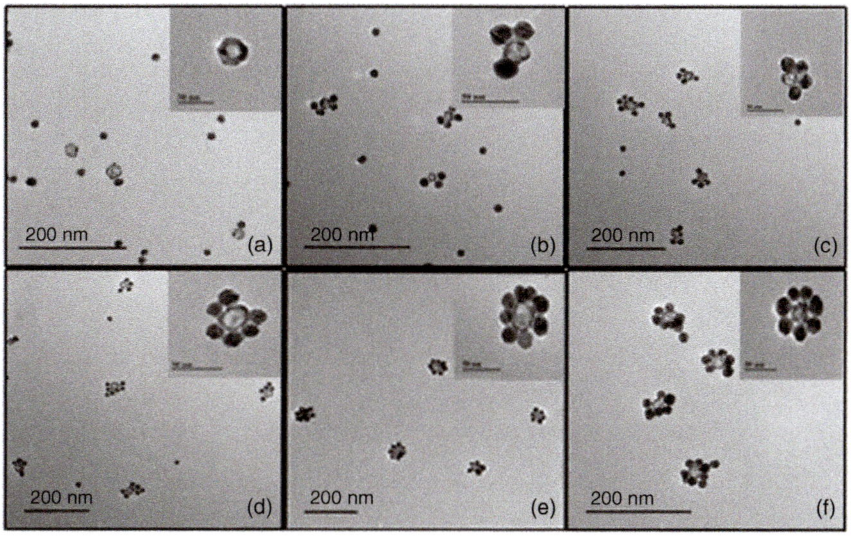


**Figure 3.29** Representative TEM images of shell–core Au satellite nanostructures at different incubation times. (a) 0.5 hours, (b) 3 hours, (c) 6 hours, (d) 9 hours, (e) 12 hours, and (f) 15 hours. Source: Cai et al. [91]/with permission of American Chemical Society.

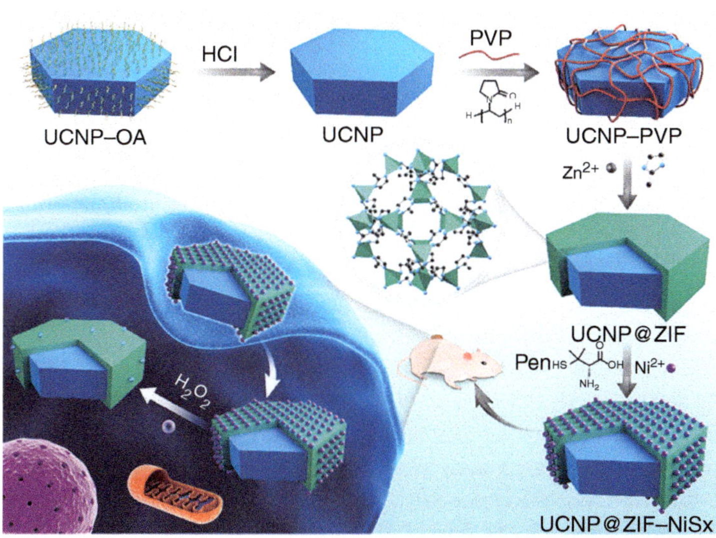


**Figure 3.30** Chiral core–shell upconversion nanoparticle@MOF nanoassemblies. Source: Hao et al. [92], figure 1 (p. 19374)/with permission of American Chemical Society.

and a layer of metal–organic framework (MOF), zeolitic imidazolate framework-8 (ZIF-8) (Figure 3.30). Subsequently, by adding nickel ions (as the nickel source) and chiral L-/D-Pen ligands (as the sulfide source) into the prepared core–shell structures, the UCNP@ZIF-NiSx-L/D nanoassemblies were finally established in the solution system [92].

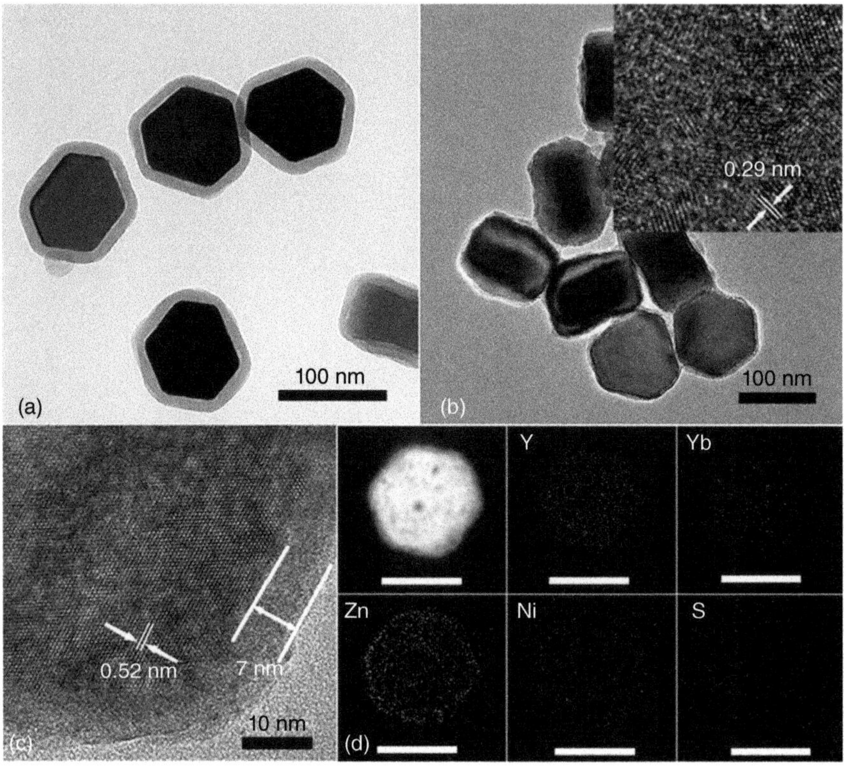


**Figure 3.31** TEM images of the (a) UCNP@ZIF assemblies and (b) UCNP@ZIF-NiSx assemblies. (c) High-resolution TEM image of UCNP@ZIF-NiSx. (d) Dark-field TEM image of one single UCNP@ZIF-NiSx and the corresponding energy-dispersive X-ray spectroscopy element mapping. Source: Hao et al. [92], figure 1 (p. 19374)/with permission of American Chemical Society.

As shown from the results of the TEM and high resolution transmission electron microscope (HRTEM) images, the core–shell geometry was obviously obtained (Figure 3.31a,b). The thickness of ZIF-8 shell was about 7 nm (Figure 3.31c). The results of elemental mapping further confirmed the successful assemblies, where all expected elements were included such as Y, Yb, Zn, Ni, and S in the assemblies (Figure 3.31d). Moreover, porous geometry can be observed from the images. The Brunauer–Emmett–Teller (BET) surface area of the UCNP@ZIF-NiSx was 416.1 $m^2\ g^{-1}$, smaller than that of the UCNP@ZIF (966.8 $m^2\ g^{-1}$).

The UCNP@ZIF-NiSx assemblies exhibited intense CD activity at 440 and 530 nm (Figure 3.32). Similar to the UCNP@ZIF-NiSx-L, CD response was observed for chiral NiSx-L NPs, further confirming the existence of chiral NiSx-L NPs in the assemblies. Mirror image CD lines of assemblies (based on the chiral penicillamine template, L-Pen and D-Pen) were observed. Concurrently, dramatically quenched UCL at 540 nm of UCNP@ZIF-$NiS_x$-L was measured. Based on the CD signal and UCL response, the constructed core–shell nanoassemblies were also employed for

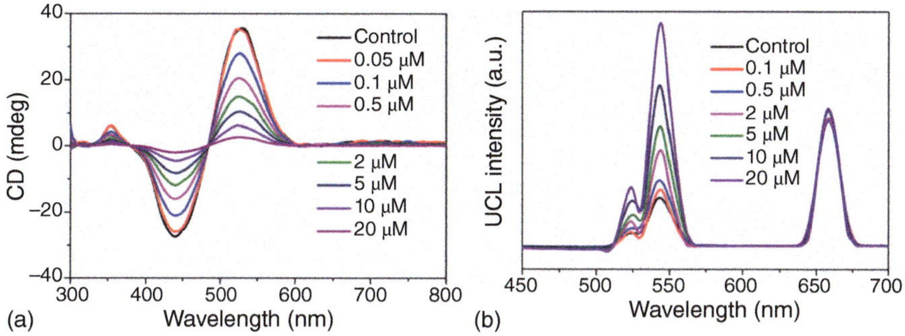


**Figure 3.32** (a) CD and (b) UCL spectra of the chiral assemblies responded to different concentrations of $H_2O_2$ in vitro. Source: Hao et al. [92]/American Chemical Society.

the subsequent ultrasensitive and selective detection of reactive oxygen species (ROS) in live cells and in vivo.

Besides the above typical core–satellite nanoassemblies, Meng et al. reported two-layer core–satellite gold structures [103]. The double-layer core–satellite architecture was made up of 30 nm Au NPs as the core, 5 nm Au NPs as the first satellite layer, and 10 nm Au NPs as the second satellite layer ($C_{30}S_5S_{10}$ NS). Importantly, Y-DNA backbones were first designed, which included three nucleotide chains of 5′-thiolated target recognition DNA ($DNA_{TR}$) and two complementary 5′-thiolated DNAs. The 5 and 10 nm Au NPs were assembled as the first and second satellite layers, respectively (Figure 3.33).

As shown in Figure 3.34a–f, the two-layer satellite gradually assembled from an incomplete to a complete structure over time. The $C_{30}S_5S_{10}$ NS became the dominant species at 9 hours, which lasted until 15 hours, enabling the control of the structural dynamics solely by time. As time proceeded with more yield of $C_{30}S_5S_{10}$ NS, intense CD and SERS signals were observed, where the maximum signal was observed at 12 hours (Figure 3.34g,h). The observed CD signal showed a positive band at 536 nm, in well agreement with the characteristic absorption peak at 536 nm. This strong collective interaction gave rise to a redshift trend of the chiroptical spectra and also resulted in increased CD signals. In the case of shorter and longer Y-DNA, CD signal was relatively reduced, indicating better chirality enhancement with a suitable gap distance.

Generally, the asymmetric collective interactions between the NPs produced the CD response of satellite superstructures. The Au NP constituents were not perfectly spherical and were slightly elongated. Lower potential energy drives the two NPs to form a scissor-like cross-configuration in space among the basic dimeric unit. Multiple collectively interacting NP dimers with similar geometries (dihedral angle) formed a more complex chiral superstructure, under the balanced repulsive and attractive forces between the core and satellite NPs. These asymmetric collective interactions produced strong CD activities.

Recently, Qu et al. reported a dual core–satellite NP probes with allosteric effect for the stimulation of the neural stem cell (NSC) differentiation under CPL [104]. The NP probes were chiroptically active reconfigurable assemblies. They were composed

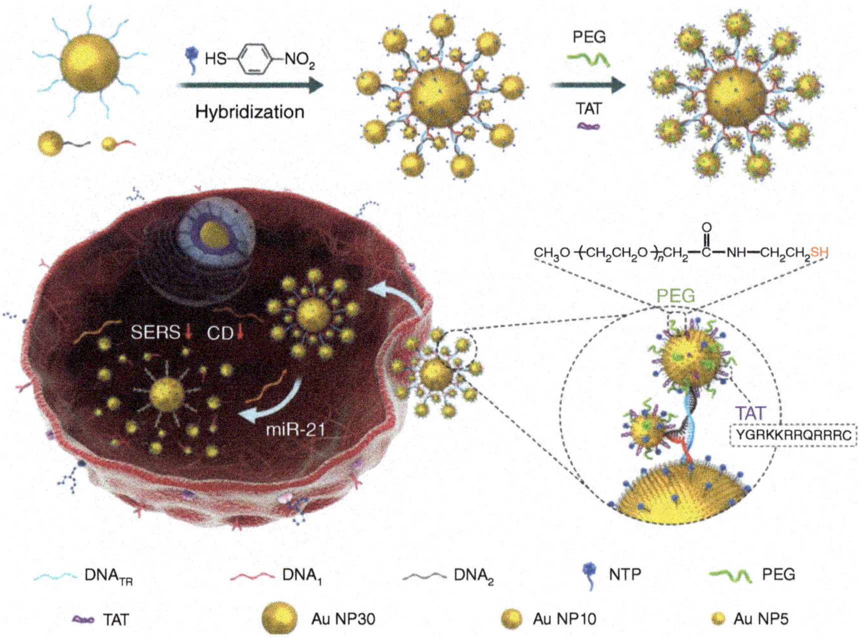


**Figure 3.33** Schemes of Y-DNA-driven construction of Au NP chiral $C_{30}S_5S_{10}$ NS; the $C_{30}S_5S_{10}$ NSs with PEG and TAT modifications was used for ultrasensitive detection of miRNA in living cells. Source: Meng et al. [103]/John Wiley & Sons.

of three different sized Au NPs (30, 20, and 5 nm). The Au NPs were coated with different chiral Cys, where D-Cys for 30 nm Au NPs and L-Cys for 20 nm Au NPs. Then, these two Au NPs were functionalized with different ssDNA to bridge together as the branched NP assemblies (Figure 3.35a). Meanwhile, 5 nm NPs were coated with different ssDNA, partly complementary with 30 and 20 nm Au NPs. The functionalized 5 nm NPs were dangling and were connected to either 30 or 20 nm NPs, forming assemblies denoted as $C_{30}S_5$-$C_{20}$ and $C_{30}$-$C_{20}S_5$, respectively. The chiral Cys coated on the 30 and 20 nm Au NPs determined the geometry and CD activity of the assemblies. The geometry and its transformation of the assemblies were further confirmed by tomographic reconstructions of cryo-TEM images (Figure 3.35b).

The DNA branches connecting 5 and 30 nm NPs were a stem-loop sequences (21 bp), which were complementary to Fox3 mRNA expressed during the NSC differentiation into neurons. Therefore, these assemblies enabled the reconfiguration inside the cells to entangle with nanofilaments of cytoskeleton. The chiroptical activity of the $C_{30}S_5$-$C_{20}$ assemblies displayed a negative peak at 524 nm ($-63.5 \pm 2.7$ mdeg), while the $C_{30}$-$C_{20}S_5$ assemblies showed a positive CD signal of $37.4 \pm 1.4$ mdeg (Figure 3.36). The CD activity of the assemblies in the visible range was attributed to the mirror asymmetry of the entire superstructure and a minor contribution of the local plasmonic enhancement of the biomolecular chirality.

Besides Au NPs, other nanomaterials are used as the building blocks for core–satellite assemblies. For example, Zhao et al. fabricated Au-QDs core–satellite

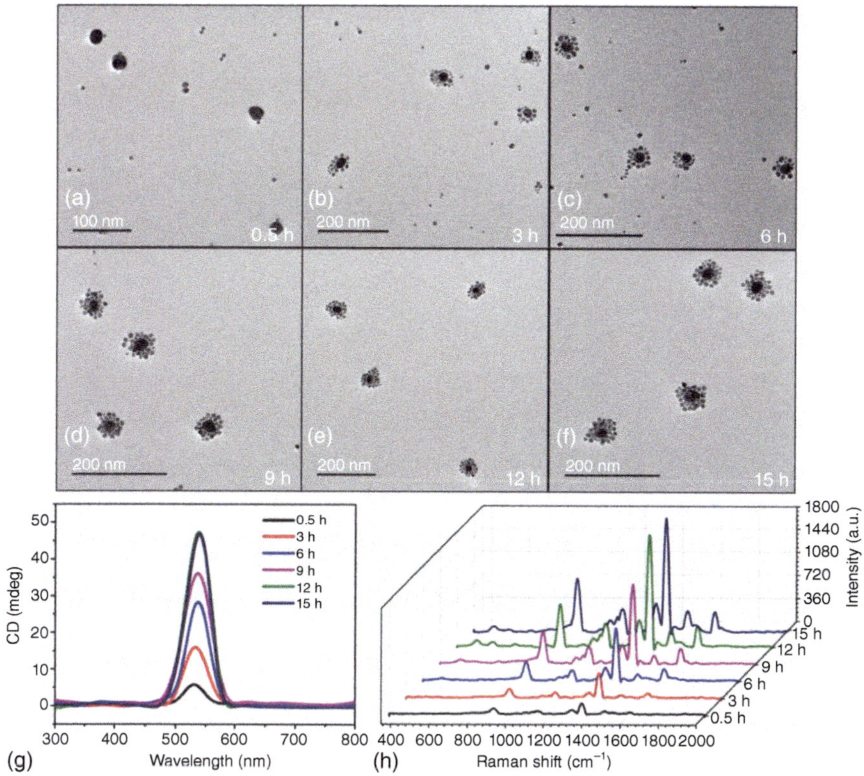


**Figure 3.34** (a–f) TEM images of dynamic satellite assemblies. (g) CD spectra and (h) SERS spectra of time-dependent dynamic assembly of satellite superstructures. Source: Meng et al. [103]/with permission of Wiley.

nanoassemblies for lighting up microRNA in vitro and in vivo [105]. The core–satellite nanostructure was fabricated using partly complementary-DNA-functionalized Au NPs of $35 \pm 2$ nm and QDs of $6 \pm 1$ nm. The hydrodynamic diameter of the assemblies was $78 \pm 3$ nm. The core–satellite assembly exhibited a slight redshift. Meanwhile, a significant quenching effect between the Au NP and QD was observed, and the quenching efficiency was ~93.8%. With the target miR, the hairpin sequence of the core–satellite assembly was dispersed by DNA hybridization, leading to restoration of fluorescence.

Some other types of chiral core–satellite nanostructures can also be established by replacing the Au NP core with other geometry of Au nanomaterials. Au NR assemblies have exhibited unexpected optical characteristics of chirality and Raman spectra. Recently, Xu et al. reported a novel plasmonic core–satellite nanostructure assemblies with Au NR core surrounded by Au NPs [106]. DNA-1 with a sequence of TAGGAATAGTTATAA-$A_{10}C_{(6)}$SH was deposited on the surface of Au NRs with an aspect ratio of 3.5 (14 nm × 48 nm). Simultaneously, Au NPs (15 nm) were functionalized with the complementary oligonucleotides DNA-2 (TTATAACTATTCCTA-$A_{10}C_{(6)}$SH). After mixing the two probes (Au NR–DNA1

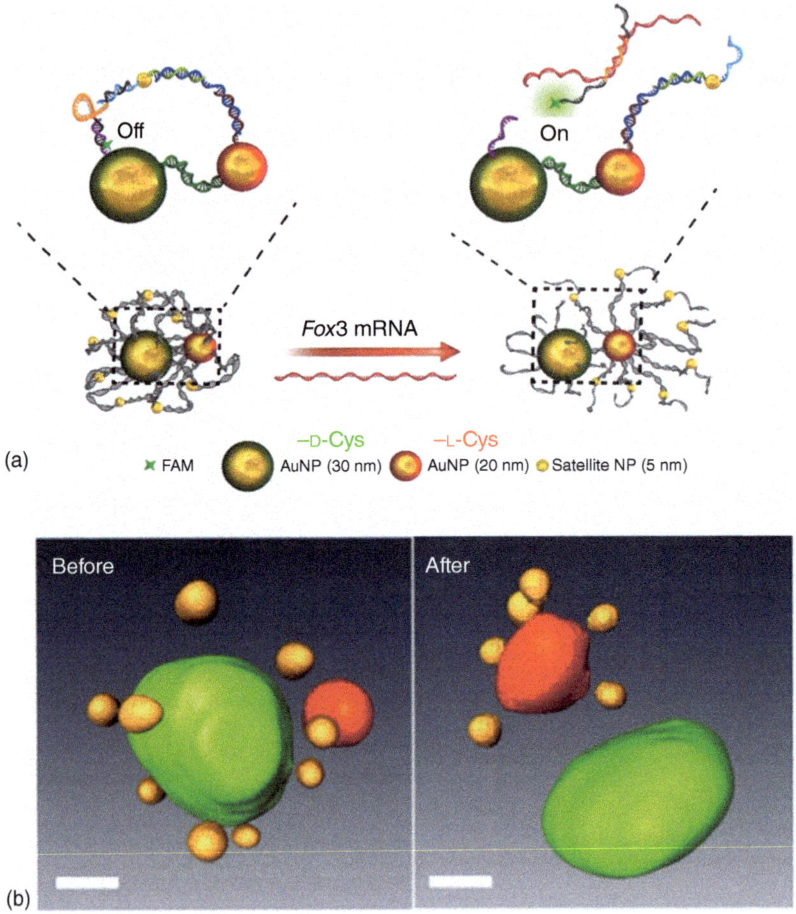


**Figure 3.35** Dual core–satellite Au NP assemblies with allosteric effect. (a) Schemes of configuration transformation of $C_{30(D)}S_5$-$C_{20(L)}$ Au assemblies induced by special mRNA. (b) 3D tomography of $C_{30(D)}S_5$-$C_{20(L)}$ before and after the transformation. Source: Qu et al. [104]/Springer Nature.

and Au NP–DNA2) together at room temperature for 12 hours, Au NR-Au NP core–satellite nanoassemblies can be observed through TEM images (Figure 3.37). The authors also used the electron tomographic analysis for further confirmation of the Au NPs organization on the assemblies. Tilted projections were aligned by using a 3D reconstruction of the nanoassemblies. By collecting a series of tilted images from −50° to +50° with 2° tilt intervals, Au NRs were spatially arranged around the Au NPs in 3D space, and it demonstrated the formation of core–satellite nanoassemblies.

In the hybridization process, Xu and coauthors measured the optical properties of the assembled core–satellites structures at various hybridization periods. As shown in Figure 3.38, the longitudinal absorption band of Au NRs shifted to 714 nm, and

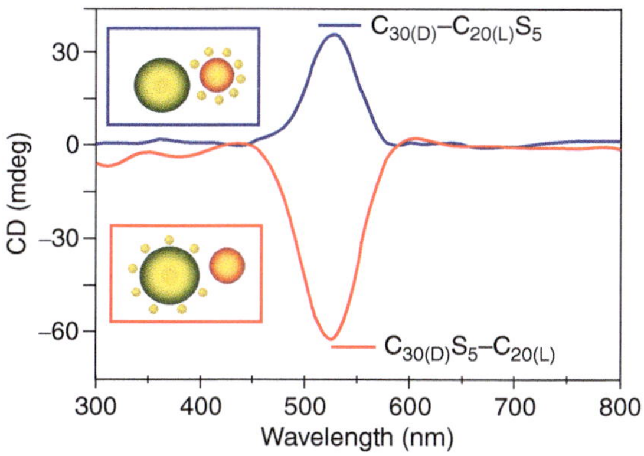


**Figure 3.36** CD spectra of $C_{30}S_5$-$C_{20}$ and $C_{30}$-$C_{20}S$ assemblies. Source: Qu et al. [104]/Springer Nature.

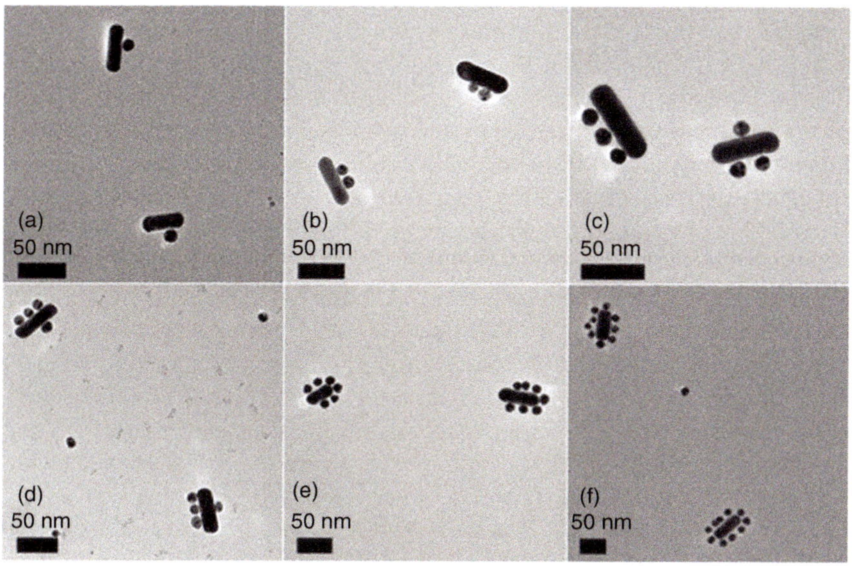


**Figure 3.37** Representative TEM images of nanoassemblies for various hybridization periods. (a) 5 minutes, (b) 30 minutes, (c) 1 hour, (d) 3 hours, (e) 6 hours, and (f) 12 hours. Source: Xu et al. [106]/with permission of American Chemical Society.

the superposition band shifted from 515 to 523 nm. These spectral position changes were typical for multiple plasmon resonance modes, resulting in a split in the energy level of charge oscillations. It was caused by both symmetric and antisymmetric hybridization between the Au NRs and the surrounding Au NPs in core–satellites assemblies. In the case of the CD activity, two new peaks at 520 and 715 nm were observed in the visible range from 400 to 800 nm. The CD band at 715 nm was ascribed to the longitudinal localized surface plasmon resonance (LSPR) of Au NRs. The CD at 520 nm was assigned to the coupling between transverse LSPR of the

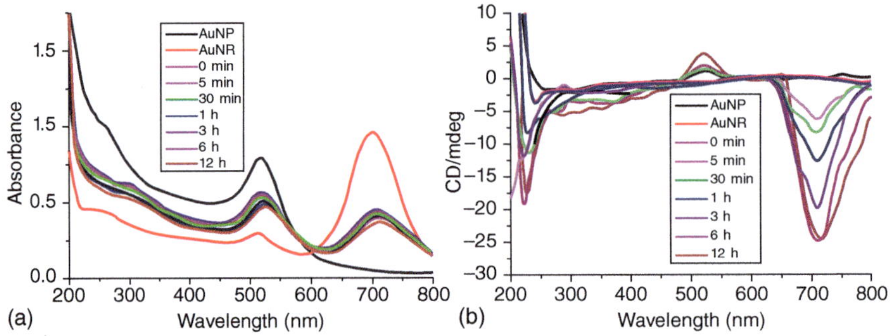


**Figure 3.38** UV–vis (a) and the corresponding CD spectra (b) of the core–satellite assemblies for different hybridization periods. Source: Xu et al. [106]/American Chemical Society.

Au NRs and the surrounding Au NPs. Importantly, the close proximity of satellites induced the coupling of their plasmon oscillations.

The author also investigated their chiral origin at the Au plasmon frequency. They ruled out the chiral NCs origin, DNA chiral arrangement, or DNA chirality transfer to the achiral metal nanoclusters. They claimed that chirality of the assemblies should have been originated from the plasmon-enhanced CD of chiral DNAs located in the hot-spot region, as well as the collective plasmon interactions between the building blocks arranged in the 3D superstructures.

Moreover, Hao et al. reported Au@AgAu yolk–shell NRs (YSNRs) as the core to fabricate hierarchical core–satellite nanostructures [107]. Chiral D/L-penicillamine (D/L-Pen) coated Au NRs as the core to establish the dispersed chiral YSNRs. The YSNRs, hollow structures with a solid Au NR core, showed intense plasmon CD signals in the band of 600–800 nm. By tuning the amount of D/L-Pen, nanogap size, and the aspect ratio (AR) of the Au NRs, their intensity and wavelength of the plasmonic circular dichroism (PCD) signal were manipulated. The prepared YSNRs were then assembled into YSNR@NP core-satellite (CS) structure. And they showed a giant optical effect. Simply, the Au@AgAu YSNR was fabricated by coating a silver shell onto the Au NR core under the reduction of $AgNO_3$ by L-ascorbic acid. The silver shell was a sacrifice layer and was then etched with $HAuCl_4$ in a galvanic replacement reaction to form interior nanogaps (Figure 3.39).

The D/L-Pen coated on the surface of the Au NR was critical for the chiroptical activity of the superstructure. Therefore, the authors optimized the concentration of D/L-Pen. Before the formation of silver shell, they decorated with different concentrations of D/L-Pen ($10–50 \times 10^{-6}$ M) on the surface of the Au NRs by adding $AgNO_3$. The silver shell was 4.5 nm with an average thickness. After the galvanic replacement reaction with $HAuCl_4$, a small nanogap (1.78 nm) formed in all the YSNR nanostructures (Figure 3.40a,b). The Au@Ag NRs exhibited an LSPR at 499 nm and a longitudinal LSPR at 639 nm. After the galvanic reaction, the YSNRs displayed a redshift in the transverse LSPR to 513 and 70 nm redshift of the longitudinal LSPR. The absorption of YSNRs displayed no observed differences. Notably, the PCD responses of YSNR-D/L displayed two distinct features. One was a bisignate spectral line shape

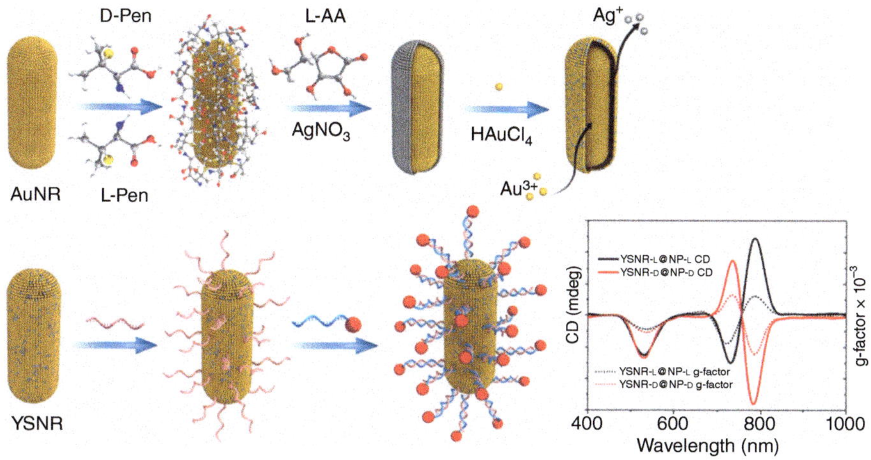


**Figure 3.39** Schemes of the prepared chiral Au@AgAu yolk-shell nanorods (YSNRs) and DNA-triggered YSNR@NP core–satellite superstructure self-assembly. Source: Hao et al. [107]/John Wiley & Sons.

with the zero-crossing point located at 750 nm. The other was a good mirror image of the PCD signals (Figure 3.40c).

As L-Pen increased to $50 \times 10^{-6}$ M, the PCD intensity showed no observed enhancement, due to the saturation of Au NR surface with L-Pen molecules. The g-factor value for YSNR-D/L synthesized in the presence of $50 \times 10^{-6}$ M D/L-Pen reached 0.0069 (Figure 3.40c). The maximum PCD signal of YSNR-L increased from 40 to 85 mdeg as L-Pen ranged from 10 to $40 \times 10^{-6}$ M, (Figure 3.40d,e). With no additional D/L-Pen, no PCD signal was observed. Thus, the intense PCD activity of YSNR-D/L was closely corresponding to the presence of D/L-Pen molecules between the Au NR core and the Au–Ag alloy shell. The authors concluded that the PCD response of YSNR-D/L was induced by the electromagnetic coupling between the D/L-Pen layer and the YSNR nanostructure.

The author examined the relationship between the interior gap size and the PCD response of the YSNRs. When the $AgNO_3$ concentration was $2 \times 10^{-3}$ M, YSNR-D/L showed the highest PCD response (±110 mdeg at 770 nm), whereas the PCD response decreased in the case of the $4 \times 10^{-3}$ M $AgNO_3$ concentration. It should be ascribed to the larger nanogap and the incomplete alloy shell at above $AgNO_3$ concentration. For $1 \times 10^{-3}$ M $AgNO_3$, the PCD signal of YSNR-D/L at 780 nm was ±76 mdeg, attributed to the fewer interior nanogaps with less hot-spot effect [108]. The g-factor for YSNR-D/L prepared with $4 \times 10^{-3}$ M $AgNO_3$ in the presence of $40 \times 10^{-6}$ M D/L-Pen was 0.005.

They further investigated the influence of shape anisotropy on the PCD response of the YSNRs using Au NRs with different ARs. As shown in Figure 3.41d–f, the chirality spectra showed that the bisignate chiral signal exhibited a redshift and intensity as the AR of Au NR increased. They caused promotion of the chiral current inside the YSNRs by the dipole of chiral Pens. The PCD signal for YSNR-2.1/2.6/4.1-L

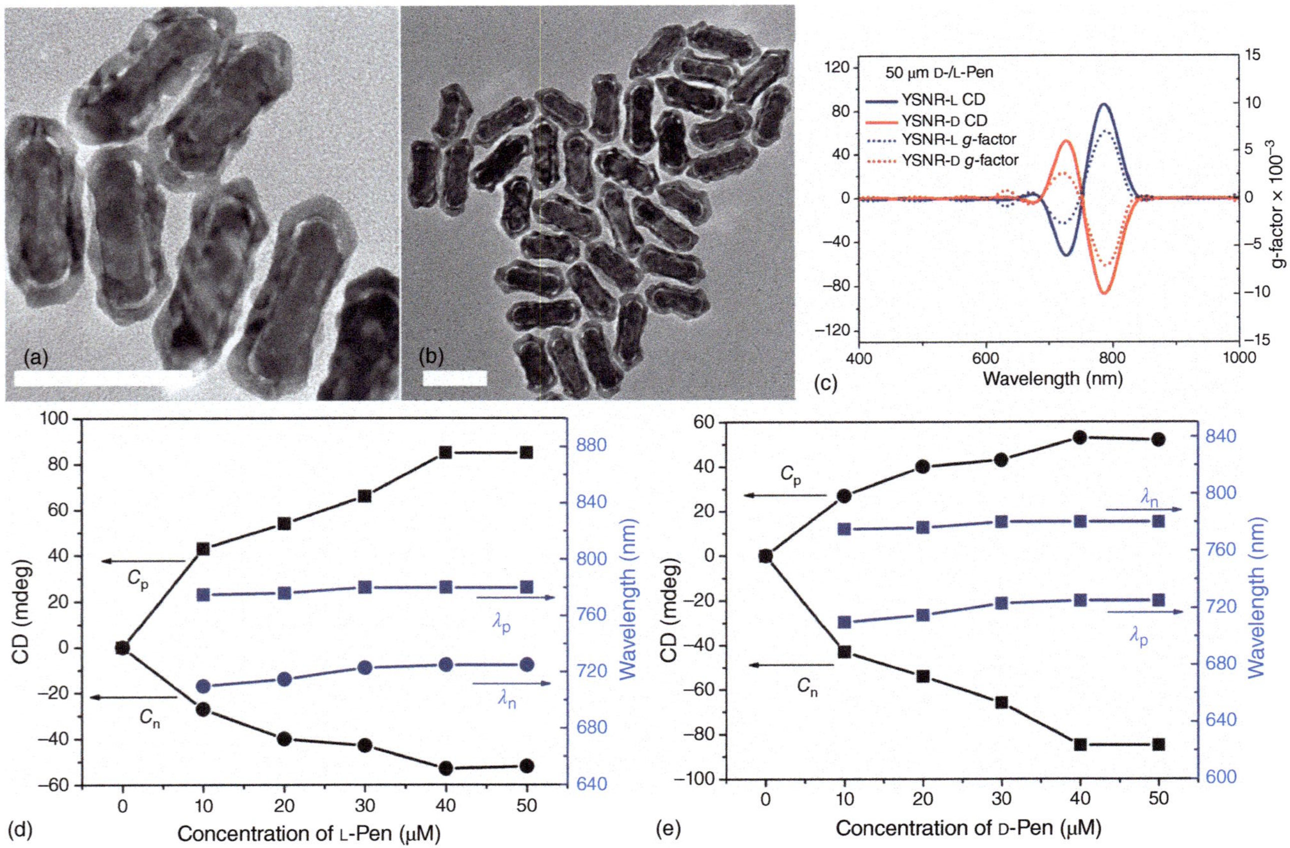


**Figure 3.40** (a, b) Typical TEM image of YSNRs; (c–e) corresponding CD spectra (solid line) and *g*-factor of YSNR-D/L. Source: Hao et al. [107]/with permission of Wiley.

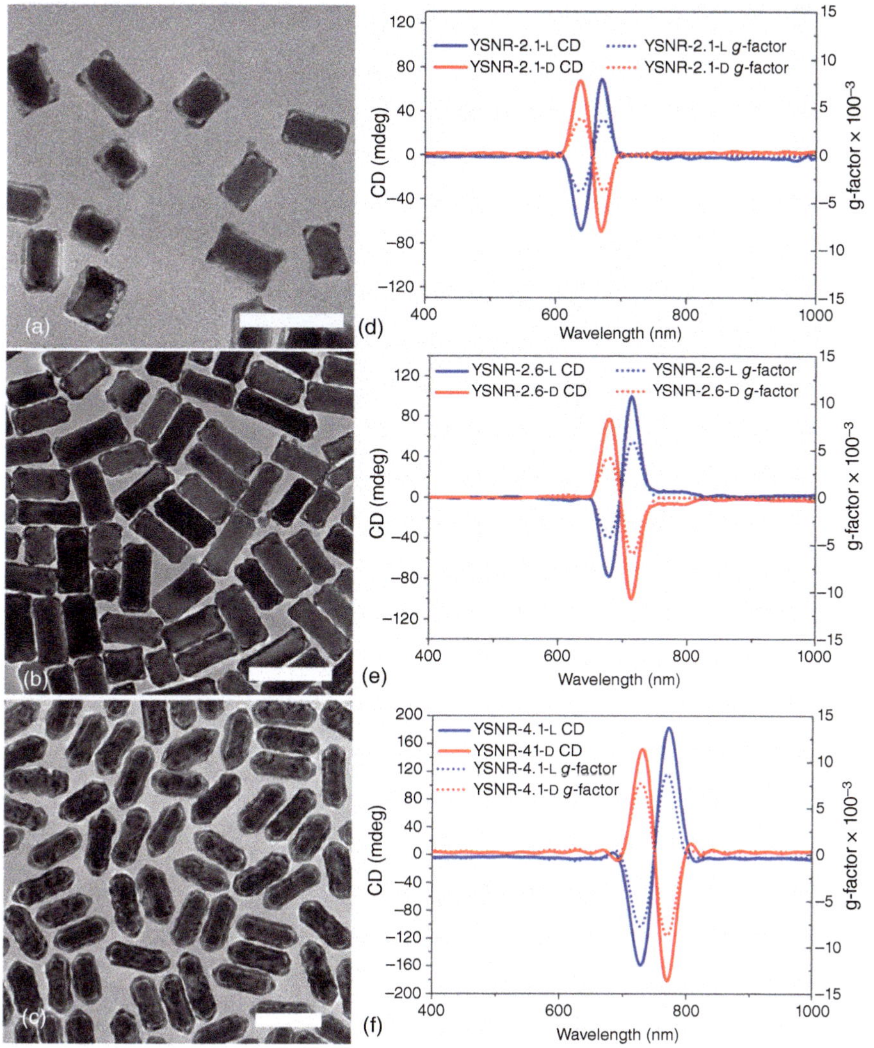


**Figure 3.41** Typical TEM image of chiral YSNRs prepared using Au NRs with different aspect ratios in the presence of $40 \times 10^{-6}$ M D-Pen and $2 \times 10^{-3}$ M $AgNO_3$, (a) YSNR-2.1-L, (b) YSNR-2.6-L, (c) YSNR-4.1-L, and the corresponding CD spectra and *g*-factor of (d) YSNR-2.1-D/L, (e) YSNR-2.6-D/L, (f) YSNR-4.1-D/L. Source: Hao et al. [107]/with permission of Wiley.

was 68 mdeg at 670 nm, 98 mdeg at 710 nm, and 180 mdeg at 770 nm. The *g*-factor for YSNR-2.1/2.6/4.1-D/L in the presence of $2 \times 10^{-3}$ M $AgNO_3$ and $40 \times 10^{-6}$ M D/L-Pen was 0.0038, 0.0061, and 0.009, respectively (Figure 3.41). Thus, various YSNRs were achieved and manipulated the intensity and wavelength of their PCD signals by tuning the AR of the Au NRs. Compared with the original chiral Pens, the chiral response in the UV range was enhanced, due to the electromagnetic field generated by the plasmonic YSNR nanostructure [109].

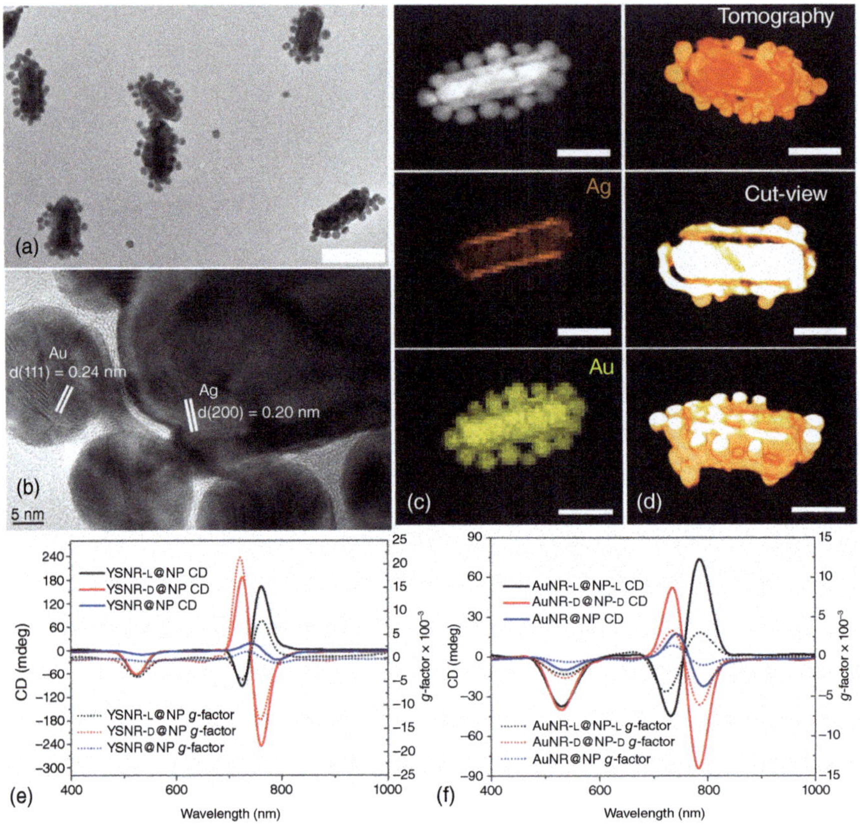


**Figure 3.42** (a) Typical TEM image, (b) HRTEM image, and (c) scanning tunneling microscope (STEM)-EDS elemental mapping images of the YSNR-L@NP core–satellite (CS) superstructure, and (d) 3D electron tomography reconstruction of the YSNR-L@NP CS, and the corresponding CD spectra (solid line) and *g*-factor (dotted line) of (e) YSNR@NP and YSNR-L/D@NP CS assemblies, (F) Au NR@NP, Au NR-L@NP-L, and Au NR-D@NP-D CS assemblies. The scale bars in (a) and (b) are 100 and 50 nm, respectively. Source: Hao et al. [107]/with permission of Wiley.

Two complementary ssDNA were firstly coated on YSNR-4.1-D/L and Au NPs. The core–satellite nanoassemblies were then achieved by DNA hybridization, which were designated as YSNR-D/L@NP CS (Figure 3.42). The corresponding energy dispersive X-ray spectroscopy (EDX) spectroscopy mapping demonstrated different compositional distributions of Au and Ag in the CS assemblies. An electron tomographic method was used to confirm the Au NPs organization in the CS assemblies. Figure 3.42d shows that Au NPs were spatially arranged around the YSNRs in 3D space. There was a clear nanogap between the Au NR core and the Au–Ag alloy shell in a cutaway view of the electron tomographic image. As shown in Figure 3.42e–f, a new band at 520 nm was observed in the visible range of 400–800 nm, which was due to the coupling between the transverse LSPR of YSNR-D/L and the surrounding Au NPs.

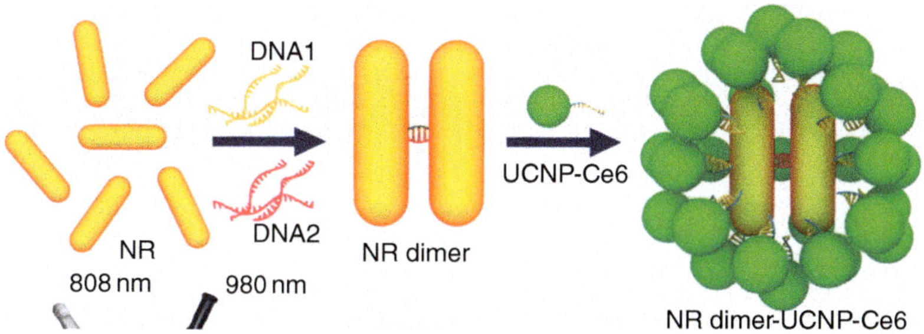


**Figure 3.43** Schemes of DNA-based NR dimer and UCNP core–satellite assembly Source: Sun et al. [110]/John Wiley & Sons.

More recently, Sun and et al. fabricated hierarchical core–satellite nanostructures with plasmonic Au NR dimers as the core and upconversion nanoparticles as the satellite (Figure 3.43) [110]. Firstly, Au NR with an aspect ratio of 4 was assembled into NR dimers through hybridization of two complementary DNA sequences (Figure 3.44a), followed by an amphiphilic diblock copolymer polystyrene-*block*-poly (acrylic acid) (PS-*b*-PAA) layer encapsulation. Then, by the 1-(3-dimethylaminopropyl)-3-ethylcarbodiimide hydrochloride technique, amino-terminated DNA was coupled to the carboxy group of the PS-*b*-PAA shell. Triggered by the "click" reaction, Yb, Er-doped $NaGdF_4$ UCNPs ($20 \pm 2$ nm) was coated with a mixture of thiolated oligonucleotides and heterobifunctional poly(ethylene glycol) molecules. Then they were grafted onto the NR dimer along with a classic photodynamic therapy (PDT) agent (Ce6) by DNA hybridization, and finally formed the NR dimer-UCNP core–satellite structure (Figure 3.44b). As a comparison, the authors also constructed single NR-UCNP core-satellite structure (Figure 3.44c).

Importantly, the layer of the dense polymer not only ensured the stability of the dimer assembly but also prevented UCNPs luminance quenching from the plasmonic metal. Under irradiation by 980 nm laser, Ce6 was excited through energy transfer from the UCNPs, where emission wavelength exactly matched the absorption peak of Ce6 (Figure 3.44d). The ultraviolet/visible (UV–vis) spectra of the NR dimer-UCNP core–satellite nanoassemblies showed a blueshift peak at 808 nm, demonstrating the suitability of the NR dimer-UCNP for PTT (Figure 3.44e). The emerging characteristic bands at 404 and 660 nm for NR dimer-UCNP-Ce6 indicated successful Ce6 loading. Due to the energy transfer between UCNP and Ce6, the UCL was significantly lower upon irradiation at 980 nm (Figure 3.44f). The UCL intensity was partially quenched by the plasmonic metal surface, while without the PS-*b*-PAA coating they showed no obvious signal.

Similarly, Gao et al. constructed a handedness-dependent CPL-activated chiral satellite assemblies from DNAzymes and spiny platinum modified with Au NRs and upconversion nanoparticles [17]. The chiral satellite probes were made up of spiny platinum coated with Au NR (Au NR@Pt) dimers (NR dimers) and upconversion nanoparticles (UCNPs) with intense absorption in the range of 600–1000 nm. An obvious CD signal was observed from the satellite assembly within 12 hours

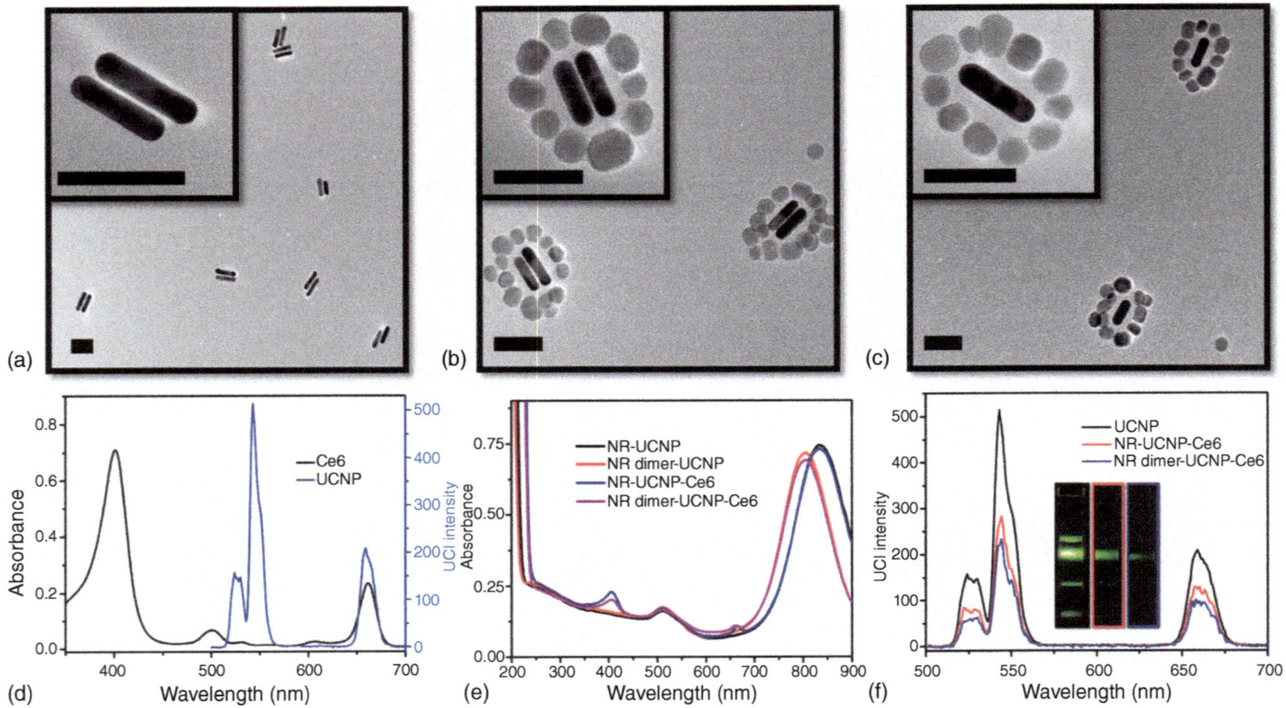


**Figure 3.44** NR dimer and UCNP self-assembly and characterization. TEM of self-assembled (a) NR dimer, (b) NR dimer-UCNP-Ce6, and (c) NR-UCNP-Ce6. (d) UCL intensity of UCNP and UV–vis absorption of Ce6. (e) UV–vis absorption spectra of nanoassemblies before and after Ce6 coupling. (f) UCL intensity variations before and after nanoassembly. Source: Sun et al. [110]/with permission of Wiley.

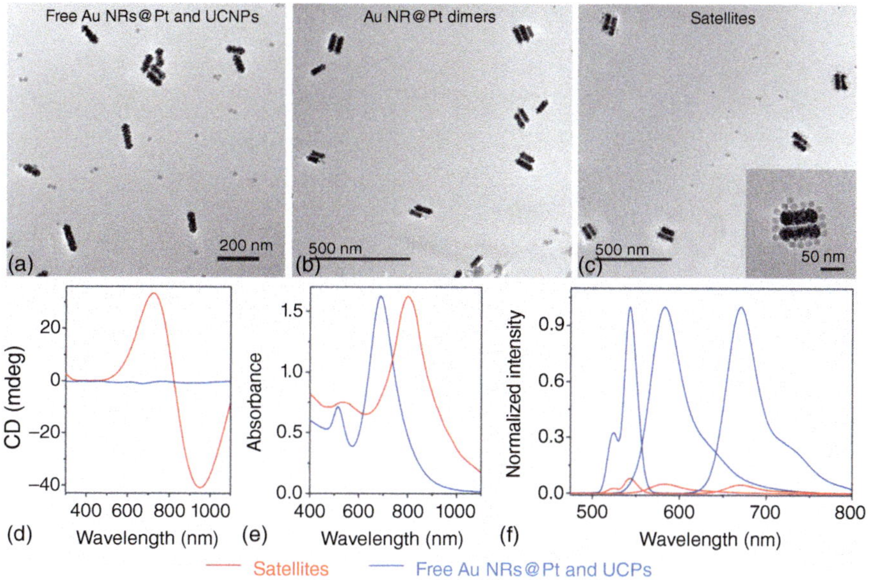


**Figure 3.45** Characterization of Au NR@Pt dimer-UCNP satellite nanoassemblies. TEM images of (a) Pt-decorated Au NRs mixed with UCNPs, (b) Pt-decorated Au NR dimers, and (c) satellite assemblies made up of Au NR@Pt dimers and UCNPs in buffer. (d) CD, (e) UV/vis absorption, and (f) luminescence spectra of dimer-UCNP satellite assemblies and nonassemblies. Source: Gao et al. [17]/with permission of Wiley.

(Figure 3.45). The Au absorbance band of the satellite probe was shifted. The chiral probes, in coordination with their corresponding divalent metal ions under 980 nm LCP light illumination, served as an in situ confocal bioimaging platform for the quantitation of the given intracellular metal ions.

## 3.7 Nanoplatelets and Chiral Films

Recently, a series of well-designed hierarchically programmed chiral nanoassemblies were reported (Figure 3.46) [111]. By orderly stacking of chiral amino-acid-stabilized nanoplatelets, twisted spike-like geometries were established. Particularly, the electrostatic repulsion among the particles decreased efficiently by the cetyl trimethyl ammonium bromide (CTAB) surfactant incorporation, and their negative charge changed to a positive charge. It allowed their staggered pattern to be edge to edge instead of face to face. Furthermore, the complex chiral nanoarchitecture formation was determined by the competitive interaction of anisotropic electrostatic and elastic forces. Therefore, enhanced optical activity and hierarchical structure were observed; their superior physicochemical properties were beneficial to subsequent biocatalysis. Thus, chiral assemblies may allow for more biotransformation processes based on the flexible controllable methods and various conformations.

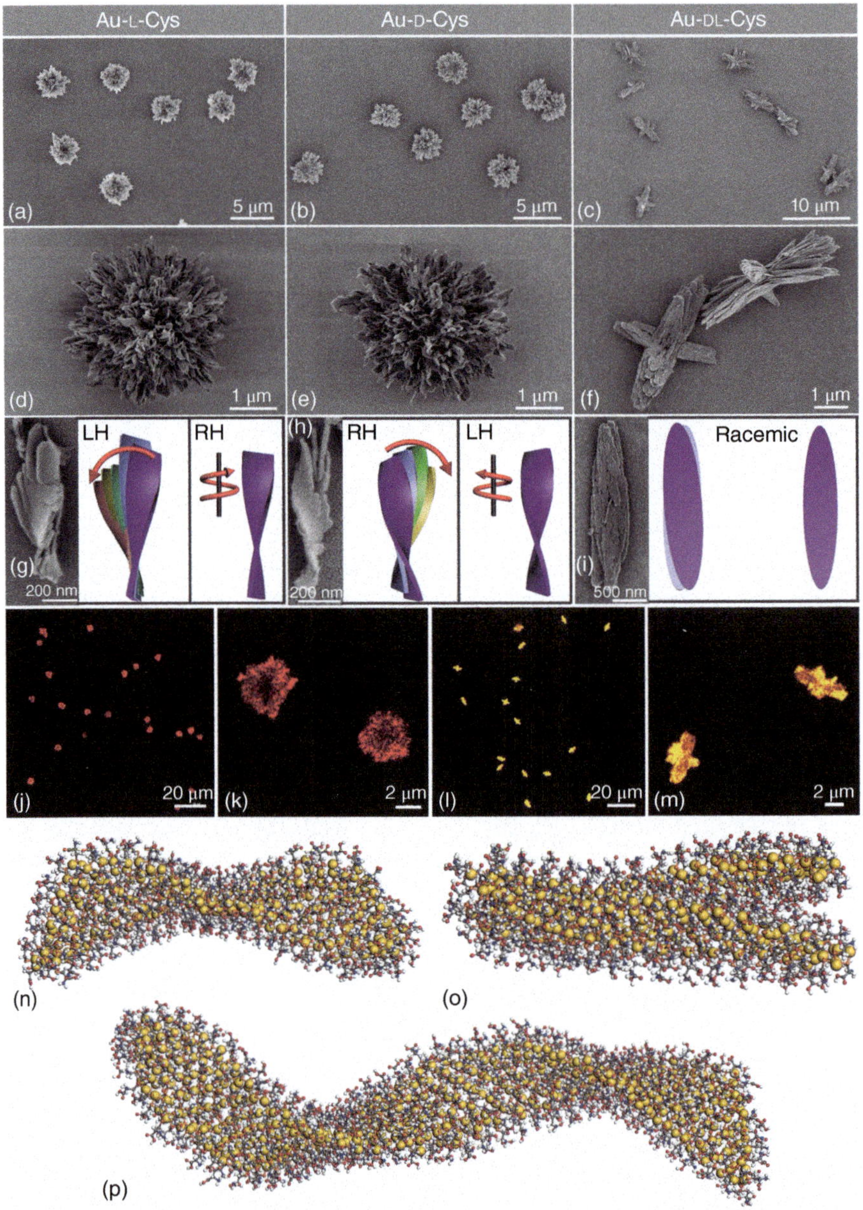


**Figure 3.46** Gold thiolate hierarchically self-organized particles (HOPs) with cysteine surface ligands. (a–c) Representative SEM images of Au-L-Cys and Au-D-Cys coccolith-like particles (CLIPs). (d–f) Enlarged SEM images of Au-L-Cys and Au-D-Cys CLIPs (g–i) SEM images and relevant schemes of segments of Au-L-Cys (g), Au-D-Cys (h), and Au-DL-Cys (I). (j–m) Confocal microscopy images of Au-L-Cys CLIPs (n–p). Atomistic molecular dynamics (MD) simulation of twisted conformation of Au-L-Cys single sheets with different lengths [(n) and (p)] and their stacks (o). The X-ray diffraction (XRD), X-ray photoelectron spectroscopy (XPS), and thermogravimetry experiment results for distinct HOPs. Source: Jiang et al. [111]/with permission of American Chemical Society.

The composition and their configuration of molecules influence electronic properties, particularly for the number, type, and spatial arrangement of the constituent atoms. Both properties determine the mix and hybridization of the electronic wave functions and lead to the production of new collective orbitals. The formation of the shared orbitals can give rise to the charge redistribution. Thus, it will produce intense interactions and binding between the atoms in the molecules. Importantly, resembling the organic molecules, hybridization will be transferred to the closely packed metallic NPs nanoassemblies, which can be named as "plasmonic molecules." Associating with the plasmon resonances, the strongly enhanced local electric fields can work as the electron wave functions. Therefore, by simply altering the size of the constituents and appropriate selected metal, clusters of individual NPs make it possible that the chiroptical bands can be transferred to the expected large wavelength range from the visible and infrared regions.

Chiral nanofilms can be simply found in nature, especially in biosystems, and exhibit unique optical properties and structure–biological function relationships. There are also some studies about the preparation of chiral film assemblies. One of the representative studies was reported by Professor Che's group [112]. They fabricated silver films with hierarchical chirality (Figure 3.47). Three levels of chirality were identified from the whole film assemblies. The primary chiral structure was from twisted nanoflakes with atomic crystal lattices. The primary distorted crystal lattices formed left-handed twisted nanoflakes. They displayed a negative scattering-based CD band (centered at 380 nm) and split-type Cotton effect (centered at 340 nm). The secondary-order subchirality should be attributed to the formation of nanoplates from the helical stacking of those nanoflakes. The secondary chiral nanoplates exhibited a positive band (centered at 700 nm) and Cotton effect (centered at 400 nm). The tertiary one was ascribed to the chiral arrangement of nanoplates from the micrometer-sized circinates. The tertiary chirality may not contribute to plasmonic absorption-based CD activity because of its large scale out of the interaction scope.

Another representative study of chiral plasmonic film was reported by Xu and coworkers [113]. The chiral plasmonic films were assembled by the Au NPs modified with L-penicillamine or D-penicillamine. The large area of monolayer Au NP films was formed at the water–hexane interface. As shown from the TEM images, closely packed uniform monolayer films were achieved. The monolayer films exhibited intense CD activity, with CD intensity of 3.5 mdeg at 550 nm and 26.8 mdeg at 775 nm (Figure 3.48). Interestingly, the CD values of the L-Pen-NP and D-Pen-NP films were equal in intensity but opposite in sign. Along with the switching of the angle between the film and light, the CD intensity varied. The maximum signal was observed if the angle rotated to 45°. The plasmonic chirality can be probably attributed to the dihedral angle between the adjacent ellipsoidal Au NPs in the monolayer film. It was induced by the balance between electrostatic repulsion and van der Waals forces, as well as the plasmonic hot spot. The bioactive films with high chirality were used to interact with cells and collect cells noninvasively, which demonstrated a deep influence on cell growth and differentiation.

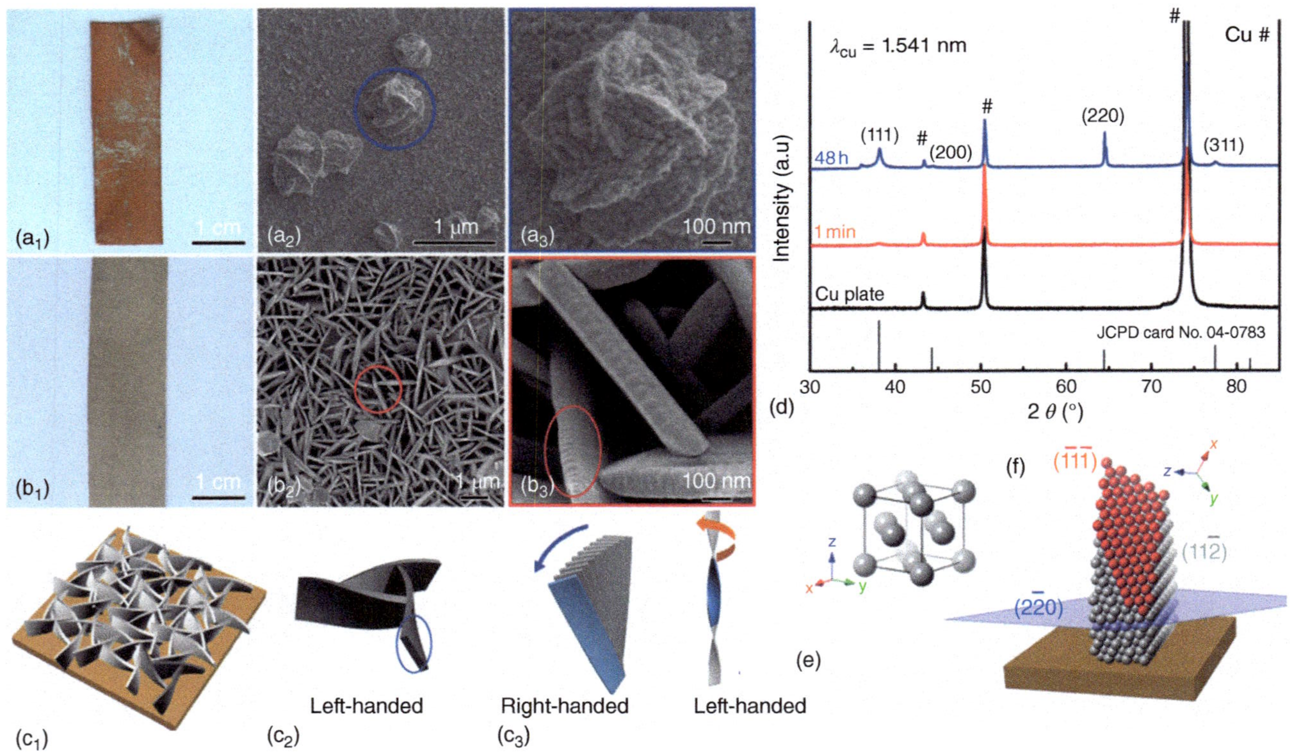


**Figure 3.47** Geometries and structures of the chiral Ag films. ($a_1$–$a_3$) SEM images at different magnifications. ($b_1$–$b_3$) Dense R-Ag film assembled for 48 hours. ($c_1$–$c_3$) Schematic illustrations of the hierarchical R-Ag film. (d) XRD patterns. (e) Crystal structure of the Ag nanoflakes. (f) Growing direction of the Ag nanoflakes along the substrate. Source: Ma et al. [112]/with permission of Wiley.

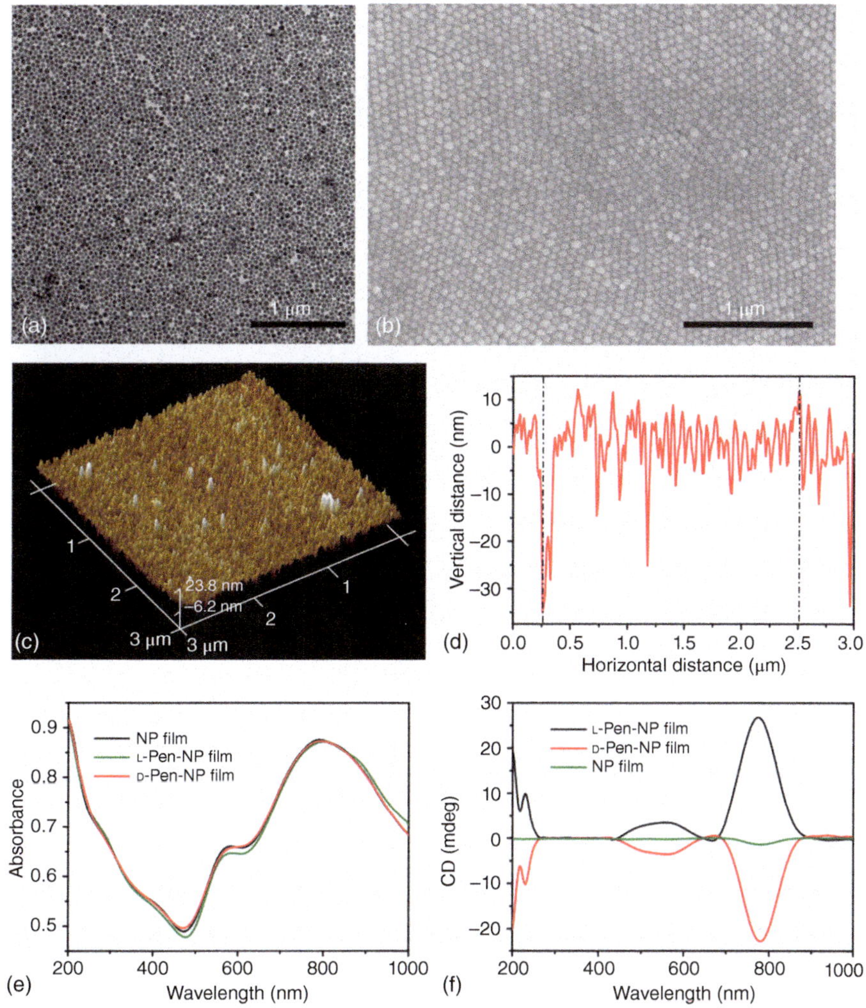


**Figure 3.48** Structural and optical characteristics of Au NP nanofilms. (a) TEM images of Au NP film. (b) SEM images of Au NP nanofilm. (c) AFM image of 3D Au NP nanofilm. (d) Vertical distance of Au NP film. (e) UV–vis absorption of Au NP film and L/D-Pen-modified Au NP nanofilm. (f) CD spectra of Au NP film and L/D-Pen-functionalized Au NP nanofilm. Source: Zhao et al. [113]/with permission of Springer Nature.

Simultaneously, Tang and coworkers reported Au nanowire chiral ultrathin nanofilms [114]. Long-range ordered chiral nanofilms were prepared based on a facile bottom-up assembly method. Different from the above studies, two nanowire layers with a designed angle were achieved, which were left- or right-handed chiral ultrathin nanofilms (Figure 3.49). The specific angle can be tailored by clockwise or anticlockwise rotation of the quartz substrate. Significantly, the prepared ultrathin films exhibited huge chiroptical activity in a broad wavelength range, which were covering the visible and near infrared regions. The anisotropic factor reached up to 0.285, which was the record value for bottom-up techniques. The strong

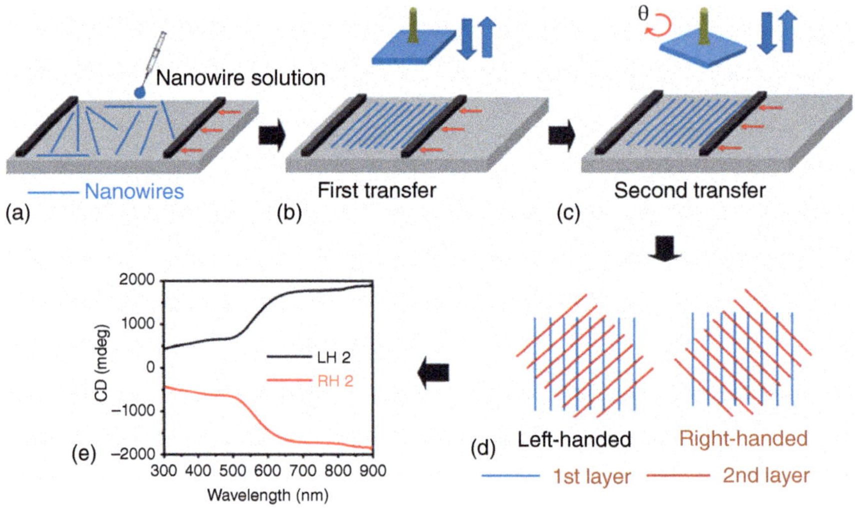


**Figure 3.49** Fabrication process of chiral Au nanowire ultrathin films. (a) Dispersed Au nanowire solution at the air–liquid interface. (b) Compressing the barrier to obtain aligned Au nanowire film at the air–liquid interface and then transferring the assembled film onto a quartz substrate with a horizontal transfer approach. (c) Rotating the substrate with angle for a second transfer. (d) Representative chiral films with opposite handedness. (e) Optical activity of the resulting chiral films. Source: Lv et al. [114]/John Wiley & Sons.

optical activity was probably originated from birefringence and dichroism of the well-aligned Au nanowire layers in the ultrathin films.

## 3.8 Chiral Supraparticles

Terminal supraparticles (SPs) are assembled from individual building blocks and are widely present in the living systems, such as lipid bilayer membranes, carboxysomes, exosomes, viruses, and endosomes [115, 116]. By taking advantage of their specific morphology and components, these natural nanoassemblies play significant roles in sustaining the normal operation of biological systems, such as signal transduction, controllable synthesis, biological catalysis, and cargo protection [115]. These superstructures displayed specific morphologies with an assembly based on characteristic dimensions from the nano- to microscale.

As far as the construction mechanism of natural organelles was concerned, self-limiting method played an important role in the construction of intricate and functional nanostructures. Different from DNA base pairing, fiber-driven nanostructures, and protein templates, SPs were constructed via the balanced noncovalent interaction between repulsive and attractive forces, such as hydrogen bonding, electrostatic interaction, and hydrophobic/hydrophilic anisotropy between the ions, solvents, NPs, and stabilizers in synthetic colloids [54]. Versatility and cost-effectiveness are important for the SP construction in future applications. The terminal SPs display remarkable characteristics of complexity and uniformity, which influence their biological properties and units [117]. They are also a powerful tool for replicating biological functions of nanoassemblies composed of strong inorganic components with special physical, chemical, optical, and magnetic properties.

Based on the self-limiting self-assembly method, micelle-like inorganic supraparticles (SPs) were achieved with good flexibility in the regulation of their sizes, components, morphologies, and optical properties. The ordered building blocks arrangement and spatial subunits confinement resulted in SPs with novel and collective properties (Figure 3.50) [116], making SPs with great potential for fundamental or biological applications [118].

As for SPs, morphology is one of their most significant characteristics. Morphology of the SPs can be flexibly regulated, including vesicle-like nanoshells and capsid-like spherical metal–semiconductor nanostructures by adjusting colloidal environment (temperature, ionic strength, pH, stabilized ligands, and compositions) [117]. It is known that nanoshells can be widely found in the living systems. They play important roles in maintenance of pH, ionic strength, and pressure of biological environments (such as carboxysomes, vacuoles, vesicles). Based on a self-limiting method, Yang et al. reported a type of artificial capsid-like nanoshell from the cadmium sulfide (CdS) NPs assembly, which bridged the gap between inorganic and natural organic self-assembled nanosystems [117]. These artificial nanoshells demonstrated promising potential for substance protection and transportation, reaction control, and homogeneous catalysis in biological systems. Compared with individual building blocks of ZnS NPs and Au NPs, spherical zinc sulfide (ZnS) SPs and Au-ZnS SPs with interior pores displayed higher catalytic and enantioselective ability, indicating the important roles of the three-dimensional chiral space in terminal SPs (Figure 3.51) [116].

Later, Xu group fabricated chiral $Co_3O_4$ supraparticles (SPs) with a mean diameter of 165 nm (Figure 3.52) [119]. The CD spectra of SPs exhibited mirror-symmetry for L-/D-cysteine and L-/D-$Co_3O_4$ SPs. The nearly identical CD spectra can be observed when compared between $Co_3O_4$ NPs and $Co_3O_4$ SPs. The chirality of these SPs was ascribed to the building blocks of $Co_3O_4$ NPs with lattice distortions [120]. Similar to disperse chiral $Co_3O_4$ NPs, a hysteresis loop of $Co_3O_4$ SPs can be observed with a coercivity of 95 Oe.

When achiral Au NRs and CdTe NPs were assembled, chiral core–shell SPs could be fabricated by changing the molar ratio of the NRs and NPs. Moreover, nanoassemblies with single NR in a shell of NPs could be formed with increased NP content. In addition to the chiral optical properties, the photon luminescence properties of QDs can also be quenched by adjacent metal NPs through Förster-resonance-energy transfer or enhanced by plasmon resonance excitation of metal NPs. Due to the chiral interactions between CdTe NPs in the shell, scissor-like SPs based on NP dimers can also form chiral nanostructures with intense chiral optical activity. Numerical simulation results showed that the chiral optical band of plasmon–exciton assemblies was caused by the coupling effects between longitudinal/transverse plasmon modes and NP excitonic state [121]. The optical properties of chiral NR-NP assemblies can combine exciton and plasmon properties, showing great potential for chiral catalysis

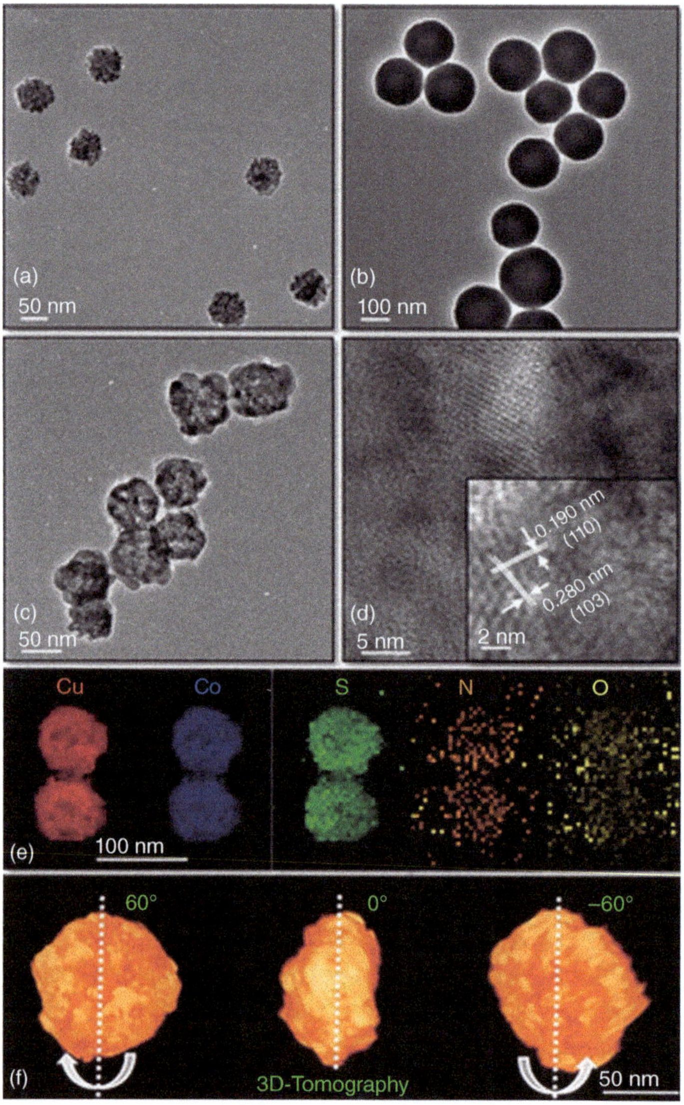


**Figure 3.50** TEM images of (a) $Cu_2S$ SPs, (b) CoS SPs, (c) $Cu_xCo_yS$ SPs. (d) HR-TEM images of $Cu_xCo_yS$ SPs. (e) Elemental mapping images of $Cu_xCo_yS$ SPs. (f) 3D electron tomography reconstruction images of $Cu_xCo_yS$ SPs. Source: Li et al. [116]/with permission of Wiley.

and chiral sensing. Single-particle fluorescence analysis showed that wurtzite zinc selenium (ZnSe) NR couples with twining structures displayed photoluminescence polarization anisotropy compared with individual NRs. The composition of NR couples was altered by the cation exchange method. Using this method, NR-couple structure families with different compositions and controllable performance can be constructed, which can be used to study electronic coupling effects between individual NRs. It can also be further used in photocatalysis, optics, and optoelectronic

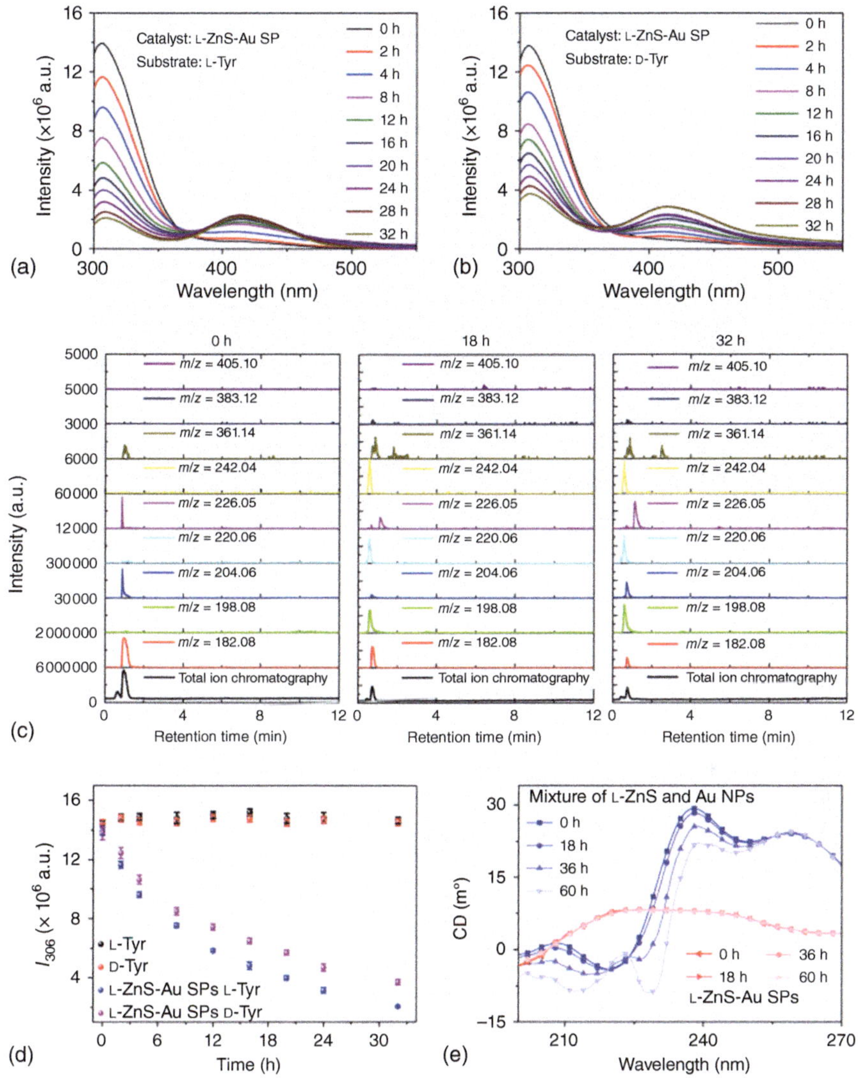


**Figure 3.51** Catalysis and stability properties of ZnS-Au SPs for photoinduced oxidative coupling of L-Tyr and D-Tyr. Fluorescence spectra of L-Tyr or D-Tyr after (a) L-Tyr with L-ZnS-Au SPs and (b) D-Tyr with L-ZnS-Au SPs under illumination with different periods with ZnS-Au SPs. (c) Extracted ion chromatography (EIC) of L-Tyr and Tyr-related products of L-ZnS-Au SP L-Tyr samples obtained at 0, 18, and 32 hours. (d) The dependence of PL intensity at 306 nm on photocatalytic reaction time with L-ZnS-Au SPs for different Tyr enantiomers. (e) CD spectra of the mixtures of ZnS and Au NPs, ZnS-Au SPs, after UV illumination for different times. Source: Li et al. [116]/Springer Nature/CC BY-4.0.

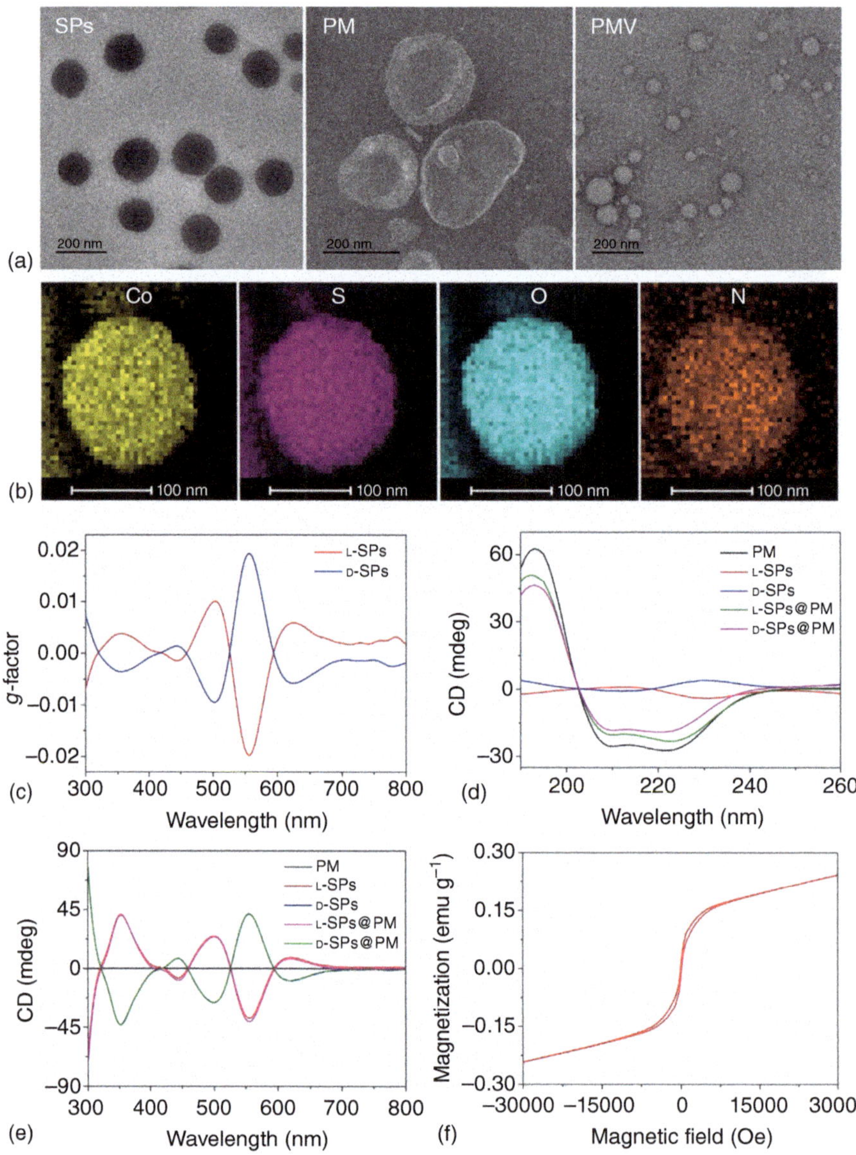


**Figure 3.52** Characterization of $Co_3O_4$ SPs@PM. (a) TEM images for D-$Co_3O_4$ SPs, platelet membrane and platelet membrane vehicles, respectively. (b) EDS mapping of D-$Co_3O_4$ SPs. (c) *g*-factor of L-/D-SPs. CD spectra of PM, L/D-SPs, and L/D-SPs@PM at (d) 190–260 nm and (e) 300–800 nm. (f) Equilibrium magnetization measurements of D-$Co_3O_4$ SPs. Source: Wang et al. [119]/with permission of Wiley.

devices. Colloidal SPs can also be fabricated with stable, broad, and tunable emissions for promising biological applications by collecting QDs together, as exemplified by the Cd(Se, ZnS) core/(Cd, Zn)S shell nanocrystals with three kinds of colors including red, green, and blue emission [122].

Importantly, SPs can serve as bio-mimicking enzymes for biological catalysis and photocatalysis, such as polymerization of amino acids and degradation of contaminants [123]. Due to the confined space, synthesized chiral multi-pores ZnS SPs and Au-ZnS SPs displayed good enantioselective catalysis for tyrosine dimerization. Compared with individual NPs, the catalytic ability of SPs can be strongly enhanced and modulated by the morphology and composition regulation of building blocks. One typical example was the SPs assembled with cytochrome C and CdTe NPs [123]. By combining inorganic NPs and biological molecules into tightly arranged components, SPs with uniform size and morphology can be constructed. Due to the effective charge and exciton transportation in SPs, catalytic efficiency can be improved [124]. These examples illustrate the potential of SPs in biological applications and, at the same time, inspire us to create new SPs with excellent catalytic performance for specific biological applications.

## References


1 Zhu, Y., Xu, L., Ma, W. et al. (2012). A one-step homogeneous plasmonic circular dichroism detection of aqueous mercury ions using nucleic acid functionalized gold nanorods. *Chemical Communications* 48 (97): 11889–11891.

2 Zhao, Y., Xu, L., Kuang, H. et al. (2012). Asymmetric and symmetric PCR of gold nanoparticles: a pathway to scaled-up self-assembly with tunable chirality. *Journal of Materials Chemistry* 22 (12): 5574–5580.

3 Xi, X., Fang, Y., Dong, T., and Cui, Y. (2011). Bottom-up assembly from a helicate to homochiral micro- and mesoporous metal–organic frameworks. *Angewandte Chemie International Edition* 50 (5): 1154–1158.

4 Wang, L., Ma, W., Xu, L. et al. (2010). Nanoparticle-based environmental sensors. *Materials Science & Engineering R: Reports* 70 (3): 265–274.

5 Wang, L., Xu, L., Kuang, H. et al. (2012). Dynamic nanoparticle assemblies. *Accounts of Chemical Research* 45 (11): 1916–1926.

6 Ma, W., Xu, L., de Moura, A.F. et al. (2017). Chiral inorganic nanostructures. *Chemical Reviews* 117 (12): 8041–8093.

7 Lim, S.I. and Zhong, C.-J. (2009). Molecularly mediated processing and assembly of nanoparticles: exploring the interparticle interactions and structures. *Accounts of Chemical Research* 42 (6): 798–808.

8 Perelson, A.S. (1978). Self-organization in nonequilibrium systems. from dissipative structures to order through fluctuations. G. Nicolis, I. Prigogine. *The Quarterly Review of Biology* 53 (3): 362–363.

9 Xu, L., Ma, W., Wang, L. et al. (2013). Nanoparticle assemblies: dimensional transformation of nanomaterials and scalability. *Chemical Society Reviews* 42 (7): 3114–3126.

**10** Guo, D., Li, Y., Zheng, X. et al. (2018). Programmed coassembly of one-dimensional binary superstructures by liquid soft confinement. *Journal of the American Chemical Society* 140 (1): 18–21.

**11** Billinge, S.J.L. and Levin, I. (2007). The problem with determining atomic structure at the nanoscale. *Science* 316 (5824): 561–565.

**12** Srivastava, S., Santos, A., Critchley, K. et al. (2010). Light-controlled self-assembly of semiconductor nanoparticles into twisted ribbons. *Science* 327 (5971): 1355–1359.

**13** Tan, S.J., Campolongo, M.J., Luo, D., and Cheng, W. (2011). Building plasmonic nanostructures with DNA. *Nature Nanotechnology* 6 (5): 268–276.

**14** Xu, L., Yan, W., Ma, W. et al. (2015). SERS encoded silver pyramids for attomolar detection of multiplexed disease biomarkers. *Advanced Materials* 27 (10): 1706–1711.

**15** Sun, M., Qu, A., Hao, C. et al. (2018). Chiral upconversion heterodimers for quantitative analysis and bioimaging of antibiotic-resistant bacteria in vivo. *Advanced Materials* 30 (50): 1804241.

**16** Li, C., Li, S., Qu, A. et al. (2020). Directing arrowhead nanorod dimers for microrna in situ raman detection in living cells. *Advanced Functional Materials* 30 (22): 2001451.

**17** Gao, R., Xu, L., Hao, C. et al. (2019). Circular polarized light activated chiral satellite nanoprobes for the imaging and analysis of multiple metal ions in living cells. *Angewandte Chemie International Edition* 58 (12): 3913–3917.

**18** Yu, R., Liz-Marzán, L.M., and García de Abajo, F.J. (2017). Universal analytical modeling of plasmonic nanoparticles. *Chemical Society Reviews* 46 (22): 6710–6724.

**19** Xu, L.G., Gao, Y.F., Kuang, H. et al. (2018). MicroRNA-directed intracellular self-assembly of chiral nanorod dimers. *Angewandte Chemie International Edition* 57 (33): 10544–10548.

**20** Auguié, B., Alonso-Gómez, J.L., Guerrero-Martínez, A., and Liz-Marzán, L.M. (2011). Fingers crossed: optical activity of a chiral dimer of plasmonic nanorods. *The Journal of Physical Chemistry Letters* 2 (8): 846–851.

**21** Kitaev, V. (2008). Chiral nanoscale building blocks – from understanding to applications. *Journal of Materials Chemistry* 18 (40): 4745–4749.

**22** Yang, M. and Kotov, N.A. (2011). Nanoscale helices from inorganic materials. *Journal of Materials Chemistry* 21 (19): 6775–6792.

**23** Kotov, N.A. (2005). *Nanoparticle Assemblies and Superstructures*. Boca Raton, New York, London, Tokyo: CRC Press.

**24** Ma, W., Kuang, H., Wang, L.B. et al. (2013). Chiral plasmonics of self-assembled nanorod dimers. *Scientific Reports* 3: 6.

**25** Dreuw, A. and Head-Gordon, M. (2004). Failure of time-dependent density functional theory for long-range charge-transfer excited states: the zincbacteriochlorin–bacteriochlorin and bacteriochlorophyll–spheroidene complexes. *Journal of the American Chemical Society* 126 (12): 4007–4016.

**26** Kuzyk, A., Schreiber, R., Fan, Z. et al. (2012). DNA-based self-assembly of chiral plasmonic nanostructures with tailored optical response. *Nature* 483 (7389): 311–314.

**27** Lan, X., Chen, Z., Dai, G. et al. (2013). Bifacial DNA origami-directed discrete, three-dimensional, anisotropic plasmonic nanoarchitectures with tailored optical chirality. *Journal of the American Chemical Society* 135 (31): 11441–11444.

**28** Kuzyk, A., Schreiber, R., Zhang, H. et al. (2014). Reconfigurable 3D plasmonic metamolecules. *Nature Materials* 13 (9): 862–866.

**29** Zhou, C., Duan, X., and Liu, N. (2015). A plasmonic nanorod that walks on DNA origami. *Nature Communications* 6: 8102–8107.

**30** Kuzyk, A., Urban, M.J., Idili, A. et al. (2017). Selective control of reconfigurable chiral plasmonic metamolecules. *Science Advances* 3 (4): 6.

**31** Kuzyk, A., Yang, Y.Y., Duan, X. et al. (2016). A light-driven three-dimensional plasmonic nanosystem that translates molecular motion into reversible chiroptical function. *Nature Communications* 7: 6.

**32** Yan, W., Xu, L., Xu, C. et al. (2012). Self-assembly of chiral nanoparticle pyramids with strong R/S optical activity. *Journal of the American Chemical Society* 134 (36): 15114–15121.

**33** Zhang, H. and Govorov, A.O. (2013). Giant circular dichroism of a molecule in a region of strong plasmon resonances between two neighboring gold nanocrystals. *Physical Review B: Condensed Matter and Materials* 87: 075410.

**34** Yan, W., Ma, W., Kuang, H. et al. (2013). Plasmonic chirogenesis from gold nanoparticles superstructures. *Journal of Physical Chemistry C* 117 (34): 17757–17765.

**35** Husu, H., Canfield, B.K., Laukkanen, J. et al. (2008). Chiral coupling in gold nanodimers. *Applied Physics Letters* 93 (18): 183115.

**36** Chandra, M., Dowgiallo, A.-M., and Knappenberger, K.L. (2012). Magnetic dipolar interactions in solid gold nanosphere dimers. *Journal of the American Chemical Society* 134 (10): 4477–4480.

**37** Ma, W., Kuang, H., Xu, L. et al. (2013). Attomolar DNA detection with chiral nanorod assemblies. *Nature Communications* 4 (1): 2689.

**38** Wu, X., Xu, L., Liu, L. et al. (2013). Unexpected chirality of nanoparticle dimers and ultrasensitive chiroplasmonic bioanalysis. *Journal of the American Chemical Society* 135 (49): 18629–18636.

**39** Yan, W., Ma, W., Kuang, H. et al. (2013). Plasmonic chirogenesis from gold nanoparticles superstructures. *The Journal of Physical Chemistry C* 117 (34): 17757–17765.

**40** Chen, W., Bian, A., Agarwal, A. et al. (2009). Nanoparticle superstructures made by polymerase chain reaction: collective interactions of nanoparticles and a new principle for chiral materials. *Nano Letters* 9 (5): 2153–2159.

**41** Guerrero-Martínez, A., Auguié, B., Alonso-Gómez, J.L. et al. (2011). Intense optical activity from three-dimensional chiral ordering of plasmonic nanoantennas. *Angewandte Chemie (International Ed. in English)* 50 (24): 5499–5503.

**42** Sone, E.D., Zubarev, E.R., and Stupp, S.I. (2005). Supramolecular templating of single and double nanohelices of cadmium sulfide. *Small* 1 (7): 694–697.

**43** Wang, X.X., Xu, L.G., Hao, C.L. et al. (2018). Circular dichroism-active interactions between fipronil and neuronal cells. *Environmental Science & Technology Letters* 5 (8): 500–507.

**44** Xu, Z., Xu, L.G., Zhu, Y.Y. et al. (2012). Chirality based sensor for bisphenol A detection. *Chemical Communications* 48 (46): 5760–5762.

**45** Xu, L.G., Xu, Z., Ma, W. et al. (2013). Highly selective recognition and ultrasensitive quantification of enantiomers. *Journal of Materials Chemistry B* 1 (35): 4478–4483.

**46** Sun, M.Z., Ma, W., Xu, L.G. et al. (2014). Chirality of self-assembled metal-semiconductor nanostructures. *Journal of Materials Chemistry C* 2 (15): 2702–2706.

**47** Tang, L.J., Li, S., Xu, L.G. et al. (2015). Chirality-based Au@Ag nanorod dimers sensor for ultrasensitive PSA detection. *ACS Applied Materials & Interfaces* 7 (23): 12708–12712.

**48** Sun, M.Z., Xu, L.G., Banhg, J.H. et al. (2017). Intracellular localization of nanoparticle dimers by chirality reversal. *Nature Communications* 8: 10.

**49** Hao, C., Xu, L., Ma, W. et al. (2015). Unusual circularly polarized photocatalytic activity in nanogapped gold–silver chiroplasmonic nanostructures. *Advanced Functional Materials* 25 (36): 5816–5822.

**50** Lin, Y.Y., Pashuck, E.T., Thomas, M.R. et al. (2017). Plasmonic Chirality imprinting on nucleobase-displaying supramolecular nanohelices by metal-nucleobase recognition. *Angewandte Chemie International Edition* 56 (9): 2361–2365.

**51** Zhang, K., Hao, L.L., Hurst, S.J., and Mirkin, C.A. (2012). Antibody-linked spherical nucleic acids for cellular targeting. *Journal of the American Chemical Society* 134 (40): 16488–16491.

**52** Gao, F.L., Sun, M.Z., Ma, W. et al. (2017). A singlet oxygen generating agent by chirality-dependent plasmonic shell-satellite nanoassembly. *Advanced Materials* 29 (18): 8.

**53** Ma, W., Fu, P., Sun, M.Z. et al. (2017). Dual quantification of micrornas and telomerase in living cells. *Journal of the American Chemical Society* 139 (34): 11752–11759.

**54** Li, S., Xu, L., Ma, W. et al. (2016). Dual-mode ultrasensitive quantification of MicroRNA in living cells by chiroplasmonic nanopyramids self-assembled from gold and upconversion nanoparticles. *Journal of the American Chemical Society* 138 (1): 306–312.

**55** Li, Z., Lv, S.W., Wang, Y.L. et al. (2015). Construction of LRET-based nanoprobe using upconversion nanoparticles with confined emitters and bared surface as luminophore. *Journal of the American Chemical Society* 137 (9): 3421–3427.

**56** Mastroianni, A.J., Claridge, S.A., and Alivisatos, A.P. (2009). Pyramidal and chiral groupings of gold nanocrystals assembled using DNA scaffolds. *Journal of the American Chemical Society* 131 (24): 8455–8459.

**57** Li, S., Xu, L., Sun, M. et al. (2017). Hybrid nanoparticle pyramids for intracellular dual MicroRNAs biosensing and bioimaging. *Advanced Materials* 29 (19): 1606086.

**58** Sun, M.Z., Hao, T.T., Li, X.Y. et al. (2018). Direct observation of selective autophagy induction in cells and tissues by self-assembled chiral nanodevice. *Nature Communications* 9: 10.

**59** Lin, Y.-X., Qiao, S.-L., Wang, Y. et al. (2017). An in situ intracellular self-assembly strategy for quantitatively and temporally monitoring autophagy. *ACS Nano* 11 (2): 1826–1839.

**60** Liu, C., Gao, Z., Zeng, J. et al. (2013). Magnetic/upconversion fluorescent NaGdF4:Yb,Er nanoparticle-based dual-modal molecular probes for imaging tiny tumors in vivo. *ACS Nano* 7 (8): 7227–7240.

**61** Gao, F., Sun, M., Ma, W. et al. (2017). A singlet oxygen generating agent by chirality-dependent plasmonic shell-satellite nanoassembly. *Advanced Materials* 29 (18): 1606864.

**62** Sun, M., Hao, T., Li, X. et al. (2018). Direct observation of selective autophagy induction in cells and tissues by self-assembled chiral nanodevice. *Nature Communications* 9 (1): 4494.

**63** Wu, X., Xu, L., Ma, W. et al. (2016). Propeller-like nanorod-upconversion nanoparticle assemblies with intense chiroptical activity and luminescence enhancement in aqueous phase. *Advanced Materials* 28 (28): 5907–5915.

**64** Viswanatha, R., Naveh, D., Chelikowsky, J.R. et al. (2012). Magnetic properties of Fe/Cu codoped ZnO nanocrystals. *The Journal of Physical Chemistry Letters* 3 (15): 2009–2014.

**65** Gibaud, T., Barry, E., Zakhary, M.J. et al. (2012). Reconfigurable self-assembly through chiral control of interfacial tension. *Nature* 481 (7381): 348–351.

**66** Caricato, M., Sharma, A.K., Coluccini, C., and Pasini, D. (2014). Nanostructuring with chirality: binaphthyl-based synthons for the production of functional oriented nanomaterials. *Nanoscale* 6 (13): 7165–7174.

**67** Caricato, M., Delforge, A., Bonifazi, D. et al. (2015). Chiral nanostructuring of multivalent macrocycles in solution and on surfaces. *Organic & Biomolecular Chemistry* 13 (12): 3593–3601.

**68** Kuang, H., Zhao, S., Chen, W. et al. (2011). Rapid DNA detection by interface PCR on nanoparticles. *Biosensors and Bioelectronics* 26 (5): 2495–2499.

**69** Kuang, H., Ma, W., Xu, L. et al. (2013). Nanoscale superstructures assembled by polymerase chain reaction (PCR): programmable construction, structural diversity, and emerging applications. *Accounts of Chemical Research* 46 (11): 2341–2354.

**70** Kuang, H., Chen, X.J., Hao, C.L. et al. (2013). Immuno-driven plasmonic oligomer sensor for the ultrasensitive detection of antibiotics. *RSC Advances* 3 (38): 17294–17299.

**71** Xie, J., Duan, Y., and Che, S. (2012). Chirality of metal nanoparticles in chiral mesoporous silica. *Advanced Functional Materials* 22 (18): 3784–3792.

**72** Shen, X., Song, C., Wang, J. et al. (2012). Rolling up gold nanoparticle-dressed DNA origami into three-dimensional plasmonic chiral nanostructures. *Journal of the American Chemical Society* 134 (1): 146–149.

**73** Chen, C.-L. and Rosi, N.L. (2010). Preparation of unique 1-D nanoparticle superstructures and tailoring their structural features. *Journal of the American Chemical Society* 132 (20): 6902–6903.

**74** Sharma, J., Chhabra, R., Cheng, A. et al. (2009). Control of self-assembly of DNA tubules through integration of gold nanoparticles. *Science* 323 (5910): 112–116.

**75** Fan, Z. and Govorov, A.O. (2010). Plasmonic circular dichroism of chiral metal nanoparticle assemblies. *Nano Letters* 10 (7): 2580–2587.

**76** Ma, W., Xu, L., Wang, L. et al. (2019). Chirality-based biosensors. 29 (1): 1805512.

**77** Korgel, B. (2005). Materials science. Nanosprings take shape. *Science (New York, N.Y.)* 309: 1683–1684.

**78** Tamoto, R., Lecomte, S., Si, S. et al. (2012). Gold nanoparticle deposition on silica nanohelices: a new controllable 3D substrate in aqueous suspension for optical sensing. *The Journal of Physical Chemistry C* 116 (43): 23143–23152.

**79** Lan, X., Lu, X., Shen, C. et al. (2015). Au nanorod helical superstructures with designed chirality. *Journal of the American Chemical Society* 137 (1): 457–462.

**80** Feng, W., Kim, J.-Y., Wang, X. et al. (2017). Assembly of mesoscale helices with near-unity enantiomeric excess and light-matter interactions for chiral semiconductors. *Science Advances* 3 (3): e1601159.

**81** Yan, J., Feng, W., Kim, J.-Y. et al. (2020). Self-assembly of chiral nanoparticles into semiconductor helices with tunable near-infrared optical activity. *Chemistry of Materials* 32 (1): 476–488.

**82** Liljestrom, V., Ora, A., Hassinen, J. et al. (2017). Cooperative colloidal self-assembly of metal-protein superlattice wires. *Nature Communications* 8: 10.

**83** Cathcart, N. and Kitaev, V. (2011). Monodisperse hexagonal silver nanoprisms: synthesis via thiolate-protected cluster precursors and chiral ligand-imprinted self-assembly. *ACS Nano* 5 (9): 7411–7425.

**84** Jana, S., de Frutos, M., Davidson, P., and Abécassis, B. (2017). Ligand-induced twisting of nanoplatelets and their self-assembly into chiral ribbons. *Science Advances* 3 (9): e1701483.

**85** Hao, C., Gao, Y., Wu, D. et al. (2019). Tailoring chiroptical activity of iron disulfide quantum dot hydrogels with circularly polarized light. *Advanced Materials* 31 (36): 1903200.

**86** Liu, M.H., Zhang, L., and Wang, T.Y. (2015). Supramolecular chirality in self-assembled systems. *Chemical Reviews* 115 (15): 7304–7397.

**87** Huo, S.W., Duan, P.F., Jiao, T.F. et al. (2017). Self-assembled luminescent quantum dots to generate full-color and white circularly polarized light. *Angewandte Chemie International Edition* 56 (40): 12174–12178.

**88** Choi, H., Cho, K.J., Seo, H. et al. (2017). Transfer and dynamic inversion of coassembled supramolecular chirality through 2d-sheet to rolled-up tubular structure. *Journal of the American Chemical Society* 139 (49): 17711–17714.

**89** Zhao, Y., Xu, L., Liz-Marzán, L.M. et al. (2013). Alternating plasmonic nanoparticle heterochains made by polymerase chain reaction and their optical properties. *The Journal of Physical Chemistry Letters* 4 (4): 641–647.

**90** Zhu, Z., Liu, W., Li, Z. et al. (2012). Manipulation of collective optical activity in one-dimensional plasmonic assembly. *ACS Nano* 6 (3): 2326–2332.

**91** Cai, J., Hao, C., Sun, M. et al. (2018). Chiral shell core–satellite nanostructures for ultrasensitive detection of mycotoxin. *Small* 14 (13): 1703931.

**92** Hao, C., Wu, X., Sun, M. et al. (2019). Chiral core–shell upconversion nanoparticle@MOF nanoassemblies for quantification and bioimaging of reactive oxygen species in vivo. *Journal of the American Chemical Society* 141 (49): 19373–19378.

**93** Jiang, X., Hao, C., Zhang, H. et al. (2021). Dual-modal fexcuyse and upconversion nanoparticle assemblies for intracellular MicroRNA-21 detection. *ACS Applied Materials & Interfaces* 13 (35): 41405–41413.

**94** Hao, T., Wu, X., Xu, L. et al. (2017). ultrasensitive detection of prostate-specific antigen and thrombin based on gold-upconversion nanoparticle assembled pyramids. *Small* 13 (19): 1603944.

**95** Li, Y., Tang, J., He, L. et al. (2015). Core–shell upconversion nanoparticle@metal–organic framework nanoprobes for luminescent/magnetic dual-mode targeted imaging. *Advanced Materials* 27 (27): 4075–4080.

**96** Chen, Q., Xie, X., Huang, B. et al. (2017). Confining excitation energy in Er3+-sensitized upconversion nanocrystals through Tm3+-mediated transient energy trapping. *Angewandte Chemie International Edition* 56 (26): 7605–7609.

**97** Wang, F., Han, Y., Lim, C.S. et al. (2010). Simultaneous phase and size control of upconversion nanocrystals through lanthanide doping. *Nature* 463 (7284): 1061–1065.

**98** Wang, F., Deng, R., and Liu, X. (2014). Preparation of core-shell NaGdF4 nanoparticles doped with luminescent lanthanide ions to be used as upconversion-based probes. *Nature Protocols* 9 (7): 1634–1644.

**99** Dong, H., Sun, L.-D., and Yan, C.-H. (2015). Energy transfer in lanthanide upconversion studies for extended optical applications. *Chemical Society Reviews* 44 (6): 1608–1634.

**100** Qu, A., Wu, X., Xu, L. et al. (2017). SERS- and luminescence-active Au-Au-UCNP trimers for attomolar detection of two cancer biomarkers. *Nanoscale* 9 (11): 3865–3872.

**101** Ge, H., Wang, D.Y., Pan, Y. et al. (2020). Sequence-dependent DNA functionalization of upconversion nanoparticles and their programmable assemblies. *Angewandte Chemie International Edition* 6: 8133–8137.

**102** Das, A., Mao, C.C., Cho, S. et al. (2018). Over 1000-fold enhancement of upconversion luminescence using water-dispersible metal-insulator-metal nanostructures. *Nature Communications* 9: 11.

**103** Meng, D., Ma, W., Wu, X.L. et al. (2020). DNA-driven two-layer core-satellite gold nanostructures for ultrasensitive MicroRNA detection in living cells. *Small* 16 (23): 9.

**104** Qu, A., Sun, M., Kim, J.-Y. et al. (2021). Stimulation of neural stem cell differentiation by circularly polarized light transduced by chiral nanoassemblies. *Nature Biomedical Engineering* 5 (1): 103+.

**105** Zhao, X.L., Xu, L.G., Sun, M.Z. et al. (2016). Gold-quantum dot core-satellite assemblies for lighting up MicroRNA in vitro and in vivo. *Small* 12 (34): 4662–4668.

**106** Xu, L., Hao, C., Yin, H. et al. (2013). Plasmonic core–satellites nanostructures with high chirality and bioproperty. *The Journal of Physical Chemistry Letters* 4 (14): 2379–2384.

**107** Hao, C., Xu, L., Sun, M. et al. (2018). Chirality on hierarchical self-assembly of Au@AuAg yolk–shell nanorods into core–satellite superstructures for biosensing in human cells. *Advanced Functional Materials* 28 (33): 1802372.

**108** Wang, W., Satyavolu, N.S.R., Wu, Z. et al. (2017). Near-infrared photothermally activated DNAzyme–gold nanoshells for imaging metal ions in living cells. *Angewandte Chemie International Edition* 56 (24): 6798–6802.

**109** Zhu, Z.N., Liu, W.J., Li, Z.T. et al. (2012). Manipulation of collective optical activity in one-dimensional plasmonic assembly. *ACS Nano* 6 (3): 2326–2332.

**110** Sun, M., Xu, L., Ma, W. et al. (2016). Hierarchical plasmonic nanorods and upconversion core–satellite nanoassemblies for multimodal imaging-guided combination phototherapy. *Advanced Materials* 28 (5): 898–904.

**111** Jiang, W., Qu, Z.-B., Kumar, P. et al. (2020). Emergence of complexity in hierarchically organized chiral particles. *Science Advances* 368 (6491): 642–648.

**112** Ma, L.G., Cao, Y.Y., Duan, Y.Y. et al. (2017). Silver films with hierarchical chirality. *Angewandte Chemie International Edition* 56 (30): 8657–8662.

**113** Zhao, X.L., Xu, L.G., Sun, M.Z. et al. (2017). Tuning the interactions between chiral plasmonic films and living cells. *Nature Communications* 8: 11.

**114** Lv, J.W., Hou, K., Ding, D.F. et al. (2017). Gold nanowire chiral ultrathin films with ultrastrong and broadband optical activity. *Angewandte Chemie International Edition* 56 (18): 5055–5060.

**115** Li, S., Guo, X., Sun, M. et al. (2021). Self-limiting self-assembly of supraparticles for potential biological applications. *Nanoscale* 13 (4): 2302–2311.

**116** Li, S., Xu, L., Hao, C. et al. (2019). Single- and multi-component chiral supraparticles as modular enantioselective catalysts. *Nature Communications* 58 (52): 19067–19072.

**117** Yang, M., Chan, H., Zhao, G. et al. (2017). Self-assembly of nanoparticles into biomimetic capsid-like nanoshells. *Nature Chemistry* 9 (3): 287–294.

**118** Nie, Z., Petukhova, A., and Kumacheva, E. (2010). Properties and emerging applications of self-assembled structures made from inorganic nanoparticles. *Nature Nanotechnology* 5 (1): 15–25.

**119** Wang, X.X., Sun, M.Z., Qu, A.H. et al. (2021). Improved reactive oxygen species generation by chiral $Co_3O_4$ supraparticles under electromagnetic fields. *Angewandte Chemie International Edition* 60 (33): 18240–18246.

**120** Li, S., Xu, L., Lu, M. et al. (2021). Metabolic profile of chiral cobalt oxide nanoparticles in vitro and in vivo. *Nano Research* 14 (7): 2451–2455.

**121** Kang, H., Buchman, J.T., Rodriguez, R.S. et al. (2019). Stabilization of silver and gold nanoparticles: preservation and improvement of plasmonic functionalities. *Chemical Reviews* 119 (1): 664–699.

**122** Fujii, M., Fujii, R., Takada, M., and Sugimoto, H. (2020). Silicon quantum dot supraparticles for fluorescence bioimaging. *ACS Applied Nano Materials* 3 (6): 6099–6107.

**123** Park, J.I., Nguyen, T.D., de Queirós Silveira, G. et al. (2014). Terminal supraparticle assemblies from similarly charged protein molecules and nanoparticles. *Nature Communications* 5 (1): 3593.

**124** Silveira, G.d.Q., Ramesar, N.S., Nguyen, T.D. et al. (2019). Supraparticle nanoassemblies with enzymes. *Chemistry of Materials* 31 (18): 7493–7500.

# 4

# Chiral Nanostructures for Biorecognition and Bioanalysis

*Chuanlai Xu[1], Yuan Zhao[2], and Xiaoling Wu[1]*

[1] *International Joint Research Laboratory for Biointerface and Biodetection, State Key Lab of Food Science and Technology, School of Food Science and Technology, Jiangnan University, Wuxi, Jiangsu Province, 214122, China*
[2] *Key Laboratory of Synthetic and Biological Colloids, Ministry of Education, School of Chemical and Material Engineering, Jiangnan University, Wuxi, Jiangsu Province, 214122, China*

## 4.1 Chiral Analysis Strategy

Chiral nanomaterials featured with tunable and strong chiral optical properties are widely used as detection probes for the fabrication of optically active sensors for biorecognition and bioanalysis [1]. Chiral nanomaterials generally exhibit adjustable optical responses with a defined peak position and intensity peak intensity by adjusting sizes, components, and shapes of NPs and even the assembled configuration. This is beneficial to specific bioapplications by choosing designed chiral nanomaterials [2, 3]. In particular, some NPs, such as Au NPs, possess good biocompatibility for a wide range of safe bioapplications. However, the chiral responses of individual NPs are weak and the signal intensity is not strong enough. Individual NPs can be assembled to form chiral nanostructures by interaction forces (hydrogen bonds, van der Waals forces, Coulomb forces, and hydrophobic interactions), biomolecules (DNA, proteins/peptides), polymers, etc. [4, 5]. Assemblies can have special collective properties due to the different spatial arrangement [6, 7]. Chiral nanomaterials display stronger chiral signals than individual NPs, which further improves the sensitivity of biorecognition and bioanalysis.

Biomolecules (such as DNA [8], proteins, and peptides [9]) usually serve as templates for the assembly of chiral nanomaterials. Importantly, the specific biorecognition between DNA and complementary, affinity for aptamers and targets, and better selectivity between the antibody and antigen provided biomolecule-engineered chiral nanostructures with unique selectivity, which further increase the accuracy and specificity of bioanalysis. Biomolecule-engineered chiral nanostructures are the alternative candidates to single chiral NPs in potential bioapplications [10]. There are many chiral molecules in a complex detection system, which would interfere with detection signals and affect the accuracy of results. Fortunately, chiral nanostructures exhibit strong chiral responses in the

*Chiral Nanoprobes for Biological Applications*, First Edition. Edited by Chuanlai Xu.

visible region, whereas these co-existing chiral molecules generally display weak chiral signals, and their chiral peaks are usually located in the ultraviolet region. The chiral signals of nanostructures are transformed from ultraviolet to visible light. These interfering signals from biomolecules would not affect the appearance of detection signals of chiral nanostructures, which further ensure the accuracy and reliability of the detection results in biorecognition and bioanalysis applications. Based on these strategies, chiral nanostructures have been widely used in the analysis and identification of abiotic and biological materials, including extracellular and intracellular analysis, such as DNA detection [11], microRNA monitoring [12], metal ion testing [13], toxin checking [14], chiral identification of amino acids and proteins [15], etc. The applications of chiral materials in biorecognition and bioanalysis are detailed in the following sections.

## 4.2 DNA Detection

DNA, which carries the genetic information necessary for the synthesis of RNA and proteins, is an essential biomacromolecule for the development and normal functioning of organisms. DNA detection has provided some key information in early disease diagnosis, pharmacogenomics, transgenic technology, and species identification [16]. DNA is usually extracted and analyzed by the traditional polymerase chain reaction (PCR) and RT-PCR techniques. The amplified DNA products are then monitored by agarose gel electrophoresis or detected by the fluorescence signals from the RT-PCR method [17]. Agarose gel electrophoresis is complex, time-consuming, and harmful. Although the traditional techniques are widely used for DNA detection, there exist still some limitations such as complex operation, inaccurate results.

DNA detection still mainly relies on fluorescence signals. There are increasing demands for sensitive, accurate, and fast optical detection in clinical diagnostics and epidemic prevention. Alternatively, the resulting DNA-driven chiral assemblies show CD signals in the visible region, which is due to the transfer of DNA molecules with chiral properties and the chiral geometry of the assembly [18]. Because the biological matrix is achiral or shows weak chiral signals in the ultraviolet region, the chiral signals produced by DNA assemblies are more resistant to interference and more sensitive than other signals [19]. Alternatively, chiral signal–dependent PCR technique was firstly proposed by Xu group for sensitive and non-interfering DNA detection (Figure 4.1a). Primers were modified on the surface of NPs, which were assembled into chiral nanomaterials by using target DNA molecules as templates. For example, forward primer–modified Au NRs and reverse primer–functionalized Au NRs were assembled into chiral Au NR assemblies. Through adjusting the modified position of the primers on the side and the end of Au NRs, side-by-side and end-to-end Au NR assemblies were obtained. In particular, the side-by-side Au NR assemblies exhibited intensive chiral signals between 500 and 800 nm, where the 620–800 nm spectral window was negative and the 500–620 nm wavelength was positive. The chiroplasmonic activity of the side-by-side assembled patterns was attributed to a 7°–9° twist between the axes of Au NRs. With the increasing

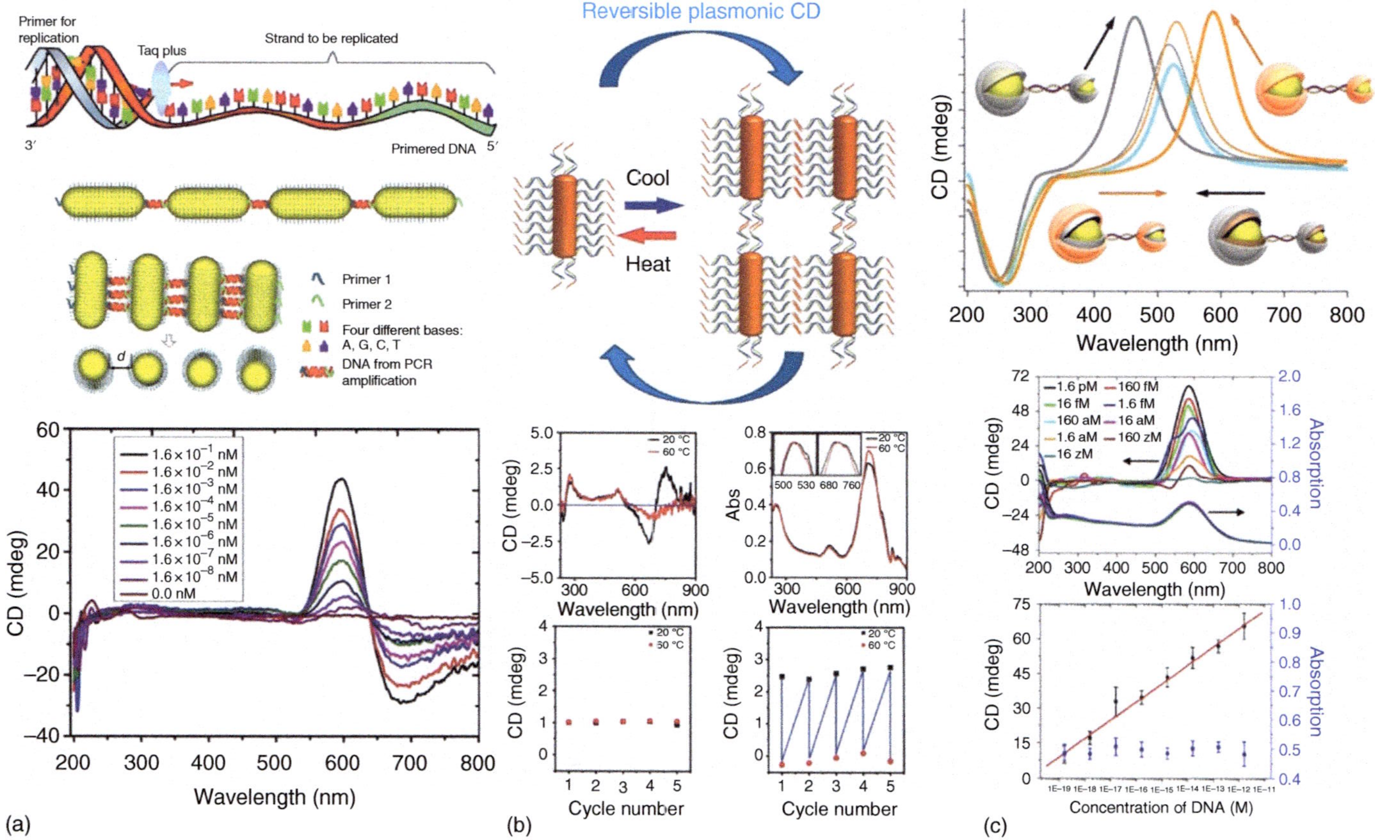


**Figure 4.1** (a) Chiral side-by-side and end-to-end Au NR assemblies for the detection of target DNA. (b) DNA detection by chiral side-by-side Au NR assemblies. (c) PCR-driven chiral core–shell NP heterodimers for DNA detection. Source: (a) Reproduced with permission Ma et al. [11]. Copyright 2013, Springer Nature; (b) Reproduced with permission Li et al. [20]. Copyright 2012, American Chemical Society; (c) Reproduced with permission Zhao et al. [21]. Copyright 2014, American Chemical Society.

concentration of DNA templates, the yields of the side-by-side Au NR assemblies improved, generating strong CD signals. The amplitude of circular dichroism showed a good linear relationship with the target DNA, and the limit of detection (LOD) was as low as 3.7 aM [11]. Reaching the attomolar limit of detection for simple and reliable bioanalysis of oligonucleotides may play a crucial role in DNA biomarker detection for early diagnosis of different diseases, forensics, and environmental monitoring. Chiral side-by-side Au NR assemblies were also prepared by controlling the assembly and disassembly of dsDNA modified on the sides of Au NRs (Figure 4.1b). The CD signal was generated in the visible region. When the temperature was higher than 50 °C, dsDNA would be disassembled, resulting in the free dispersion state of Au NRs with decreased CD responses. When the temperature was lower than 20 °C, the sticky ends of DNA could be further complementary, resulting in spontaneous assembly of side-by-side Au NR assemblies with enhanced CD signals. The CD signal changes were reversible by controlling the temperature. As-prepared chiral side-by-side Au NR assemblies achieved DNA detection with an LOD of 75 nM [20]. Alternatively, chiral side-by-side Au NR assemblies fabricated by the PCR technique achieved ultrasensitive DNA detection, which was attributed to the strong chiral signal and the exponential amplification characteristics of the PCR technique.

PCR-driven chiral assemblies showed potential applications for accurate and sensitive DNA detection. By changing the used plasmonic building blocks, distinct structures of chiral assemblies with different CD responses were obtained. Primer-modified Au NPs with different sizes were also successfully assembled into different chiral nanostructures, by controlling the density of the modified primers on NPs and the PCR cycle number. For example, forward primer–modified small Au NPs and reverse primer–functionalized large Au NPs were assembled into chiral Au NP heterodimers by controlling only one primer modified on the surface of Au NPs. With the increasing number of PCR, the yield of Au NP heterodimers increased, generating enhanced CD responses for sensitive DNA detection. When the number of primers was controlled to two, adjacent two kinds of Au NPs would be connected. With the increasing number of PCR, more Au NPs would be connected one by one, and chiral Au NP chains were also obtained [22]. Therefore, with the increasing number of PCR, the yield of heterodimers increased and the chains had a long length. Heterochains displayed strong plasmonic chirality at 500–600 nm, and the chiral activity revealed nonmonotonous dependence on the length of heterochains. Furthermore, through controlling the asymmetrical and symmetrical modification of primers on different sized Au NPs, chiral supernanostructures were also obtained [18]. The configuration of assemblies was complex and closely related to the chiral responses. In comparison to other chiral nanostructures, PCR-driven chiral heterodimers possessed simple assembly structures and high yield as well as intensive CD responses and, therefore, became an alternative chiral candidate for DNA detection. Au NP heterodimers showed chiral signals due to their imperfect oval and scissor shapes, which could also be used for specific DNA detection. To increase the detection sensitivity, amplified CD responses of chiral Au NR heterodimers were obtained by depositing another metal shell through the reduction of metal precursors. Chiral Au NP heterodimers showed blueshifted chiral signals from 525 to 418 nm after the covering of the Ag shell due to shorter local surface plasmon resonance (LSPR) absorption of Ag, but displayed redshift from 525 to 586 nm and amplified CD intensity after the deposition of the Au shell owing to longer LSPR absorption of Au (Figure 4.1c). Different shell types, thicknesses, and layer sequences made it possible to tailor their chiroptical activity with respect to the spectral position and intensity. The strongest CD peak was obtained at 586 nm. Combined with the exponential amplification characteristics of PCR technique and enhanced CD intensity, shell-deposited chiral Au NP heterodimers improved detection sensitivity and achieved zeptomolar DNA detection [21]. By changing the deposited material components, sequence, and thickness of the metal shell, the chiral peak position and intensity of the chiral signals were successfully adjusted, so that multiplex and defined DNA sequence detection can be achieved by using distinct chiral nanomaterials with designed CD responses. In comparison to methods using fluorescent labels, the chiroplasmonic method circumvented some sources of variability related to the use of photolabile fluorophores and used the CD signal in a part of the spectrum that had contribution and therefore noise from other components.

Based on the principle of base complementary pairing, a single strand of DNA can hybridize and bind to its complementary strand, which is very sensitive and specific, even to the difference of a single base. Target DNA can be sensitively detected by carefully designing the base sequences of DNA fragments that bind specifically in the fabrication of chiral nanostructures [23]. For example, DNA-engineered NP pyramid structures were fabricated through the hybridization between complementary strands and target DNA and were used for DNA detection. 10, 20 nm Au NPs, and 10, 20 nm Ag NPs were assembled to pyramids by using DNA as linkers. NP pyramids exhibited obvious chiral signals at 414 and 522 nm, attributing to the plasmonic interaction between Au and Ag NPs and the asymmetrical pyramid structures. The sum intensity of two peaks at 414 and 522 nm reached up to $-51.91 \pm 1.4$ mdeg. When the target DNA was added, the frame structures would undergo recombination/decomposition depending on the off–on and on–off principles, respectively. In the first off–on structure, four identical Au NPs were connected by four ssDNA to form a pyramidal asymmetrical framework. After hybridization with the target DNA, an asymmetrical framework with chiral activity was formed. With the increasing amounts of target DNA, more pyramids were assembled, generating higher CD signals. An off–on switch chiral sensor was developed for DNA detection with a detection range of 50–5000 aM and an LOD of 20.3 aM. The second on–off structure was also attempted to fabricate through connecting four different NPs with four ssDNA. A chiral pyramidal symmetrical framework was formed (Figure 4.2a). After the addition of target DNA, the pyramidal structures were decomposed into heterotrimers, which made the chiral signals to disappear. A on–off switch chiral sensor was developed for DNA detection with a detection range of 10–5000 aM. The LOD was as low as 3.4 aM [24]. A small amount of DNA induced changes in the structures of NP assemblies, resulting in changes in CD responses. Such high sensitivity could achieve the detection of other DNA-recognized molecules with much

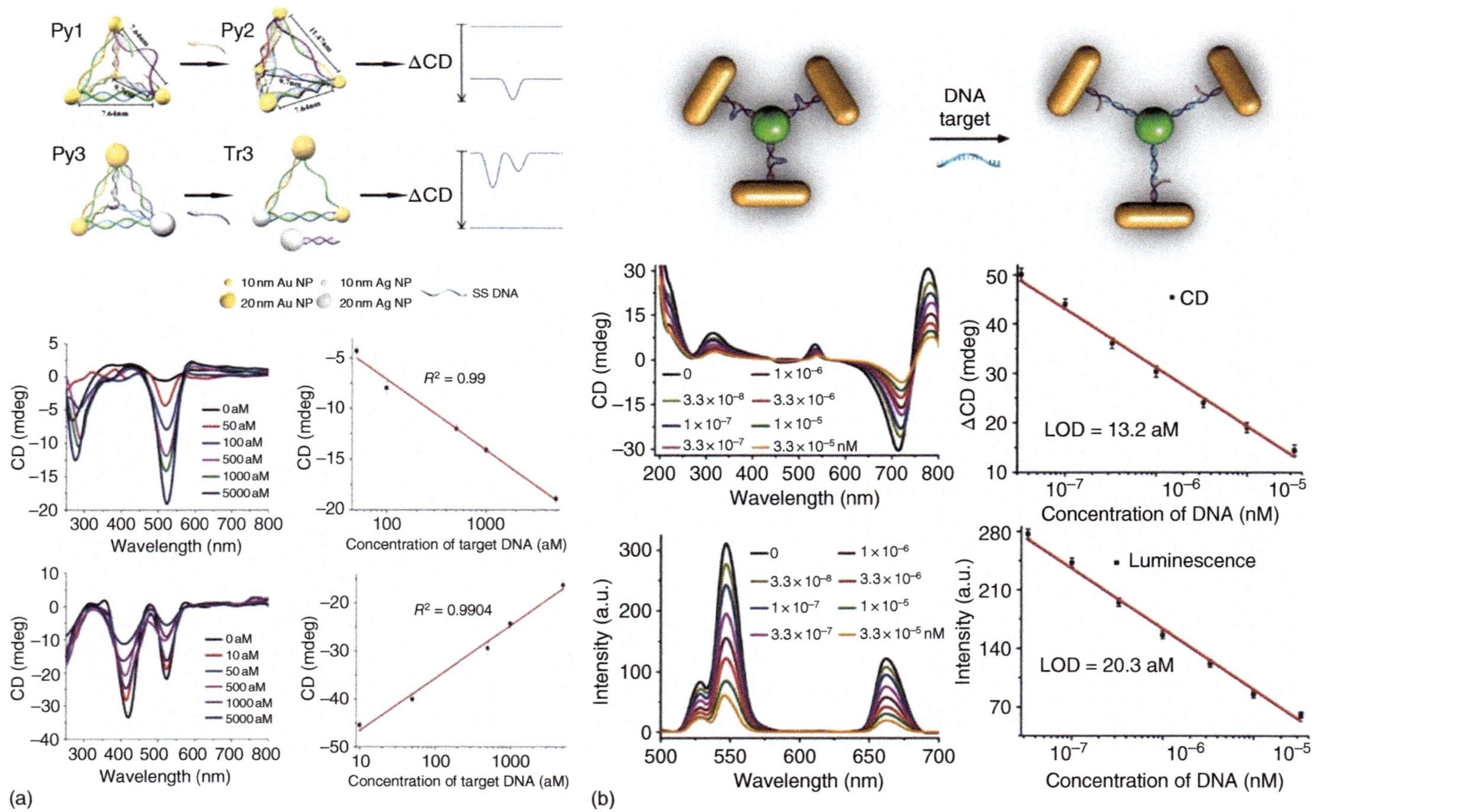


**Figure 4.2** (a) DNA-engineered chiral NP pyramidal structures for DNA detection. (b) Chiral and fluorescent UCNP-Au NR propellers for dual-mode DNA detection. Source: (a) Reproduced with permission Yan et al. [24]. Copyright 2014, Wiley-VCH; (b) Reproduced with permission Wu et al. [25]. Copyright 2016, Wiley-VCH.

low LODs using a similar mechanism. Furthermore, the strong CD responses of the pyramids in the visible region and the minimal interference of background signal in the ultraviolet region ensured DNA detection with high sensitivity and better accuracy.

Although chiral signal–dependent optical sensors have been reported for DNA detection, dual-mode optical sensors by combining CD signals with other optical signals will display better reliability for accurate DNA detection. For example, fluorescence signal–dependent optical sensors were fabricated for multiple real-time fluorescent DNA detection. Nanomaterials with unique fluorescence-quenching ability were also achieved for sensitive DNA detection, such as $MoS_2$, graphene oxide, and graphdiyne. Such single-layer 2D nanosheets showed higher affinity toward ssDNA than dsDNA. Dye-labeled ssDNA was quenched after the combination with single-layer 2D nanosheets. The target DNA would hybrid with dye-labeled ssDNA, forming dsDNA and retaining the fluorescence signals. Sensitive DNA detection was achieved by fluorescence signals. It was reported that few-layer graphdiyne nanosheets used as a sensing platform showed best performance for real-time detection of DNA with low background and high signal-to-noise ratio, in comparison to graphene oxide and $MoS_2$. The LOD was as low as 25 fM [26]. It was because of higher noncovalent binding ability of graphdiyne for nucleobases and amino acids compared to $MoS_2$ and graphene oxide. Benefiting from the better detection ability of fluorescence signals, dual signals combining chiral signals with fluorescence signals were designed for DNA detection. For example, UCNP-Au NR propeller-like tetramers were studied (Figure 4.2b). Strong chiral signals were originated from chiral configurations with enhanced polarization rotation, and an amplified PL signal was observed due to the plasmon-enhanced PL mechanism. By adjusting the length of the DNA sequence, the length–width ratio of Au NRs and the particle sizes of UCNPs, chiral activity and PL intensity were adjusted. In the presence of target DNA, the designed hairpin-like dsDNA structures were extended to longer dsDNA. This would largely increase the interparticle distances between Au NRs and UCNPs, which generated weak chiral intensity and decreased PL peaks. A switch-off dual-mode optical sensor was fabricated for accurate DNA detection, with LODs of 13.2 aM for the chiral signal and 20.3 aM for the PL signal, respectively [25]. The sensitivity of DNA detection based on the chiral signal was higher than that of the fluorescence signal. Dual-mode optical sensors by combining chiral signals with other optical signals will largely improve the reliability of detection and should be further developed for bioanalysis, such as chiral and fluorescence signals, chiral and surface-enhanced Raman scattering (SERS) signals [27], and so on. Although most chiral nanomaterials were fabricated for DNA detection, the used building blocks mainly focused on 1D nanodots, NPs, and NRs. Alternatively, 2D nanostructures with superior electron-capturing ability need to be further explored for their potential chiroptical properties. Furthermore, 2D materials also possess better fluorescence-quenching ability. The introduction of 2D nanomaterials will generate distinct CD signals and fluorescence responses. Furthermore, biosensors based on fluorescence signals and Raman signals may achieve simultaneous DNA detection, but it is still difficult to achieve multiple and even high-throughput DNA detection using chiral signals.

## 4.3 Biomarker Bioanalysis

In recent years, cancer has become the leading cause of death, and its morbidity and mortality have been increasing every year. The detection of tumor biomarkers can be used for the early diagnosis and treatment of malignant tumors, thus reducing the fatality rate and improving the quality of life of patients with malignant tumors [28]. A large number of tumor biomarkers have been widely reported, such as human epidermal growth factor receptor-2 (HER2), α-fetoprotein (AFP), vascular endothelial growth factor (VEGF), prostate-specific antigen (PSA), thrombin, mucin-1, which can be used to detect breast cancer, lung cancer, colorectal cancer, prostate cancer, and other diseases [29]. Besides protein biomarkers, miRNA, micromolecules, and circulating tumor cells (CTCs) are regarded as biomarkers for human cancers. At the same time, as a global health problem, the incidence of food allergy has increased significantly with the process of modern industrialization. Testing for allergens in food is an effective way to avoid food allergies. Eggs, peanuts, milk, soybeans, wheat, tree nuts, fish, and crustaceans contain allergens. Many chiral sensors have been used to detect tumor markers and food allergens [30].

Due to the sensitive responses and non-interfering characteristics of chiral signals, chiral biosensors were firstly fabricated for monitoring biomarkers in serum samples and intracellular systems. AFP is a critical biomarker for liver cancer [31]. Chiral Au NP heterodimers were fabricated by using AFP aptamers and the complementary sequences as linkers (Figure 4.4a). AFP aptamer sequences and its complementary sequences were fixed on the $20 \pm 2$ nm and $25 \pm 3$ nm Au NPs, respectively. Au NP heterodimers were formed and showed a strong chiral signal at 528 nm. When AFP was added, AFP would bind with the aptamer sequence, resulting in disintegration of Au NP heterodimers and a decrease in the CD signal. The intensity of the CD signal was linearly related to the AFP concentration, and the sensitivity of AFP detection by this method reached up to 11 pg ml$^{-1}$ [32]. Chiral Au NP heterodimers achieved AFP detection by using the chiral signal as the sensitive detection signal and aptamers as the identification unit, which provided the established sensors with high sensitivity and better selectivity. By adjusting the aptamers used, chiral Au NP heterodimers could also be used for the detection of other biomarkers. For example, chiral Au NP heterodimers were applied for the monitoring of 8-hydroxy-2′-deoxyguanosine (8-OHdG) in human serum samples (Figure 4.3a). 8-OHdG is a product of hydroxylation of the guanosine moiety of a nucleic acid base and has been considered a biomarker that represents cumulative oxidative DNA damage in animals and humans exposed to mutagenic agents. Aptamers for 8-OHdG were reported and showed high affinity toward 8-OHdG and can be used as recognition units for selective detection of 8-OHdG. Probe-1-modified 25 nm Au NPs were assembled with probe-2-functionalized 20 nm Au NPs in the presence of aptamers, and Au NP heterodimers were formed and showed strong

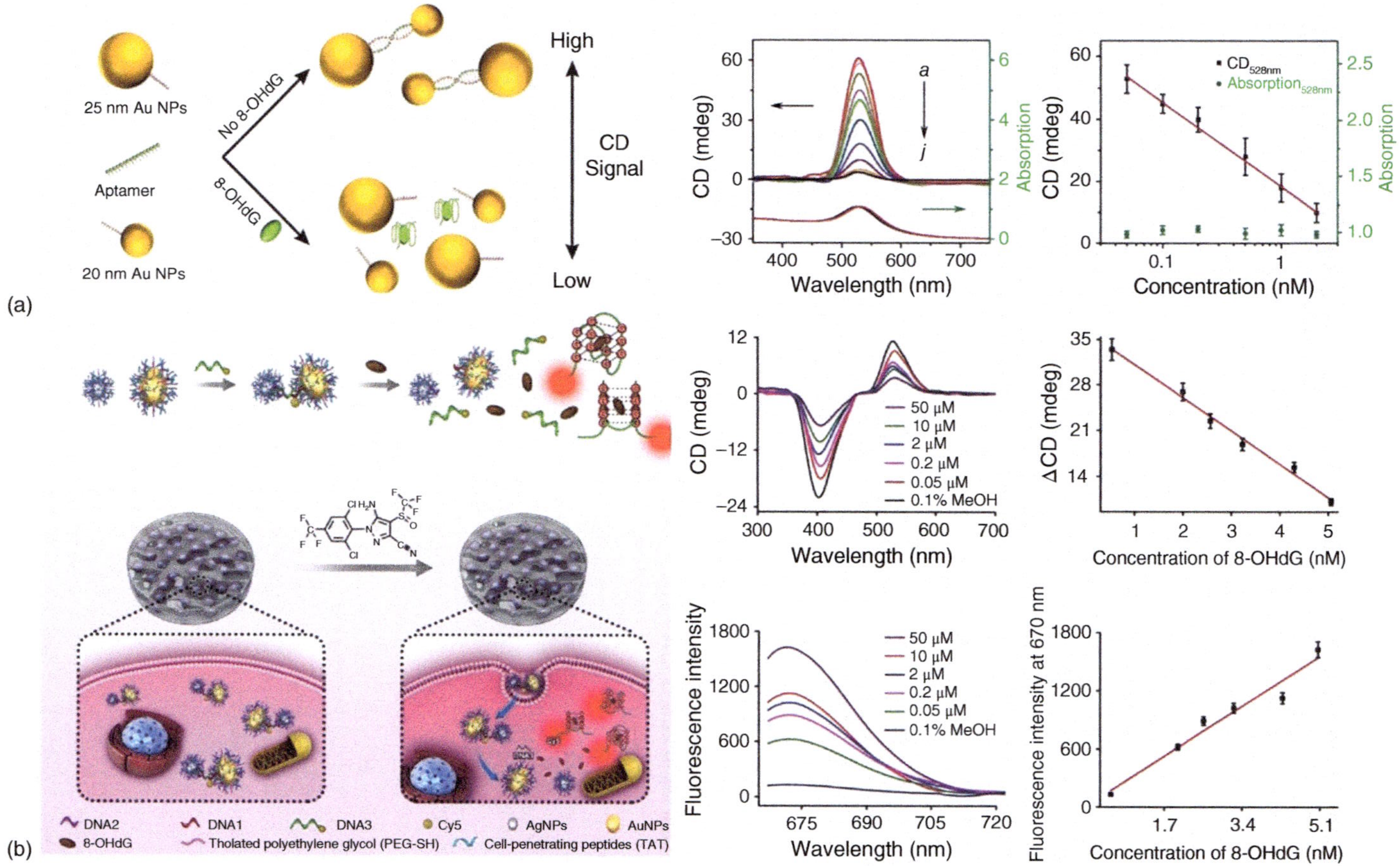


**Figure 4.3** (a) Chiral Au NP heterodimers were applied for the monitoring of 8-OHdG in the human serum sample. (b) Au NP-Ag NP dimers were also fabricated for the detection of 8-OHdG. Source: (a) Reproduced with permission Liu et al. [35]. Copyright 2016, American Chemical Society; (b) Reproduced with permission Wang et al. [36]. Copyright 2018, American Chemical Society.

CD responses at 528 nm. When 8-OHdG existed in the system, the strong affinity of aptamers and 8-OHdG destroyed hybridization of 8-OHdG aptamers and their complementary sequence, resulting in the disassembly of Au NP heterodimers and decreasing CD signals. The CD intensity was in log-linear correlation with the concentration of 8-OHdG ranging from 0.05 to 2 nM. The LOD was 33 pM. Thus, 8-OHdG could be sensitively and specifically detected in low concentrations based on the changes in CD intensity. The LOD of this method was better than or comparable to that of most of the previously reported methods, and it also exhibited higher sensitivity, a wider linear range for 8-OHdG detection. Recoveries ranged from 92.5% to 107%, and the relative standard derivations were in the range of 4.89–7.27%. On the basis of this principle, a simple, highly sensitive, and selective CD method for detection of 8-OHdG was conducted. Good accuracy, precision, and sensitivity were obtained when the method was applied in a complex matrix such as human serum samples, which was significant for clinical diagnosis. In addition, the detection processes were conducted at room temperature and ambient conditions, and the operation of experiments was relatively simple and convenient [35]. In comparison to Au NP dimers, Au NP-Ag NP dimers showed amplified CD responses for ultrasensitive biomarker detection. Au NP-Ag NP dimers were also fabricated for the detection of 8-OHdG (Figure 4.3b). By using 8-OHdG aptamers as linkers, $20 \pm 5$ nm Au NPs and $15 \pm 5$ nm Ag NPs were modified by DNA1 and DNA2, respectively. DNA1 and DNA2 could hybridize with the aptamers of 8-OHdG, inducing the formation of Au NP-Ag NP dimers. Au NP-Ag NP dimers showed strong plasmonic chiral signals at 526 and 405 nm, and Au NP-Ag NP dimers quenched the fluorescence signal from Cy5 on the end of the aptamers. The presence of 8-OHdG induced the disassembly of Au NP-Ag NP dimers, due to the higher affinity between 8-OHdG and aptamers than the combination of the aptamers with the complementary sequences. This change in structures induced decreased CD signals and recovered fluorescence signals. Intracellular 8-OHdG not only could be detected by the plasmonic CD response of Au NP-Ag NP dimers but also could be imaged through the changes in fluorescence signals of Cy5. The LOD was as low as 0.1 nM for the CD signal and 0.3 nM for the fluorescence signal. The CD signal showed higher sensitivity than the fluorescence signal. This proposed dual optical sensor (chiral and fluorescence signals) was applied to measure the oxidative DNA damage caused by nonylphenol and alternariol in cells. Chiroplasmonic platform was a promising and applicable method for detecting 8-OHdG in cells but also for assessing the capacity of chemicals to induce intracellular oxidative DNA damage [36].

Not only aptamers could be used as bridges for the assembly of Au NP-Ag NP dimers, but antibodies and antigens could also be used as linkers for the fabrication of chiral Au NP-Ag NP dimers. For example, antibody-functionalized Au NPs and Ag NPs were assembled into sandwich structures in the presence of PSA, resulting in the formation of Au NP-Ag NP heterodimers (Figure 4.4b). Increased serum PSA levels indicated pathological changes or injury to the prostate. PSA levels were elevated in the vast majority of patients with prostate cancer [37]. Sensitive PSA detection in serum samples was achieved by using chiral signals as detection signals

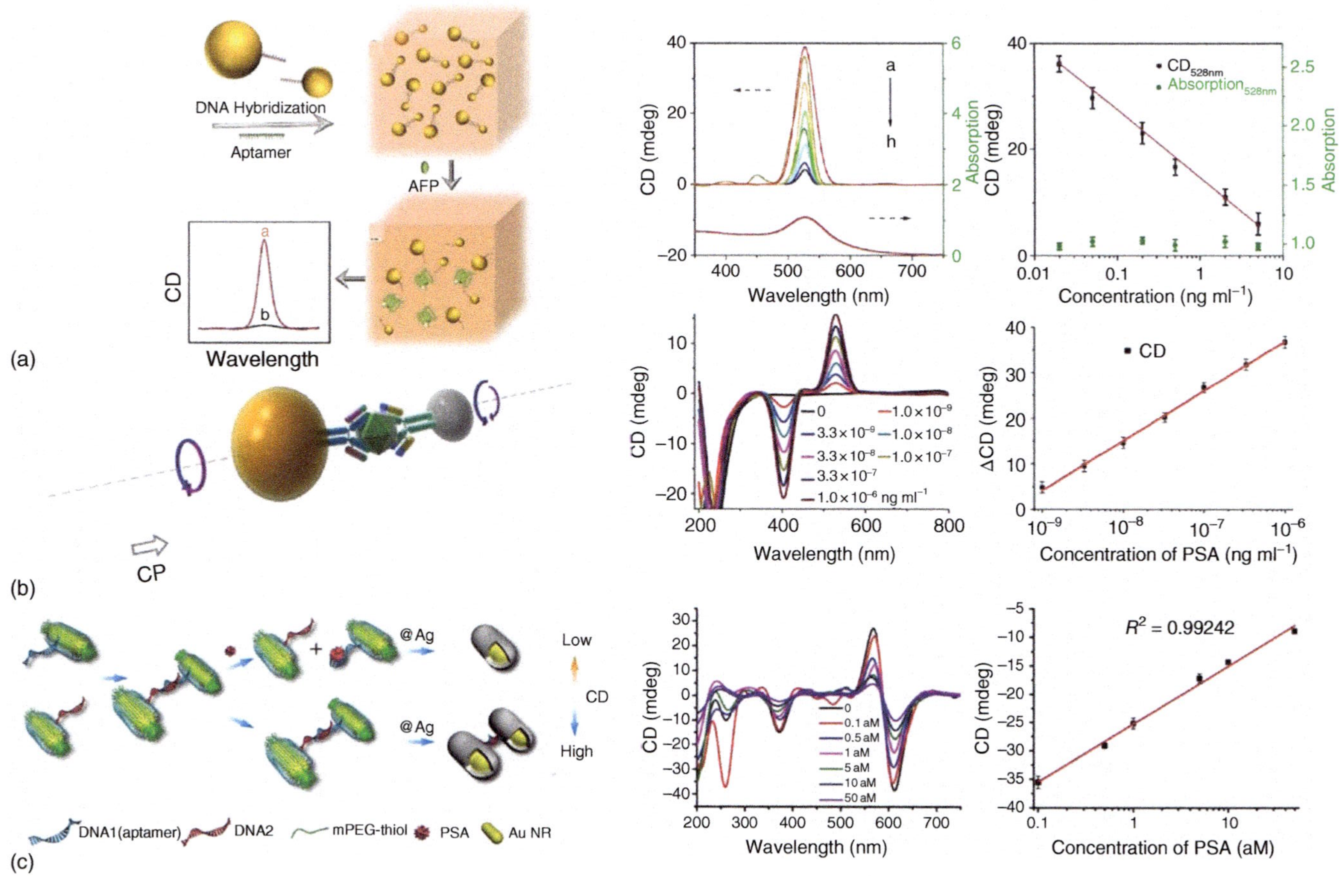


**Figure 4.4** (a) Aptamer-engineered chiral Au NP heterodimers for AFP detection. (b) Antibody-driven chiral Au NP-Ag NP heterodimers for PSA detection. (c) Aptamer-driven chiral Au NR dimers@Ag core–shell nanostructures for PSA detection. Source: (a) Reproduced with permission Zhao et al. [32]. Copyright 2017, Springer Nature; (b) Reproduced with permission Wu et al. [33]. Copyright 2013, American Chemical Society; (c) Reproduced with permission Tang et al. [34]. Copyright 2015, American Chemical Society.

and antibody as identification units. Au NP-Ag NP heterodimers showed intensive CD signals, due to the scissor-like geometry with an angle of 9° between the long axes of NPs. With the increasing amount of PSA, more chiral Au NP-Ag NP heterodimers were assembled. There was a positive correlation between the PSA concentration and chiral signals, with the detection range from $10^{-9}$ to $10^{-6}$ ng ml and an LOD of $5 \times 10^{-10}$ ng ml$^{-1}$ ($1.5 \times 10^{-20}$ M) [33]. Owing to the specific recognition between antigens and antibodies, the proposed chiral immunoassay showed better selectivity for PSA detection. Besides antigens and antibodies as bridges, aptamer-driven chiral sensors were developed for PSA detection. PSA aptamer-modified Au NRs and the complementary-sequence-modified Au NRs were assembled into side-by-side Au NR dimers. Au NR dimers exhibited CD signals in the visible region (Figure 4.4c). To amplify the chiral intensity and improve the detection sensitivity, Au NR dimers were further deposited on an Ag shell with a different thickness by reducing various concentrations of $AgNO_3$ solution. Au NR dimers@Ag core–shell nanostructures demonstrated a blueshifted and amplified CD peak, in which the longitudinal LSPR shifted from 710 to 533 nm and the CD signals were increased fourfold than that of bare Au NR dimers. The presence of PSA could dissociate the assembly and reduce the CD signals [34]. Au NR dimers@Ag core–shell nanostructures engineered with chiral aptasensors achieved sensitive and selective PSA detection, owing to the amplified chiral responses after shell deposition and unique specificity of aptamers.

Besides heterodimer structures, Au NP heterotrimers exhibited intensive CD signals in the visible region, and can be used as detection probes for detection. Fu's team used different sized Au NPs to construct a trimer structure for detection of shellfish allergen tropomyosin (TROP) (Figure 4.5). TROP is the allergen of most shellfish-induced allergic reactions, and its sequence conserved is highly homologous, so it can be used as a biomarker to realize authenticity identification of shellfish products and tracking of shellfish allergens. Using 16-mercaptohexadecanoic acid (16-MHDA) as ligands, 20 and 30 nm Au NPs were prepared. 16-MHDA is a nonionic surfactant that can bind the antibody to the surface of Au NPs by covalent bonding, thus improving the stability and detection sensitivity of Au NPs. Under the action of 16-MHDA, the TROP was connected to 20 nm Au NPs (Au NPs-TROP). Meanwhile, the anti-TROP monoclonal antibodies (mAb) were covalently modified to the 30 nm Au NPs (Au NPs-mAb). Au NP heterotrimers were formed between two Au NPs-TROP and one Au NPs-mAb, owing to the specific binding of antigens and antibodies. Such Au NP heterotrimers displayed strong CD responses at 525 nm. The presence of free TROP antigen in the samples would bind on the surface of Au NPs-mAb, inducing the disassembly of Au NP heterotrimers. The number of TROP in the system increased and the CD signal decreased with a weak redshift. This was because the dimers had a smaller dipole moment and a smaller electromagnetic field enhancement than the trimers, resulting in a weaker CD signal under the same conditions. The sensitivity of this method was high, and there was a good linear relationship between the CD signal and TROP concentration in the range of 0.1–15 ng ml$^{-1}$. The LOD was less than 21 pg ml$^{-1}$, and the quantification limit was 70 pg ml$^{-1}$. It could be applied to the practical detection of seven shellfish samples. At

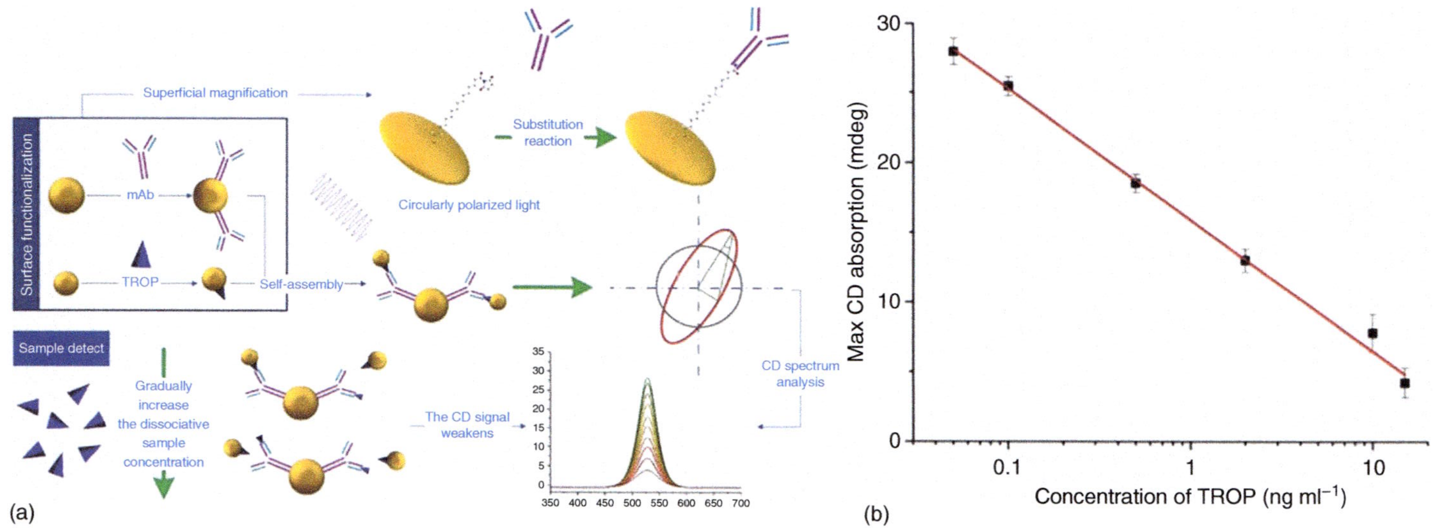


**Figure 4.5** (a) Schematic of a trimer structure constructed using different sizes of Au NPs for the detection of TROP in shellfish. (b) The linear curve of CD signal and TROP concentration in the range of 0.1–15 ng ml$^{-1}$. Source: (a, b) Reproduced with permission Wang et al. [38]. Copyright 2019, Elsevier.

the same time, specificity was strong, and the nonallergic protein would not interfere with the actual sample detection [38].

As mentioned above, Ag shell deposition not only provided a way to adjust the position of CD peaks but also largely amplified the CD responses of chiral Au NP nanostructures for specific applications. Ag displayed strongest SPR properties in the visible region but was limited by poor stability. It was known that the deposition of Au shell on Ag NPs would induce the displacement reaction, resulting in the formation of hollow nanostructures. To inhibit the displacement reaction, Ag@Au NPs were successfully synthesized by adjusting the pH of solution (Figure 4.6a). Chiral Ag@Au NP assemblies were prepared by using HER2 aptamers as bridges for the detection of HER2. HER2 is a proto-oncogene of SK-BR-3 cells, and its oncogene and protein product are overexpressed and amplified in a variety of tumors. The positive expression of HER2 can be used as an independent indicator for the prognosis of breast cancer. Zhao et al. deposited Au shells on the surface of Ag NPs, which adjusted and reversed the spectral position, as well as amplified the intensity of the CD signal of Ag@Au NP assemblies. The affinity between of HER2 and aptamers induced the disassembly of Ag@Au NP assemblies, resulting in decreased chiral intensity. There was a negative relationship between the HER2 concentration and chiral signals in the range from 50 to $10^5$ cells $ml^{-1}$. The LOD was calculated to be $10 \pm 6$ cells $ml^{-1}$. The chiral aptasensor realized sensitive and specific detection of cancer cells with HER2 overexpression on the surface [39].

Chiral biosensors not only are widely applied for sensitive and accurate detection of different biomarkers in serum samples by changing the chiral nanostructures and recognition biomolecules but also have achieved non-interfering monitoring of biomarkers in extracellular and intracellular systems. For example, epithelial cell-adhesion molecule (EpCAM) is a glycosylated membrane protein and is overexpressed on the surface of cancer cells. Chiral aptasensor by using GO-Au NP assemblies was developed for the detection of EpCAM on MCF-7 cell membranes (Figure 4.6b) [12]. Two different DNA-probe-modified GO and Au NPs were assembled with the aid of the nucleotide sequence against EpCAM. Au NPs were uniformly assembled on the surface of large-scale GO. The plasmonic coupling between elliptical Au NPs and the enhanced electronic oscillation coupling of Au NPs in the presence of conductive GO both induced the generation of CD responses. The affinity between the aptamers and EpCAM induced the separation of Au NPs from GO surface, resulting in decreased chiroptical properties. Chiral GO-Au NP assemblies achieved the extracellular detection of EpCAM on MCF-7 cell membranes, with the detection range of 8.47–74.78 pg $ml^{-1}$ and an LOD of 3.63 pg $ml^{-1}$. Furthermore, microRNA (miRNA) are endogenous short single-stranded non-protein and non-coding RNA molecules. They play an important role in the generation, prognosis, and treatment of human diseases and tumors. However, due to their low abundance, degradability, small size, and sequence similarity, it is difficult to detect. Traditional miRNA detection methods mainly include PCR and microarray methods but are limited by high cost, complex operation of cell lysis and RNA extraction. Chiral nanostructures overcome these shortcomings and have high sensitivity and specificity and have been applied for

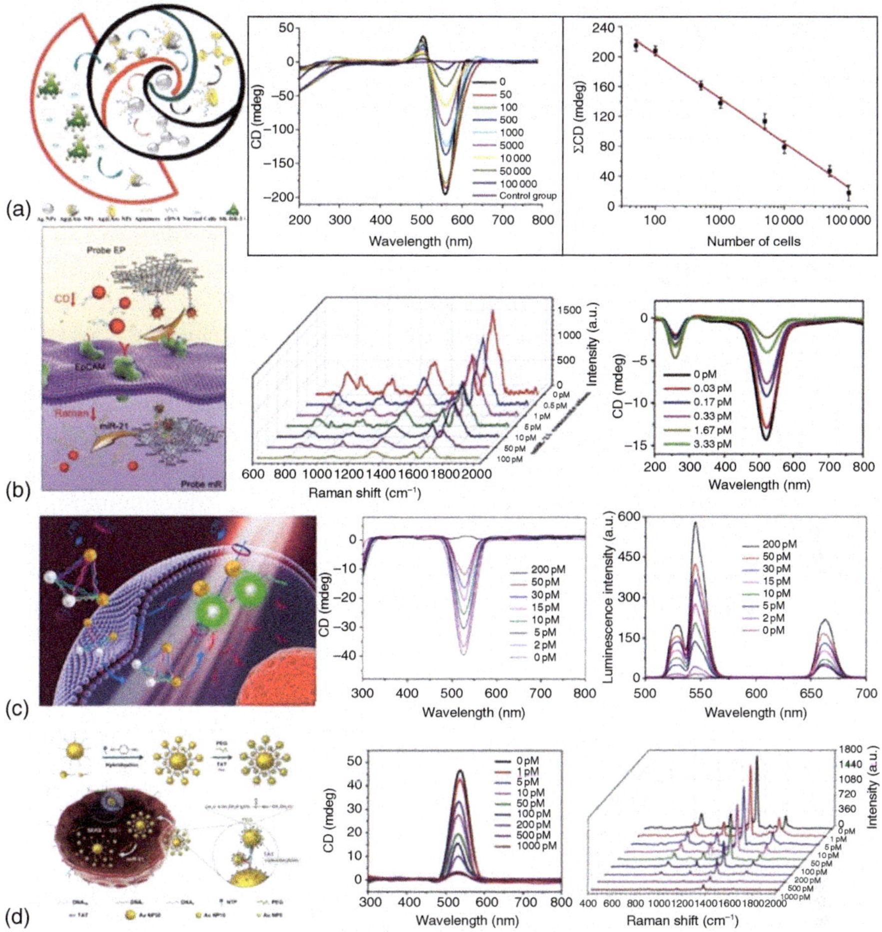

**Figure 4.6** (a) Chiral Ag@Au NP assemblies for HER2 detection expressed on SK-BR-3 cells. (b) Chiral GO-Au NP assemblies for the detection of extracellular EpCAM on the membrane of MCF-7. (c) Detection of miR-21 in living cells by using the CD and PL signals from the DNA-driven Au NP-Au NP-UCNP-UCNP pyramid structure. (d) CD- and SERS-dependent two-layer core–satellite Au NP assemblies for the detection of miR-21 in living cells. Source: (a) Reproduced with permission Zhao et al. [39]. Copyright 2016, Wiley-VCH; (b) Reproduced with permission Ma et al. [12]. Copyright 2017, Wiley-VCH; (c) Reproduced with permission Li et al. [40]. Copyright 2015, American Chemical Society; (d) Reproduced with permission Meng et al. [41]. Copyright 2020, Wiley-VCH.

the monitoring of miRNAs in living cells. By using similar GO-Au NP assemblies, miR-21 was monitored in living cells. The GO-Au NP assemblies also showed intensive SERS signals due to collective coupling of GO and Au NPs. The intracellular miR-21 induced the disassembly of GO and Au NPs, decreasing the SERS signals. The detection ranges were from 0.07 to 13.68 amol $ng^{-1}{}_{RNA}$ and the LOD was as low as 0.03 amol $ng^{-1}{}_{RNA}$. Extracellular and intracellular different biomarkers could be simultaneously monitored by using similar nanostructures with different recognition units and distinct optical signals.

In comparison to one kind of optical signal, dual-mode optical sensors by employing chiral with other different optical signals was proposed for the accurate miRNA detection. Xu and coworkers achieved miR-21 detection in living cells by using CD and PL signals in the visible and near-infrared regions (Figure 4.6c) [40]. DNA-driven Au NP-Au NP-UCNP-UCNP pyramids showed intensive chiroptical signals at 521 nm owing to unique chiral configurations and quenched PL signals at 500–600 nm due to luminescence resonance energy transfer (LRET) of UCNPs and Au NPs. The appearance of miR-21 would bind with the designed complementary DNA sequence on each side of the DNA frame, leading to dissociation of pyramids. The changes in nanostructures led to the "on-to-off" switch of CD signals and "off-to-on" switch of PL properties. Therefore, a negative correlation curve was established between chiral signals and miR-21 concentration and a positive relationship curve between PL signals and miR-21 concentration. The LODs were 0.03 fmol/10 $\mu g_{RNA}$ for CD signals and 0.12 fmol/10 $\mu g_{RNA}$ for PL signals. Alternatively, CD signals showed higher sensitivity than PL signals. Dual-mode optical signals provided biosensors with better accuracy for the monitoring of intracellular miRNA, especially in living cells. Furthermore, dual CD and SERS signals were proposed for the detection of miR-21 in living cells (Figure 4.6d). "Y"-shaped DNA skeleton was designed for the assembly of three different sized Au NPs. $DNA_1$-modified 5 nm Au NPs and $DNA_2$-capped 10 nm Au NPs were assembled with $DNA_{TR}$-modified 20 nm Au NPs, forming two-layer core–satellite nanostructures. This fabricated two-layer core–satellite nanostructures exhibited strong CD responses and SERS activity. The existing miR-21 in living cells would combine with $DNA_{TR}$ on 20 nm Au NPs, releasing 5 nm Au NPs and 10 nm Au NPs from the assemblies. Therefore, decreased CD intensity and SERS signals were both observed. There was a good relationship between CD signals and the amount of miR-21 ranging from 0.011 to 20.94 amol $ng^{-1}{}_{RNA}$, and a linear correlation was obtained between SERS intensity and the concentration of miR-21 in the range of 0.052–34.98 amol $ng^{-1}{}_{RNA}$. The LODs were 0.0051 amol $ng^{-1}{}_{RNA}$ for CD signals and 0.0281 amol $ng^{-1}{}_{RNA}$ for SERS signals [41]. This dual-mode detection can be applied for accurate and sensitive monitoring miRNA in living cells.

## 4.4 Enzyme Bioanalysis

Enzymes are generally dissolved in the cytoplasm, and are combined to various membrane structures, or are located at specific sites in other cellular structures (a product of the cell). Common intracellular enzymes include DNA endonuclease, DNA methyltransferases, telomerase, etc. DNA endonuclease is used to recognize specific deoxyribonucleotide sequences and cleaves phosphodiester bonds between two deoxyribonucleotide sequences at specific locations in each strand [42]. DNA methyltransferases cause DNA methylation by covalent bonding to a methyl group at the carbon site of cytosine 5 of the CpG dinucleotide in the genome. A large number of studies have shown that DNA methylation can lead to changes in chromatin structure, DNA conformation, DNA stability, and DNA interaction with proteins

[43]. Telomerase, a basic nucleoprotein reverse transcriptase responsible for telomere elongation in cells, can add telomeres to the end of chromosomes in eukaryotic cells and replace the telomeres lost in DNA replication. Telomerase plays an important role in maintaining telomere stability, genome integrity, long-term cell activity, and potential continuing proliferation [44].

DNA-driven chiral nanomaterials have been widely fabricated and show intensive chiroptical properties. In particular, endonucleases are specific enzymes capable of breaking the phosphodiester bonds between nucleic acids in a polynucleotide chain. Deoxyribonuclease I (DNase I) can cleave the DNA skeleton, which will disassemble the chiral configuration and generate decreased chiral signals. Chiral signal–based biosensors were developed for sensitive detection of DNase I. For example, DNA-engineered Au NR-Au NP core–satellite assemblies and Au NP heteropyramids were prepared (Figure 4.8a) [45, 47]. These nanostructures generated intensive chiral signals, in which Au NR-Au NP core–satellite assemblies exhibited two CD peaks at 520 and 715 nm, and Au NP heteropyramids displayed chiral signals at 525 nm. The intensity of the chiral peaks was not changed without DNase I, but obviously decreased in the presence of DNase I. With the increasing amount of DNase I, more DNA strands were cleaved by DNase I, more chiral nanostructures were decomposed, and the chiral signals were generally reduced. There was a good relationship between the concentration of DNase I and the CD peak intensity. The detection ranges of DNase I by Au NP heteropyramid-dependent chiral bioanalysis were from 0.005 to 1 U $ml^{-1}$, and the LOD was 0.0036 U $ml^{-1}$. Based on this principle, DNA-engineered chiral assemblies could also be used to screen DNase I inhibitors. DNA methyltransferases generally transfer a methyl group from the methyl group donor of *S*-adenosylmethionine to the C5-position of cytosine or the 6-amino group of adenines at a specific base sequence. DNA methyltransferases participate in many physiological processes such as cell development, cell differentiation, aging, memory formation, and paternal and maternal imprinting. A variety of human diseases have a close relationship with the abnormal DNA methyltransferase activities, such as leukemia, Alzheimer's disease, autoimmune disease, and multiple types of cancers including lung, colon, prostate, breast, and endometrium cancers [48]. DNA methyltransferase activity is closely related to the DNA methylation level and is considered a potential biomarker of diseases. The assessment of DNA methyltransferase activity and screening of inhibitors are critical for the regulation of human diseases. Plasmonic NP-dependent sensors have been widely reported for monitoring the activities of DNA methyltransferases. For example, Au NPs were widely used for the fabrication of colorimetric sensors, chiral sensors, and fluorescence sensors for DNA methyltransferase detection. Qu and Ren et al. applied Au NPs for the establishment of a simple and universal colorimetric assay of DNA methyltransferase activity (Figure 4.7) [49]. Red color 13 nm Au NPs were prepared and modified by probe DNA (5′-CAC GAG TTG ACA-3′) and then were assembled together in the presence of DNA-1 linkers. The color of Au NP solution changed from red to purple. DNA-1 linkers contained 18-mer base pair overlap, which could form a stable duplex with 12-mer dangling ends, and the restriction site 5′-GATC-3′ was located in the duplex region. When

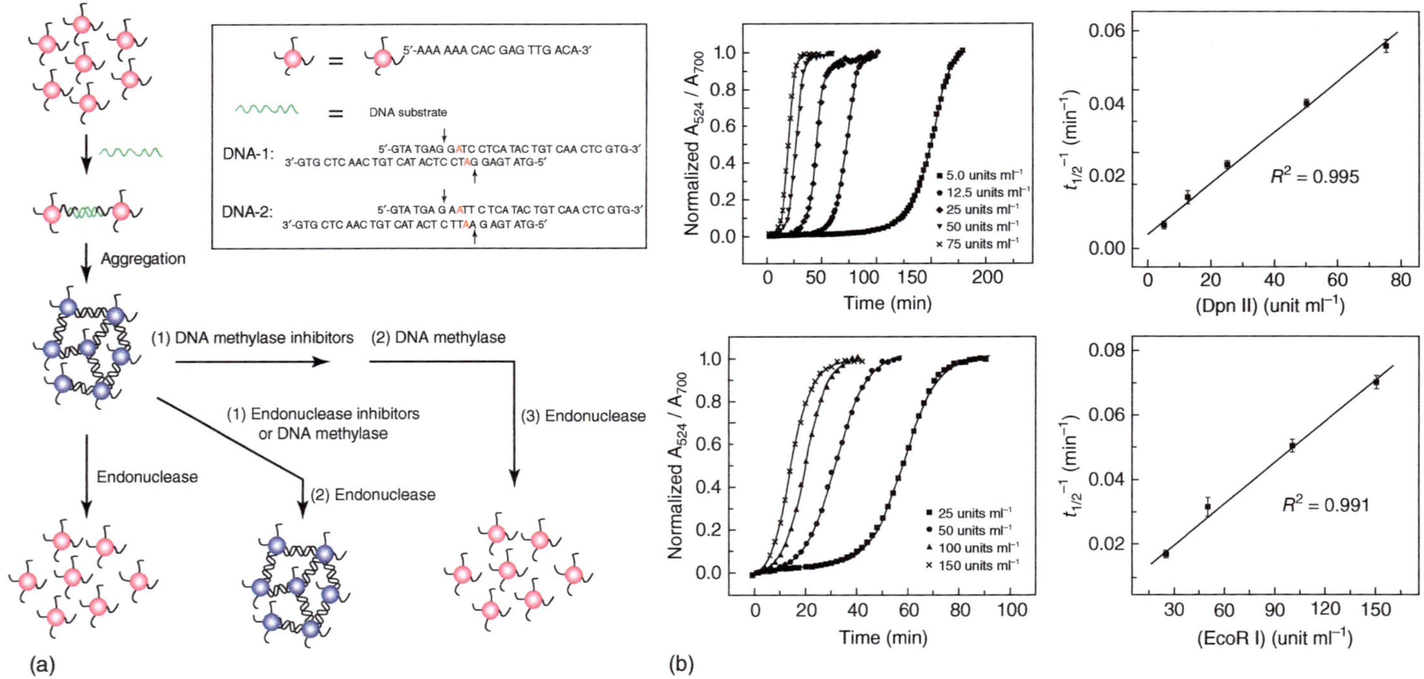


**Figure 4.7** (a) Schematic diagram of a colorimetric sensor for the detection of DNA methyltransferases using Au NPs. (b) The normalized resolution curves and linear curves of DPN II and EcoR I were detected using DNA-Au NP aggregates. Source: (a, b) Reproduced with permission Song et al. [49]. Copyright 2009, American Chemical Society.

Dpn II restriction endonuclease was present, Dpn II restriction endonuclease would recognize the restriction site 5′-GATC-3′ and further degraded the DNA–duplex interconnections, inducing the disassembly of Au NP nanostructures. The color of NP solution changed from purple to red, and the absorbance intensity at 700 nm for Au NP assemblies decreased and the absorbance intensity at 524 nm for released Au NPs increased. At the same time, Dam MTase could catalyze the methylation reaction at the recognition sequence to yield the methylation duplex DNA 5′-GAmTC-3′, which could not be cleaved by Dpn II restriction endonuclease. The Au NP assemblies were stable and still exhibited purple color. With the increasing amount of Dam MTase, there was a small change in the ratios of the absorbance intensity between 524 and 700 nm, which indicated the trend of methylation of DNA substrate by Dam MTase. Based on a similar principle, Au NP–based colorimetric assay could also be applied for monitoring the activity of EcoR I MTase by using the restriction site 5′-GAATTC-3′-embedded DNA-2 linkers in the presence of EcoR I.

Au NP assemblies not only possessed colorimetric properties but also chiral properties. Chiral signal–dependent sensors by using chiral Au NP assemblies were also proposed for the assay of DNA methyltransferase activity. For example, Wei et al. used chiral Au NP dimers for the evaluation of the activity of DNA methyltransferases and even their inhibitors. In this work, DNA-engineered chiral Au NP dimers were prepared and showed the chiral peak at 528 nm, and the chiral intensity was explored by optimizing the sizes of Au NPs and the interparticle distances (Figure 4.8b). HpaII could recognize and clear 5′-CCGG-3′ sequence in the DNA frame, which induced the disassembly of Au NP dimers and decreased CD signals. The presence of DNA methyltransferases could methylate the second C base of 5′-CCGG-3′ sequence, and the sequence would not be destroyed by HpaII. With the increasing concentration of DNA methyltransferases, more and more Au NP dimers were formed, generating stronger CD peaks. A linear correlation between DNA methyltransferase concentration and CD peak intensity was established, ranging from 0.5 to 150 U $ml^{-1}$, and the LOD was as low as 0.27 U $ml^{-1}$. This method can be applied to complex matrix samples such as human serum with high accuracy and sensitivity [46]. Due to high specificity of enzymes, the developed chiral biosensors were endowed with better selectivity for enzyme bioanalysis. Besides Au NPs, anisotropic Au NRs possess unique optical properties for biosensing. Au NR–dependent chiral assemblies will show high sensitivity for monitoring DNA methyltransferase activity. Although great progress has been made for sensitive and selective detection of DNA methyltransferases by using plasmonic nanostructures, most studies still focus on the detection of bacterial DNA methyltransferases (such as M.SssI and Dam). Further research should devote to monitoring the activities of human DNA methyltransferases (such as Dnmt1, Dnmt3a, and Dnmt3b). It is difficult to monitor the activities of DNA methyltransferases in living cells due to the requirements of nuclease. It is important to develop sensitive and accurate methods for imaging DNA methyltransferases in living cells without nuclease. The development of multiple detection of DNA methyltransferases is critical to accurate evaluation of cellular functions under external stimulation from xenobiotics.

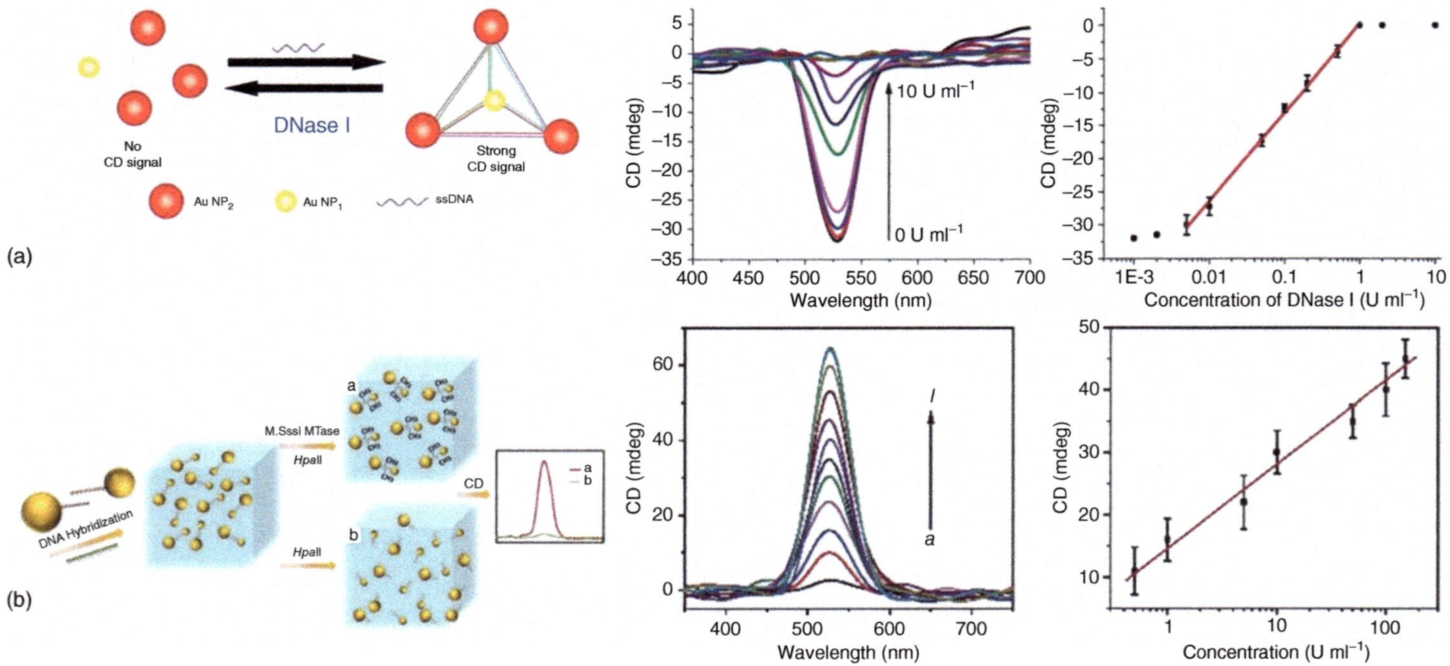


**Figure 4.8** (a) Chiral bioanalysis of DNase I enzyme by using Au NP heteropyramids. (b) Chiral bioanalysis of DNA methyltransferases by using DNA-engineered chiral Au NP dimers. Source: (a) Reproduced with permission Hao et al. [45]. Copyright 2013, Royal Society of Chemistry; (b) Reproduced with permission Liu et al. [46]. Copyright 2015, Royal Society of Chemistry.

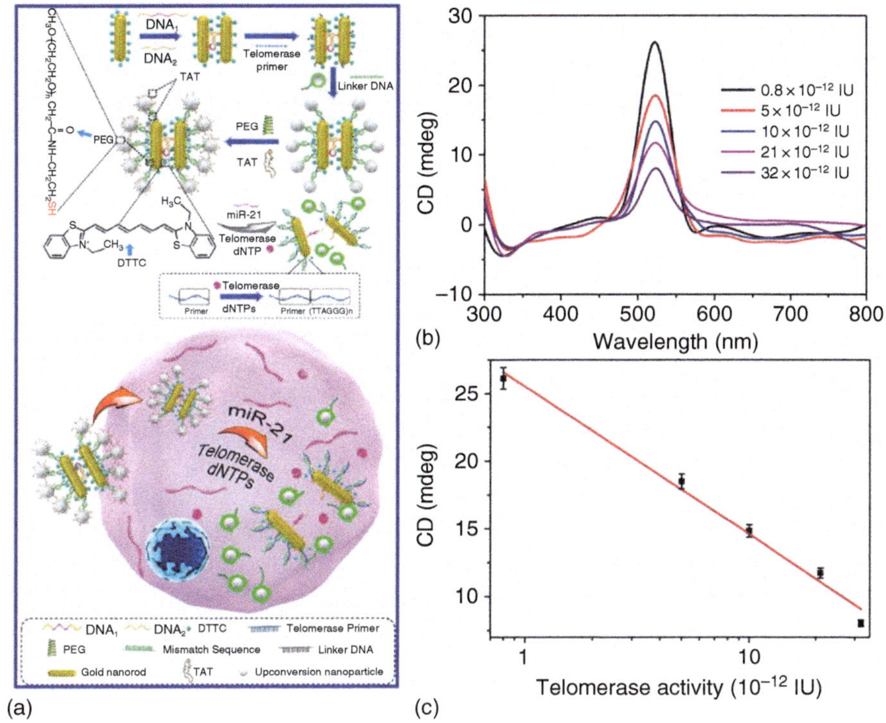


**Figure 4.9** (a) Detection of telomerase activity in HeLa cells by using chiral Au NP dimers. (b, c) CD spectra and standard linear curve for detecting intracellular telomerase activity using chiral Au NP dimers by CD signal. Source: (a–c) Reproduced with permission Sun et al. [1]. Copyright 2016, Wiley-VCH.

Although much great process has been made for enzyme detection by chiral biosensors, the reported studies are mainly limited to the applications in serum samples. Recently, chiral nanomaterials were used for probing intracellular enzyme activity. In particular, the concentration of telomerase is generally low in normal cells, and the length of telomeres is gradually shortened as the cell cycle progresses. However, most cancer cells mainly overexpress the level of telomerase and maintain the length of telomeres for cell proliferation. Therefore, telomerase is regarded as a critical indicator for cancer diagnosis and therapy. For example, intracellular telomerase activity was explored using chiral Au NP dimers (Figure 4.9). The characteristic of telomerase is that it can enable the repeated extension of TTAGGG sequence in the presence of endogenous RNA templates. Telomerase primer–modified Au NPs were assembled with Au NPs functionalized with linker complementary DNA strands in the presence of linker DNA, resulting in the formation of Au NP dimers with a yield of 82.5%. As-prepared Au NP dimers exhibited a strong chiral peak at 523 nm, attributed to the scissor-like chiral configurations. Cell-penetrating peptides (TAT) and thiolated polyethylene glycol (PEG) were modified on the surface of Au NP dimers, to overcome the intracellular membrane. The presence of telomerase in the cell induced the repeated extension of TTAGGG sequence triggered by the telomerase primer chain coupled on large Au NPs and resulted in the release of linker complementary DNA-strand-coupled small NPs. This would lead to the disintegration of the asymmetrical Au NP dimers, resulting in decreased CD signals. The LOD was $1.7 \times 10^{-15}$ IU in a single HeLa cell [1]. At the same time, the release of linker complementary DNA-strand-coupled small NPs enabled the light up of PL signals on the end of linker DNA. The decreased chiroptical spectra and turn-on PL images both indicated the intracellular telomerase activity, which showed potential applications in many types of cells. This strategy provides a new avenue for chiroplasmonic sensors in the detection of intracellular biomarkers, which hold great promise for early diagnosis of diseases in the clinic. Besides chiral and PL signals, dual-mode optical biosensors based on SERS and PL signals have been established for the simultaneous monitoring of telomerase and miR-21 in living cells. DNA-engineered Au NR dimers were fabricated and coupled with telomerase primers, and further were assembled with UCNPs in the presence of linker DNA. Au NR dimer-UCNP core–satellite assemblies were fabricated and showed intensive SERS activity as a result of EM enhancement from Au NR dimers and displayed quenched UCPL intensity due to the LRET process between Au NRs and UCNPs. TAT and PEG were also modified on the surface of Au NR dimer-UCNP core–satellite assemblies to cross cellular membranes for intercellular detection. However, the presence of telomerase induced the repeated extension of TTAGGG sequence and the release of UCNPs, leading to the recovery of PL signals of UCNPs. The presence of miR-21 resulted in the disassembly of Au NR dimers, generating decreased SERS signals [50]. Therefore, enhanced PL signals were used for the monitoring of intracellular telomerase activity, and the reduced SERS signals were applied for the detection of miR-21 in living cells. There was a good linear relationship between the PL intensity and the logarithmic concentration of intracellular telomerase in the range of $0.6 \times 10^{-12}$ to $31 \times 10^{-12}$ IU. These results were in good agreement with Raman and confocal imaging. The LOD for intracellular telomerase was calculated to be $3.2 \times 10^{-13}$ IU. At the same time, a good linear relationship was also established between the Raman intensities and the logarithmic concentration of intracellular miR-21. The detection range was between 0.021 and 22.36 amol $ng^{-1}{}_{RNA}$, and the LOD was determined to be 0.011 amol $ng^{-1}{}_{RNA}$. Double signal–dependent dual-mode optical biosensors were fabricated and they achieved simultaneous detection of intracellular enzymes and miRNA in living cells. This paves a new way to understand the chemistry within cellular systems and its application in the diagnosis of diseases.

## 4.5 Metal Ion Bioanalysis

Metal ions are widely found in nature and in the human body. They are divided into harmful heavy metal ions (such as $Pb^{2+}$, $Ag^{+}$, and $Hg^{2+}$) [51–53] and essential metal ions needed by human body (such as $K^{+}$, $Zn^{2+}$, $Cu^{2+}$, and $Mg^{2+}$). In particular, when the $Pb^{2+}$ content in drinking water is 10 $\mu g\,l^{-1}$ [54], it can cause the blood lead concentration to exceed the standard in children. $Pb^{2+}$ can accumulate in the human body, and its toxicity causes anemia, neurological disorders, and kidney damage [55]. $Ag^+$ settles in the body and causes a permanent blue-gray color to the skin, eyes, and mucous membranes. Because the water contains $Ag^+$, it can be combined with sulfur-containing components in cooking food to improve the absorption rate, so the concentration of $Ag^+$ in drinking water cannot exceed 0.05 mg $ml^{-1}$ [56]. $K^+$ can regulate appropriate osmotic pressure in cells and the acid–base balance of body fluids and participate in the metabolism of intracellular sugars and proteins [57]. The traditional methods for metal ion detection mainly include enzyme linked immunosorbent assay (ELISA), inductively coupled plasma mass spectrometry (ICP-MS), atomic fluorescence spectroscopy (AFS), and cold vapor atomic absorption spectrometry. However, these methods cannot be widely used because of time consuming, complex operation and the need for specialized instruments and skilled operators [58].

Chiral signal–dependent optical biosensors are widely studied for the detection of the level of heavy metal ions and the monitoring of essential metal ions in the complex matrix samples and even in living cells. Since $Ag^+$ can bind to cytosine (C) in DNA to form C-$Ag^+$-C structure, it is possible to assemble two Au NPs of different sizes using DNA strands with $Ag^+$ to form a chiral dimer structure (Figure 4.10a). With more $Ag^+$, more dimers were formed, and a stronger CD signal was obtained [59]. Similarly, $Hg^{2+}$ can bind to thymine (T) in DNA to form a T-$Hg^{2+}$-T structure. Using $Hg^{2+}$ and DNA strands, Au NRs were assembled into side-by-side assemblies (Figure 4.10b). $Hg^{2+}$ induced the transformation of ssDNA into dsDNA. The SPR effect between Au NRs and the DNA double strands caused the CD signal to change from the UV region to the visible region. With the increasing amount of $Hg^{2+}$, more Au NRs were assembled into side-by-side structures, generating enhanced CD peaks. There was a good relationship between the concentration of $Hg^{2+}$ and chiroptical intensity. The LOD was 0.03 ng $ml^{-1}$ [60]. Benefiting from the specific affinity of C-$Ag^+$-C and T-$Hg^{2+}$-T, DNA strands and these metal ions induced the assembly process of NPs for the production of strong chiroptical responses. A positive linear curve was generally established. Alternatively, such a positive correlation showed better accuracy and reliability than the negative one for the chiral analysis. Recently, chiral ligand–capped semiconductors were reported for $Pb^{2+}$ detection (Figure 4.10c). Cys-modified CdSe nanoplatelets were fabricated through Cd—S bonds. CdSe nanoplatelets were achiral and exhibited chiral properties after ligand exchange with cysteine, which was attributed to orbital coupling effects. L- and D-Cys-modified CdSe nanoplatelets showed a mirrored CD line shape with an anisotropy factor of $\sim 10^{-4}$. In the presence of $Pb^{2+}$, Pb—S bonds were formed between $Pb^{2+}$ and Cys, resulting in the separation of Cys from CdSe nanoplatelets, unlocking the chiral excitonic transitions and inducing decreased CD signals. With the increasing concentration of $Pb^{2+}$, the electronic state interaction between Cys and CdSe nanoplatelets weakened and the CD signals decreased. Interestingly, this method had strong specificity and sensitivity toward $Pb^{2+}$. Even though $Ca^{2+}$, $Mg^{2+}$, $Fe^{3+}$, $Cu^{2+}$, $Zn^{2+}$, and $Fe^{2+}$ could induce the decrease in CD peaks, the decreased extents were less than 15% in comparison to the control groups. However, obvious CD peak changes with a 58% decrease in intensity still occurred in the

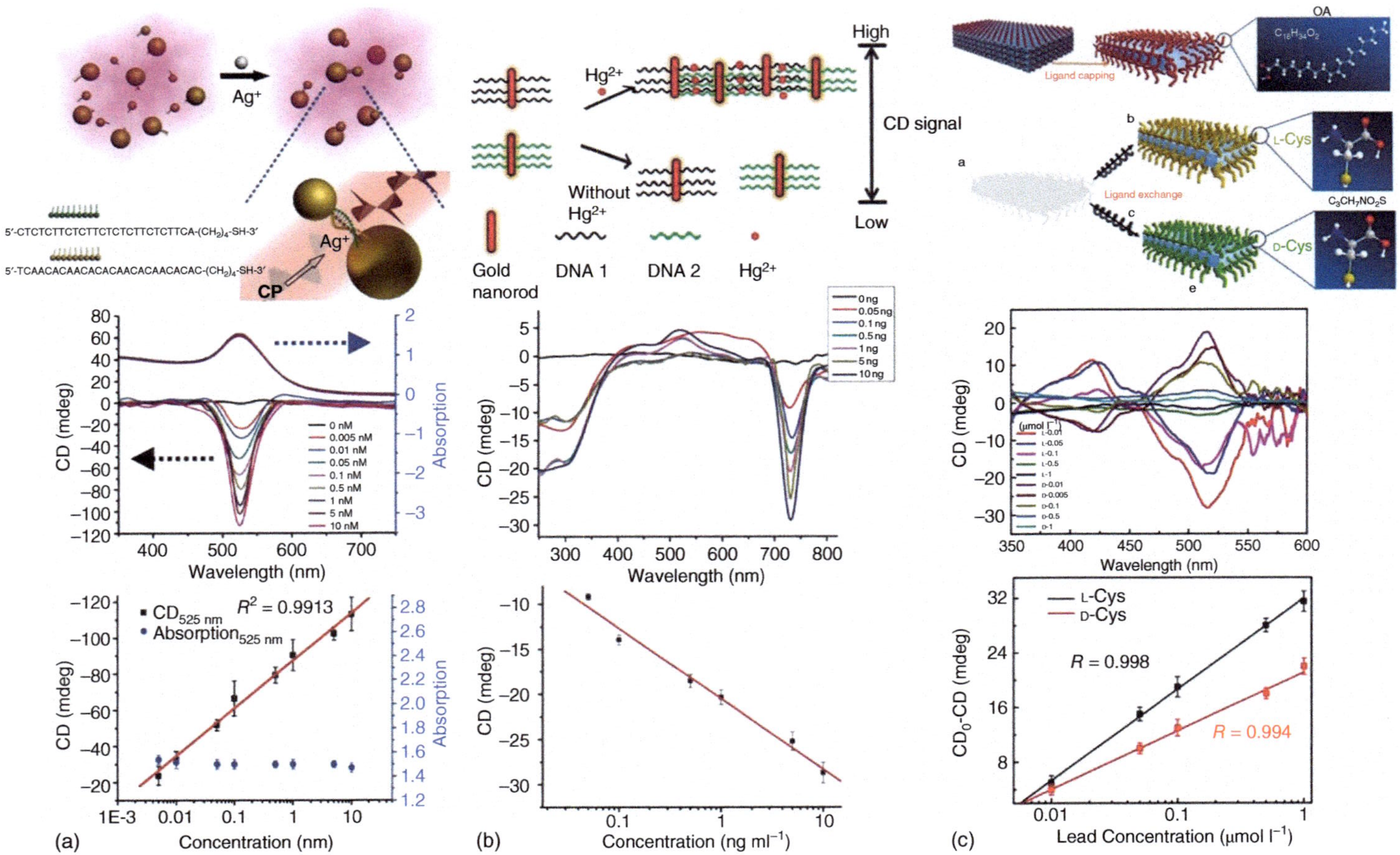


**Figure 4.10** (a) The detection of $Ag^+$ by chiral Au NPs heterodimers. (b) The detection of $Hg^{2+}$ by chiral side-by-side Au NR assemblies. (c) The detection of $Pb^{2+}$ by L- and D-Cys-modified CdSe nanoplatelets. Source: (a) Reproduced with permission Xu et al. [59]. Copyright 2013, Wiley-VCH; (b) Reproduced with permission Zhu et al. [60]. Copyright 2012, Royal Society of Chemistry; (c) Reproduced with permission Wang et al. [61]. Copyright 2019, Royal Society of Chemistry.

presence of $Pb^{2+}$. This high selectivity was due to the stronger interaction between the thiol groups and lead ions compared to other six kinds of different metal ions. The LOD was as low as 4.9 nM [61]. These results revealed that as-prepared chiral CdSe nanoplatelets achieved chiral signal-dependent heavy metal ion detection in wastewater processes, which could not be achieved for detection of heavy metal ions based on photoluminescence signals due to the non-luminous CdSe nanoplatelets formed after ligand exchange. The exploration of the chiral properties for semiconductors may provide a versatile platform for applications in chiral sensing, chiral separation, and design of chiral devices.

Besides the unique recognition ability for C-$Ag^{+}$-C and T-$Hg^{2+}$-T, DNAzymes as functional DNA strands showed high selectivity for metal ion detection. Kuang groups produced chiral dimers formed by two Ag NPs of different sizes via DNAzymes, structures that are chiral due to asymmetry (Figure 4.11a). Adding $Pb^{2+}$ to the system, triggered the DNAzymes to cut the DNA strands into pieces. Ag NP dimers were dissociated and the CD signal was weakened, so this chiral Ag NP dimer–based detection principle was designed to detect $Pb^{2+}$ in tap water. The detection range was from 0.05 to 10 ng ml$^{-1}$, and the LOD was down to 0.02 ng ml$^{-1}$ [62]. A large number of metal ion–specific DNAzymes have been widely explored, especially for essential metal ions ($Zn^{2+}$, $Cu^{2+}$, and $Mg^{2+}$). Metal ions play an important role in the study of life sciences. Trace metal elements account for 50–70% of the enzyme components in the human body. Metal ions also participate in the synthesis of certain hormones and vitamins and interact with nucleic acids, proteins, and other biological macromolecules. The emergence of metal ion–specific DNAzymes has made quantification of metal ion levels possible in living cells by using chiral nanomaterials. For example, $Zn^{2+}$ deficiency can cause loss of appetite, stagnation of growth and development in children, taste and smell function loss, fatigue, etc. Chiral biosensors were studied for the detection of $Zn^{2+}$ by using $Zn^{2+}$-specific DNAzymes. $Zn^{2+}$-specific DNA-cleaving DNAzymes were chosen as the DNA frame for the fabrication of Au@AgAu yolk–shell NRs-Au NP core–satellite assemblies (Figure 4.11b). These assemblies displayed stronger CD intensity at 530 and 740 nm compared to single Au@AgAu yolk–shell NRs, attributing to the chiral configuration and the intrinsic chirality of building blocks. By adjusting the aspect ratio of the Au NRs, the concentration of chiral molecules, and the size of nanogaps, the CD signal could be adjusted. $Zn^{2+}$ triggered the cleavage of DNAzymes for the DNA frame, destroying the Au@AgAu yolk–shell NRs-Au NP core–satellite assemblies and releasing Au NPs from Au@AgAu yolk–shell NRs. A decrease in CD peak intensity at 530 and 740 nm was observed in the presence of $Zn^{2+}$. As-fabricated chiral biosensor achieved the detection of different concentrations of $Zn^{2+}$ in SH-SY5Y cells. A good linearity was established, ranging from $76.6 \times 10^{-6}$ M/$10^6$ cells to $211.5 \times 10^{-6}$ M/$10^6$ cells, and the LOD was $38.7 \times 10^{-6}$ M/$10^6$ cells [63]. Besides $Zn^{2+}$, the lack of $Mg^{2+}$ can cause muscle spasms, rapid heart rate, and mental disorders, and $Cu^{2+}$ deficiency can cause brain tissue atrophy, loss of neurons, Alzheimer's disease, and so on. Multiple divalent metal ions ($Zn^{2+}$, $Cu^{2+}$, and $Mg^{2+}$) were simultaneously analyzed by using chiral nanomaterials (Figure 4.11c). $Zn^{2+}$-specific DNAzyme-coupled Au

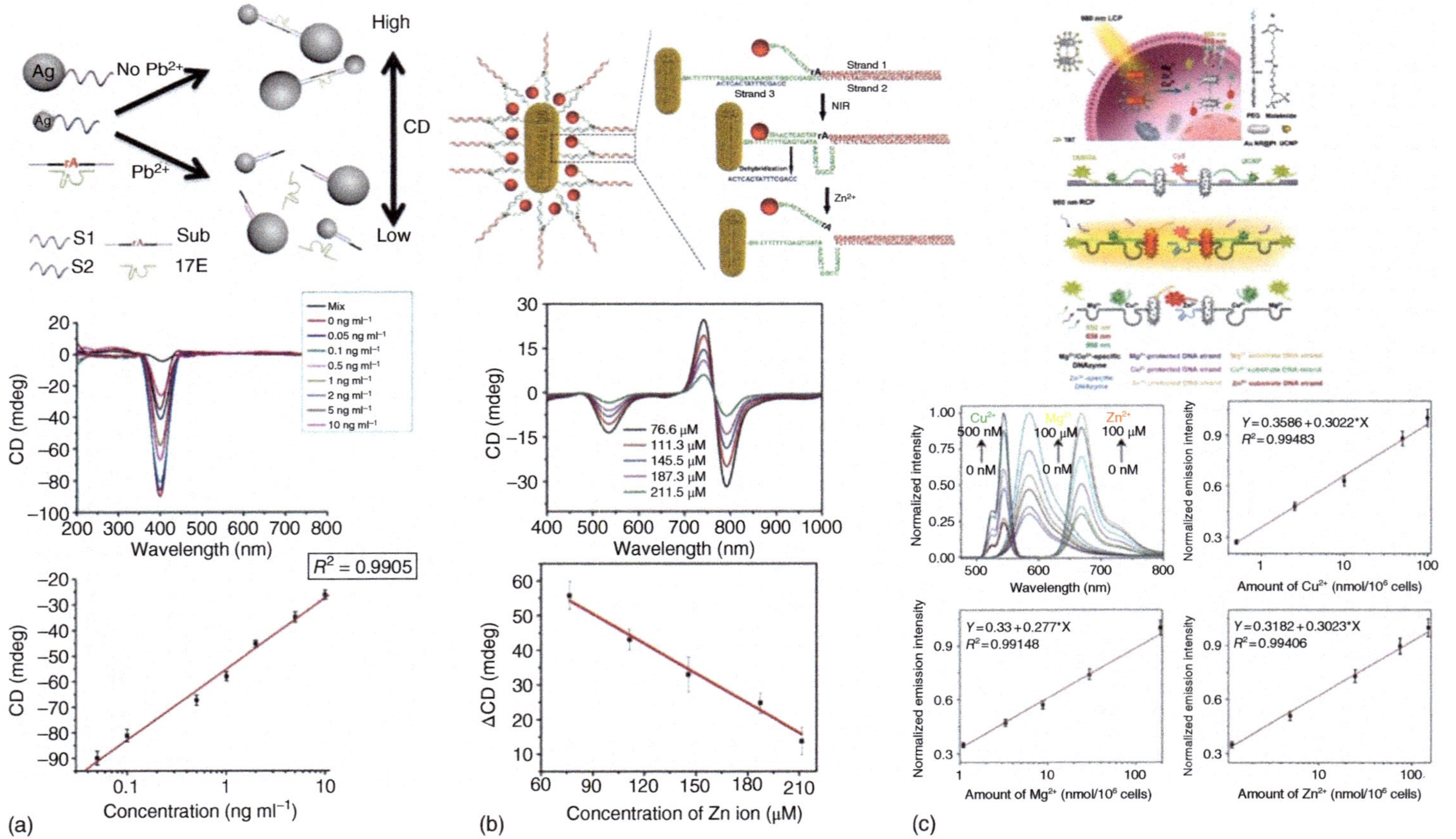


**Figure 4.11** (a) The detection of $Pb^{2+}$ by chiral Ag NP heterodimers. (b) Au@AgAu yolk–shell NR-Au NP core–satellite structures dependent chiral biosensor for the detection of different concentrations of $Zn^{2+}$ in SH-SY5Y cells. (c) $Zn^{2+}$, $Mg^{2+}$, and $Cu^{2+}$ in living cells detection by Au NRs@Pt-UCNP core–satellite assemblies under circularly polarized light at 980 nm. Source: (a) Reproduced with permission Kuang et al. [62]. Copyright 2013, MDPI; (b) Reproduced with permission Hao et al. [63]. Copyright 2018, Wiley-VCH; (c) Reproduced with permission Gao et al. [64]. Copyright 2019, Wiley-VCH.

NRs@Pt and $Zn^{2+}$-protected DNA-strand-attached Au NRs@Pt were assembled into dimers. $Mg^{2+}/Cu^{2+}$-specific DNAzymes were modified on the surface of the dimers. 5-carboxytetramethylrhod-amine (TAMRA)-attached $Mg^{2+}$ substrate DNA strand was combined on the 5′ end of $Mg^{2+}/Cu^{2+}$-specific DNAzymes, and UCNP-modified $Cu^{2+}$ substrate DNA strand was attached on the 3′ end of $Mg^{2+}/Cu^{2+}$-specific DNAzymes, and Cy5-functionalized $Zn^{2+}$ substrate DNA strand was hybridized with $Zn^{2+}$-specific DNAzymes. $Zn^{2+}/Mg^{2+}/Cu^{2+}$-protected DNA strands and DNAzymes as well as substrate DNA strands were hybridized to form metal-ion-specific DNAzyme precursors. Au NRs@Pt-UCNP core–satellite assemblies exhibited observable chiral signals. Under irradiation at 980 nm circular polarized light, chiral Au NRs@Pt-UCNP core–satellite assemblies absorbed the light with increase in in local temperature. Protected DNA strands were released from DNAzymes, and the presence of metal ions induced the cleavage of DNA strands triggered by DNAzymes, releasing the substrate DNA strands and restoring the PL signals of the corresponding dyes and UCNPs. $Cu^{2+}$, $Mg^{2+}$, and $Zn^{2+}$ triggered the recovered PL signals from UCNPs, TAMRA, and Cy5, respectively. Under circularly polarized light at 980 nm, $Zn^{2+}$, $Mg^{2+}$, and $Cu^{2+}$ in living cells were quantitatively detected. The detection limits were 1.1 nmol/$10^6$ cells, 1.02 nmol/$10^6$ cells, and 0.45 nmol/$10^6$ cells, respectively [64, 65]. Interestingly, extracellular and intracellular Au NRs@Pt-UCNP core–satellite assemblies exhibited different chiral responses and displayed distinct absorption ability toward RCP and LCP for temperature control. Circularly polarized light-active chiral nanomaterials showed potential applications for detection of specific multiplex metal ions with the aid of DNAzymes in extracellular and intracellular systems.

## 4.6 Small Molecule Detection

Macromolecules (such as proteins, enzymes, DNA) were monitored in intracellular and extracellular systems depending on specific chiral aptasensors and immunoassays [66]. Small molecules, involving toxins, microcystins, reactive oxygen species (ROS), and some organic chemicals, are also detected by using chiral nanosensors. Compared to other chiral nanostructures, chiral NP dimers are simple to fabricate and possess high assembly yields of more than 80%, as well as exhibit intensive CD peak intensity. These properties ensure good stability and high sensitivity in the detection applications. For example, immunoassays based on chiral Au NP dimers and aptasensors based on chiral Au NP dimers were successfully established for the detection of bisphenol A (BPA). BPA is an important organic compound used in the synthesis of materials (such as polycarbonate, epoxy resins), so it is widely found in metal coatings, food and beverage packaging, and water supply pipes. BPA can enter the body through the digestive tract, the respiratory tract, and skin and can have harmful effects on many systems, especially its reproductive system toxicity [67]. Studies have shown that exposure doses higher than 0.05 mg $kg^{-1}$ of body weight per day can cause health risks. Chiral Au NP heterodimers were assembled for BPA detection by using different Au NPs of 10 and 20 nm and applying antigen–antibody

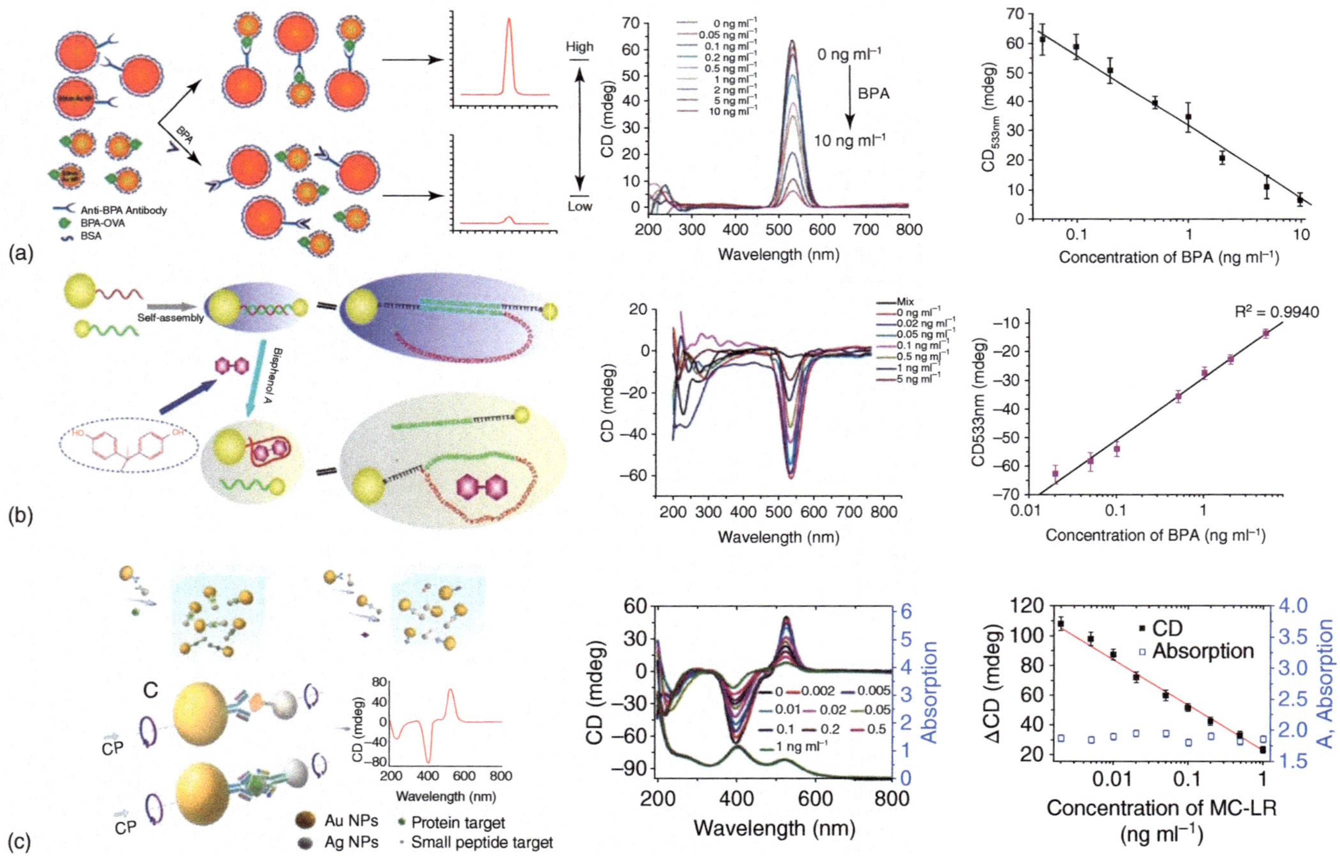


**Figure 4.12** (a) Antigen–antibody bridged chiral Au NP heterodimers for BPA detection. (b) BPA aptamers and the complementary DNA strands driven Au NP heterodimers for the detection of BPA. (c) The detection of MC-LR by chiral Au NP-Ag NP heterodimers. Source: (a) Reproduced with permission Xu et al. [14]. Copyright 2012, Royal Society of Chemistry; (b) Reproduced with permission Kuang et al. [68]. Copyright 2014, American Chemical Society; (c) Reproduced with permission Wu et al. [33]. Copyright 2013, American Chemical Society.

as bridges (Figure 4.12a). The chiral geometry of Au NP heterodimers enabled the CD signal to be converted from the UV region to the visible region. With the addition of BPA, anti-BPA antibody would competitively combine with BPA, and Au NP heterodimers were dissociated and the CD signal weakened. This method achieved BPA detection in tap water with an LOD of 0.02 ng $ml^{-1}$ [14]. Besides using antibody–antigen as recognition units, BPA aptamers and the complementary DNA strand-driven Au NP heterodimers were also fabricated and showed strong CD responses at 533 nm (Figure 4.12b). The addition of BPA would specifically be recognized by the aptamers, resulting in the disassembly of Au NP heterodimers and a decrease in CD signals. A lower LOD of 0.008 ng $ml^{-1}$ for chiral aptasensors was achieved, which was 0.02 ng $ml^{-1}$ for chiral immunoassays [68]. This may be due to the better recognition ability of aptamers for BPA than immune recognition. It is a critical way to improve the detection sensitivity by using chiral nanostructures with strong CD responses.

In comparison to Au NP heterodimers, Au NP-Ag NP heterodimers assembled by different metal NPs exhibited higher chiroptical properties, which were originated from the strong plasmonic coupling between Au NPs and Ag NPs for the generation of intensive EM field. Immune-recognition-engineered Au NP-Ag NP heterodimers were fabricated for the detection of microcystin-LR (MC-LR). Microcystins are a class of biologically active cyclic heptapeptides, which are the most widely distributed hepatotoxins. They are mainly produced by the freshwater algae Microcystis aeruginosa and have considerable stability. They strongly inhibit protein phosphatase activity and are a strong liver tumor promoter [69]. The maximum allowable concentration of MC-LR in drinking water has been determined to be 1 nM by the World Health Organization (WHO) since 1998[70]. Using antigen–antibody bridges, Au NPs and Ag NPs were assembled into Au NP-Ag NP heterodimers. Bisignate nature of CD bands occurred and possessed high g-factor of $0.5 \times 10^{-2}$ due to the nonspherical and scissor-like conformation. The addition of MC-LR induced competitive recognition with anti-MC-LR antibody, resulting in the disassembly of Au NP-Ag NP heterodimers and decreased chiral signals. There was a good linear relationship between MC-LR concentration and CD signal, and the LOD was down to 0.8 pM [33]. Aptamers of MC-LR were also screened and showed good selectivity toward MC-LR. The plasmonic coupling between bimetallic Au NPs and Ag NPs induced the generation of intensive EM field for the enhancement of SERS activity. SERS aptasensors were also established for sensitive and accurate MC-LR detection. For example, "Y"-shaped hybrid aptamer chains were designed and applied as DNA frames for the fabrication of SERS-active bimetallic Au nanoflower (Au NF)-Ag NP core–satellite heterostructures (Figure 4.12c). MC-LR aptamer-driven Au NF-Ag NP core–satellite heterostructures exhibited intensive SRES activity, owing to (i) the formation of more and strong hot spots between Ag NPs and Au NFs with multiple tips; (ii) the bimetallic plasmonic coupling between Ag NPs and Au NFs for the amplification of EM field; (iii) the utilization of Ag NPs with best plasmonic enhancement in the visible region; (iv) the reduced interparticle distances between Ag NPs and Au NFs through designing "Y"-shaped hybrid structures. It was reported that SERS activity of as-prepared Y-shaped

DNA-driven assemblies was 2.5 orders of magnitude stronger than that of dsDNA. The presence of MC-LR induced the disassembly of structures and generated decreased SERS signals. MC-LR was detected in Tai Lake water with an LOD of 8.6 pM [33]. Furthermore, the $\pi$–$\pi$ stacking interaction between MC-LR aptamers and GO sheets enabled the assembly of SERS-active bimetallic Au@gap@AgAu NPs on the surface of GO sheets. Au@gap@AgAu NPs exhibited gap- tunable amplified SERS activity and served as SERS tags. The GO/$Fe_3O_4$ sheets displayed a unique and stable Raman band from the graphitic component, making them an ideal internal standard for quantitative Raman analysis. As-prepared MC-LR aptamers driven Au@gap@AgAu NPs-GO/$Fe_3O_4$ sheets were utilized for the detection of MC-LR by using Au@gap@AgAu NPs as detection probes and GO/$Fe_3O_4$ sheets as reference probes. In the presence of MC-LR, aptamer-modified Au@gap@AgAu NPs were dissociated from GO/$Fe_3O_4$ sheets owing to the affinity between aptamers and MC-LR, leading to changes in Raman intensity of SERS tags. Reliable and accurate MC-LR in Tai Lake water was detected with an LOD of 9.82 pM [70].

Chiral biosensors based on bimetallic NPs were applied for the detection of mycotoxins. Mycotoxins mainly refer to the toxic metabolites produced by mold in the contaminated food. They can enter human and animal bodies through feed or food and cause acute or chronic poisoning, thereby damaging tissues and organs such as the liver and kidneys. Mycotoxins are secondary metabolites produced due to improper food storage, such as ochratoxin (OTA) and aflatoxin, which are widely found in grains, fruits, dried fruits, cocoa, beer, and wine. Mycotoxins can enter the human body through the food chain or can be directly ingested by humans, leading to liver and kidney toxicity, immune suppression, and even carcinogenicity and teratogenesis [71]. The traditional detection methods mainly include chromatography and ELISA. Chromatography needs expensive large equipment and complex operation time; ELISA needs sophisticated technologies and special reagents, which cause complications in detection. Therefore, it is imperative to develop a new toxin detection method. Aptasensors dependent on chiral nanostructures possess high sensitivity and good specificity and have been widely studied. OTA aptamer-coupled Au NPs were hybridized with OTA complementary-modified Ag NPs to form Au NP-Ag NP core–satellite structures. Strong CD peaks were observed at 400 and 527 nm, which were attributed to plasmon-enhanced chiral DNA and the asymmetry structures of the assemblies (Figure 4.13a). With the increasing concentration of OTA, Ag NPs would separate from the surface of Au NPs, leading to weakened CD signals at two peaks. A good linear relationship was established between the concentration of OTA and the summation of peak intensity at 400 and 527 nm. The LOD was as low as 0.16 pg $ml^{-1}$, which could be applied to the detection of OTA in red wine [72]. In order to further improve the detection sensitivity of OTA, high chiral-active nanostructures need to be designed. In comparison to solid NPs, shell NPs possess stronger electric field intensity for the enhancement of chiral signals. OTA aptamer-driven shell NP-Au NP core–satellite assemblies showed high CD signals at 521 nm, owing to the strong asymmetrical dipole–dipole interaction at the enhanced electric field (Figure 4.13b). The presence of OTA induced the disassembly of Au NPs from shell NPs, resulting in decreased

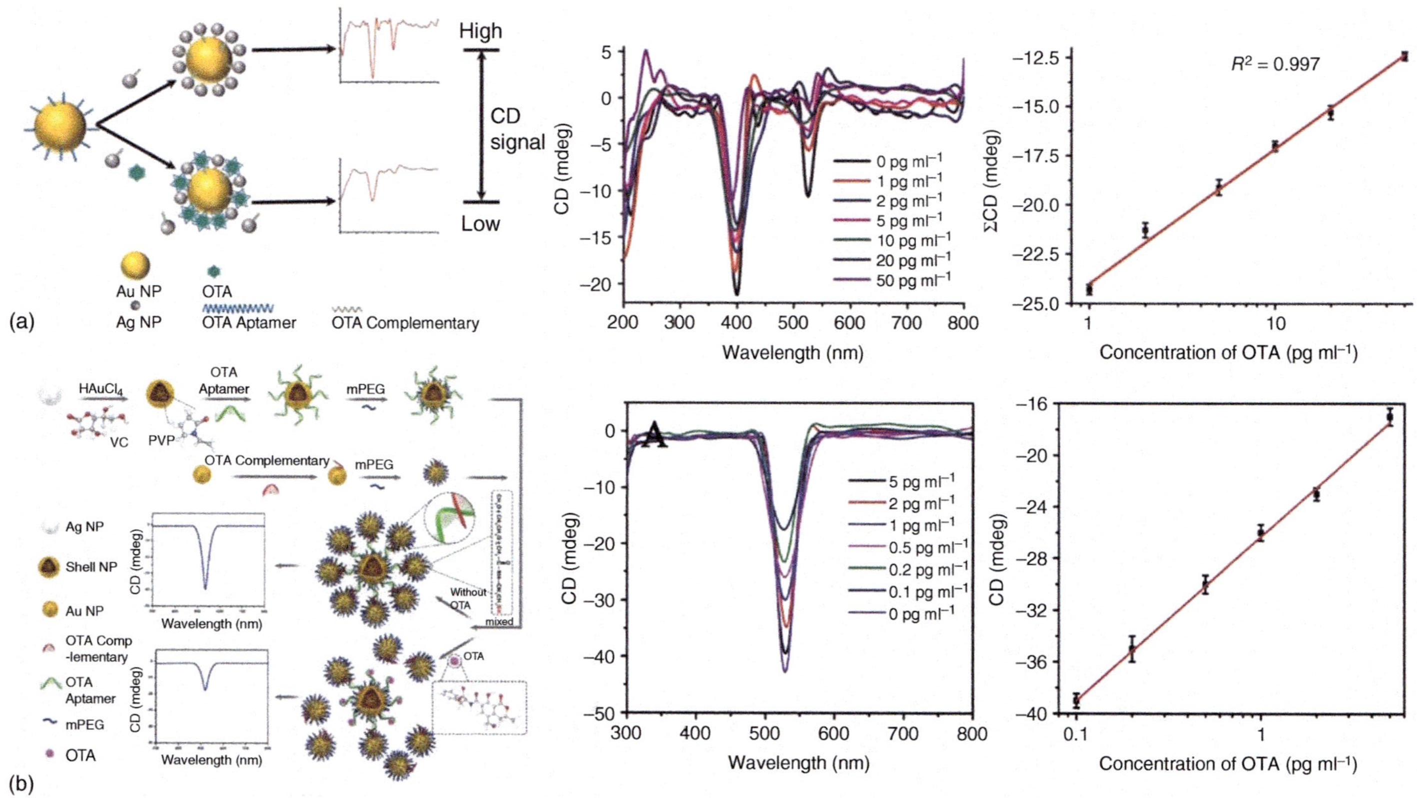


**Figure 4.13** (a) Chiral Au NP-Ag NP core–satellite structures for OTA detection in red wine. (b) OTA aptamer–driven chiral shell NP-Au NP core–satellite assemblies for OTA detection. Source: (a) Reproduced with permission Zhao et al. [72]. Copyright 2014, Elsevier; (b) Reproduced with permission Cai et al. [73]. Copyright 2018, Wiley-VCH.

chiral signals. There was a linear relationship between CD peak intensity at 521 nm and the amount of OTA; and the LOD was down to 0.037 pg $ml^{-1}$ [73]. Generally, chiral signals from chiroplasmonic nanostructures occurred in the visible region, which could not be affected by the chiral molecules with chiral signals in the UV region in the detection system. Chiral signal–dependent biosensors showed better accuracy. Besides chiral biosensors, SERS aptasensors were established for the detection of OTA. To improve the detection accuracy and stability, label-free and anti-interfering SERS probes were designed for OTA detection. For example, Zhao et al. reported a ratiometric SERS aptasensor for OTA detection by using plasmonic Au-Ag Janus NPs as detection probes and MXenes nanosheets as reference probes. Au-Ag Janus NPs were fabricated by the reduction of $AgNO_3$ on the surface of Au NPs using 2-mercaptobenzoimidazole-5-carboxylic acid (MBIA). As-prepared Au-Ag Janus NPs showed intensive SERS activity from MBIA, and the bimetallic plasmonic coupling between Au and Ag largely amplified the SERS signals of MBIA. This avoided the extra modification of Raman molecules on the surface of plasmonic NPs. As-prepared Au-Ag Janus NPs possessed better stability in the pH from 4 to 8, at 20–50 °C and within 90 days. At the same time, MXenes nanosheets also possessed a unique and stable Raman signal and could be used as an ideal internal standard for quantitative Raman analysis. OTA aptamer–modified Au-Ag Janus NPs were assembled with MXenes nanosheets depending on the hydrogen bond and the chelation interaction between MXenes nanosheets and OTA aptamers. In the presence of OTA, Au-Ag Janus NPs were dissociated from MXenes nanosheets due to the formation of aptamer/OTA complex, resulting in the attenuation of the Raman signal of Au-Ag Janus NPs; meanwhile, the signal of MXenes nanosheets remained constant. Upon correction by the IS Raman signals, sensitive and quantitative detection can be achieved with an LOD of 1.28 pM [71]. There generally existed more than one kind of mycotoxins in the complex matrix, for example OTA and aflatoxinB1 (AFB1). The European Commission sets the maximum allowable concentration of OTA in coarse grains at 5 μg $kg^{-1}$. In the United States and China, the current maximum allowable level of AFB1 in food is 20 ng $g^{-1}$ [74]. Double detection of OTA and AFB1 was achieved depending on SERS biosensors by using SERS-active Ag@Au core–shell NPs as reporters [75]. For example, by embedding Raman molecules into the junction of Ag core and Au shell, anti-interfering and strong SERS-active Ag@Au core–shell NPs were obtained. Multiple SERS-active Ag@Au core–shell NPs could be prepared by changing the types of Raman molecules. 4-ATP- and 4-NTP-embedded Ag@Au core–shell NPs were fabricated and showed different characteristic peaks at 1076 and 1335 $cm^{-1}$, respectively. Ag@4-ATP@Au core–shell NPs and Ag@4-NTP@Au core–shell NPs were used as detection probes for the double detection of OTA and AFB1, respectively. Such prepared SERS-active probes not only showed strong SERS activity owing to the plasmonic coupling between Ag core and Au shell but also possessed better stability due to the protection of Raman molecules from Au shell and provided possibility to fabricate more SERS tags for multiple detection applications. Although chiral biosensors show high sensitivity for mycotoxin detection, it is still difficult to achieve chiral signal–based double and even multiple mycotoxin detection.

Dual-mode optical biosensors are also developed for the monitoring of small molecules expressed on the surface of bacteria. Immunoassays dependent on chiral and UCPL signals achieved the quantitative analysis and bioimaging of antibiotic-resistant bacteria through the detection of polymyxin B. UCNPs and Au YSNPs were conjugated into heterodimers by polymyxin B and antibody, which showed quenched UCPL and high CD signals (Figure 4.15a). Polymyxin-B-functionalized UCNPs had low affinity for polymyxin-B-resistant lipid A radicals. Therefore, more UCNPs entered the sensitive strains than the resistant strains, and the UCL signal in the sensitive strains was stronger than that in the resistant strains. UCNPs that did not enter the cell were assembled with Au YSNPs into a dimer, so the CD signal of polymyxin-B-resistant strains was weaker than that of polymyxin-B-sensitive strains. Quantitative levels of polymyxin-B-resistant bacteria could be detected by using UCPL signals and CD signals. In addition, Au YSNPs produce singlet oxygen, which could be used as a photodynamic agent to completely eliminate the inflammation caused by the polymyxin-B-sensitive strain by photodynamic therapy [76]. Chiral and UCPL signals dependent dual-mode optical biosensors achieved the ultrasensitive determination of sulfamethazine (SMZ) by using Au@Ag@Au NP-UCNP heterodimers (Figure 4.14). SMZ is an antibacterial drug in veterinary medicine worldwide and is a growing and serious threat in animal-derived food supplies. It is necessary for the establishment of reliable and accurate approaches for the detection of SMZ in food products. It has been reported that aptamers of SMZ were selected and showed strong affinity toward SMZ. Aptamer-driven Au@Ag@Au NP-UCNP heterodimers were designed for dual-mode monitoring of SMZ. Aptamers of SMZ were modified on the surface of UCNPs, and Au@Ag@Au NPs were modified by the complementary DNA of SMZ aptamers. Depending on the unique complementary base-pairing reactions, SMZ aptamers would be hybridized with the complementary DNA sequence to form dsDNA structure, which induced the formation of Au@Ag@Au NP-UCNP heterodimers. Such Au@Ag@Au NP-UCNP heterodimers exhibited intensive CD responses owing to the chiral heterodimer structures and displayed quenched UCPL signals due to the efficient inner filter effect (IFE) from UCNPs (energy donor) to Au@Ag@Au NPs (energy acceptor) under excitation of 980 nm laser. When the target (SMZ) was present, the aptamers of SMZ would combine with SMZ, thereby decreasing the degree of the assembly. As a result, the chiral signal decreased, and the fluorescence signal recovered. SMZ was detected in the range of 0.10–100 ng $ml^{-1}$ for the fluorescence signal and 1.00–100 ng $ml^{-1}$ for the chiral signal. The recoveries were ranged from 88.05% to 108.33%. The strategy was applied to the reliable determination of SMZ in spiked milk with high recovery, better accuracy and reproducibility [78].

Chiral semiconductors, such as CdS QDs, were mainly constructed by introducing chiral ligands or constructing helical structures by means of electric dipole moment regulation and could be used to detect small molecular substances according to the change of the CD signal. Glucose plays an important role in the field of biology. It

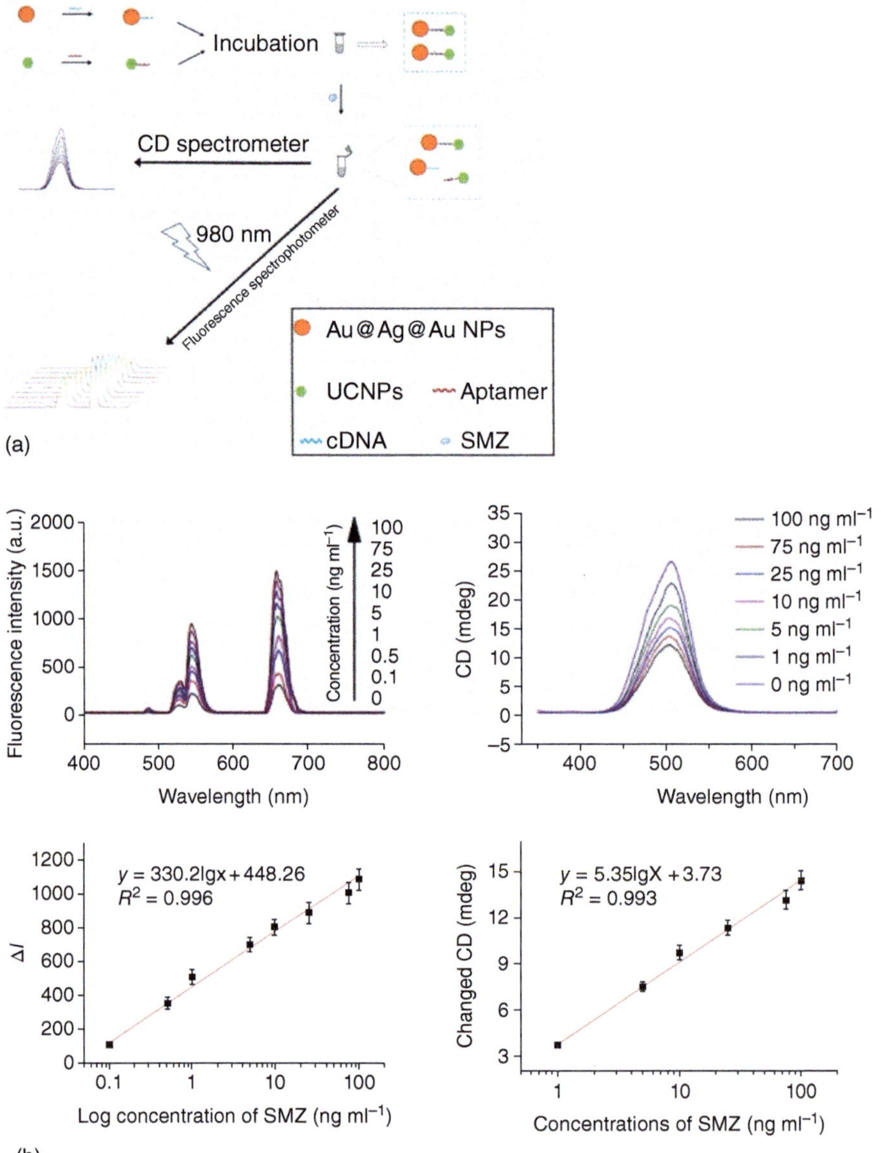


**Figure 4.14** (a) Schematic of the detection of SMZ by a chiral and UCPL signal-dependent biosensor using Au@Ag@Au NP-UCNP heterodimer. (b) UCPL and chiral signal with a standard linear curve for detecting SMZ using Au@Ag@Au NP-UCNP heterodimer. Source: (a, b) Reproduced with permission Wu and Cui [78]. Copyright 2019, Springer Nature.

is the energy source and the metabolic intermediate of living cells and the main energy supply substance of living organisms. Ngeontae et al. prepared chiral CdS QDs for the detection of glucose according to the changes of the CD signal. First, Cys-capped CdS QDs were prepared and reacted with the chiral compound DPA containing sulfhydryl group based on Cd—SH bond, in order to form DPA/Cys-CdS

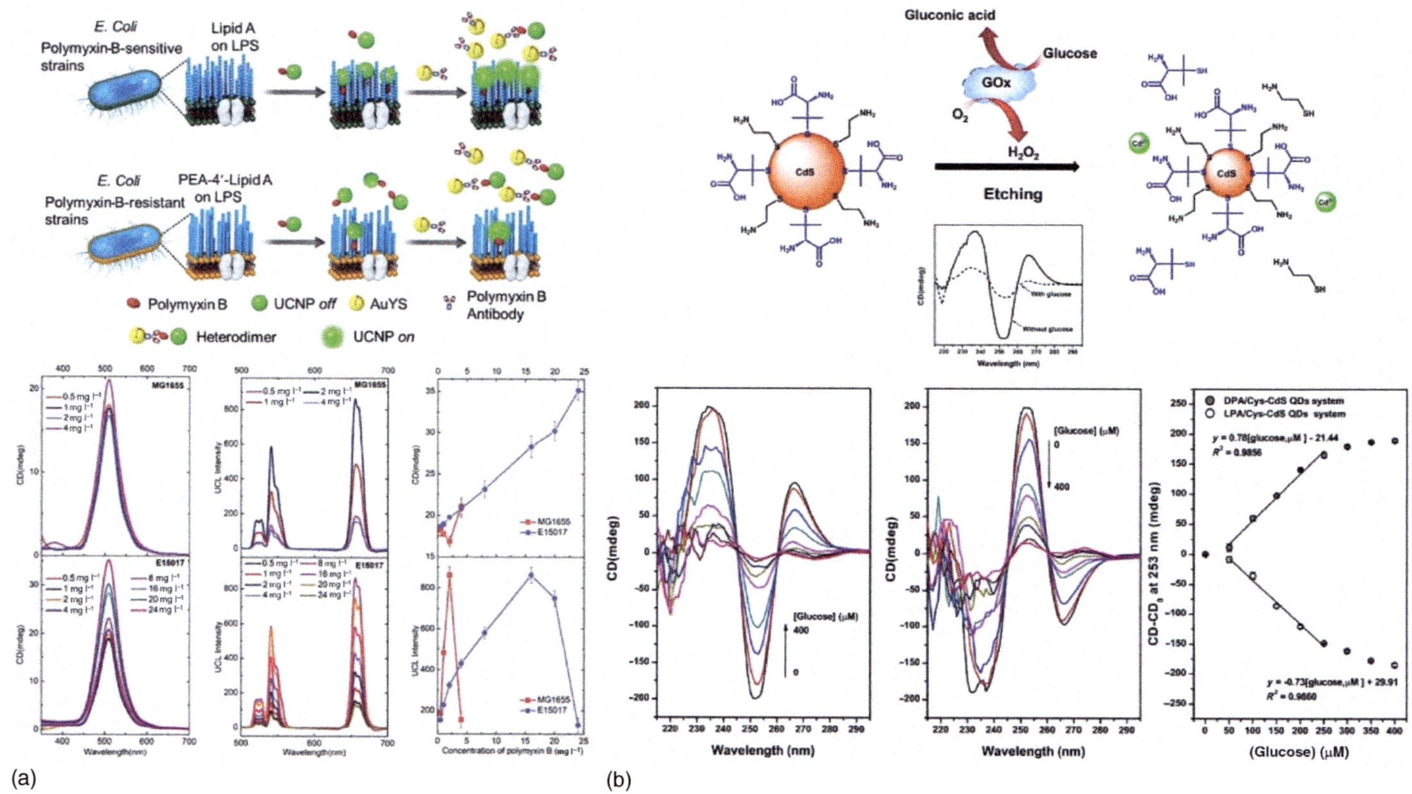


**Figure 4.15** (a) Chiral and UCPL signal–dependent immunoassay for the quantitative analysis and bioimaging of antibiotic-resistant bacteria through the detection of polymyxin B by chiral upconversion heterodimers. (b) Chiral CdS QDs for the detection of glucose. Source: (a) Reproduced with permission Sun et al. [76]. Copyright 2018, Wiley-VCH. (b) Reproduced with permission Nagamdee and Ngeontae [77]. Copyright 2018, Elsevier.

QDs (Figure 4.15b). DPA/Cys-CdS QDs showed strong CD responses in the range of 220–290 nm. Under the action of glucose oxidase, glucose and dissolved oxygen reacted with each other and produced $H_2O_2$. The produced $H_2O_2$ would further oxidize DPA/Cys-CdS QDs, inducing desorption of DPA and Cys from the surface of DPA/Cys-CdS QDs. The CD signal correspondingly decreased. This method was applied to the determination of glucose concentration in actual samples, such as coconut water, with an LOD of 31 μM [77]. Great challenges need to be overcome for simultaneous monitoring of multiple small molecules by developing multiplex chiral biosensors.

## 4.7 Chiral Recognition

Living organisms are complex and large chiral systems composed of asymmetrical compounds. Common biomolecules, such as nucleic acids, amino acids, enzymes, polysaccharides, and other important substances in life activities, mostly possess chiral characteristics. Metabolic reactions in the human body are closely related to chiral recognition, such as specific selective catalysis of proteins and the complementary pairing of nucleic acids. In the aspect of life science, physical and chemical properties of two enantiomers of chiral compounds may also be different. The exploration of chiral recognition helps us to understand the origin of life and the essence of life process and also helps to study the physiological activity and metabolic process of chiral molecules or compounds in the human body [79]. Chiral recognition is a common phenomenon that exists widely in nature, such as stereoselective specificity of enzymes, metabolism of chiral drugs in biological organs. In particular, chiral drugs have a very important research significance in medicine, life science, food safety, material synthesis, and other aspects [80]. Commonly used methods for chiral identification include chromatographic analysis, capillary electrophoresis, etc. However, most of these techniques require expensive chiral columns and complicated sample pretreatment [81]. Enantiomer molecules can participate in the formation of chiral nanostructures to construct chiral nanostructure sensors; this method has strong specificity and high sensitivity.

L/D-Cys with -SH groups can easily couple on the surface of NPs, which enable the assembly of NPs for the generation of intensive chiral signals. The distinct CD response of chiral NP assemblies has achieved chiral recognition of enantiomeric cysteines (L-Cys and D-Cys). For example, L/D-Cys were successfully modified on Au NPs (Figure 4.16a). Through controlling the modified amounts of L/D-Cys and incubation times, Au NP dimers and Au NP clusters were fabricated benefitting from the electrostatic attraction and hydrogen bonding of zwitterionic groups of Cys. In particular, Au NP dimers displayed positive CD responses in the presence of D-Cys and a negative chiral peak in the presence of L-Cys at 525 nm. Different ratios of L-/D-Cys were prepared (0 : 100, 20 : 80, 40 : 60, 50 : 50, 60 : 40, 80 : 20, and 100 : 0) and distinguished by these chiral biosensors developed with the total concentrations ranging from 0 to 5 nM. For the mixtures of L-Cys and D-Cys, the increasing amounts of D-Cys

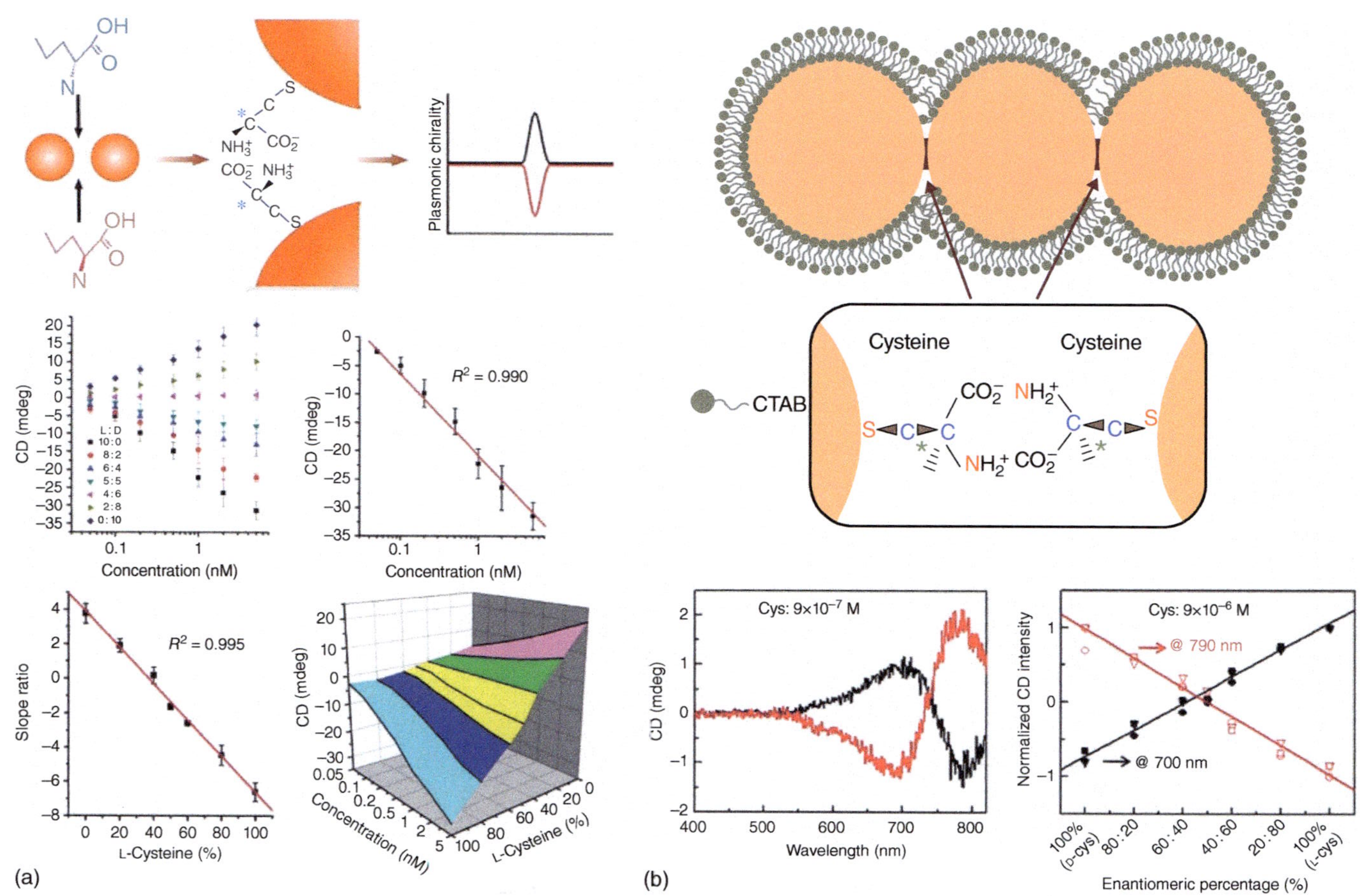


**Figure 4.16** (a) L/D-Cys-modified Au NP dimers for the chiral recognition of L- and D-Cys. (b) L/D-Cys-driven Au NP clusters for the chiral recognition of D- and L-Cys. Source: (a) Reproduced with permission Xu et al. [82]. Copyright 2013, Royal Society of Chemistry; (b) Reproduced with permission Wang et al. [83]. Copyright 2014, American Chemical Society.

induced a decreased negative CD signal at 525 nm from L-Cys. According to the chiral recognition mechanism, the detection range was from 0.05 to 5 nM, and the LOD of L-Cys obtained by this method was 20 pM [82]. Different from Au NP dimers, L/D-Cys-driven Au NP clusters were fabricated (Figure 4.16b). Cysteine enantiomers can assemble Au NPs into dimers, trimers, and tetramers. L-Cys-modified Au NP clusters showed a positive CD peak at 700 nm and a negative chiral signal at 790 nm, whereas D-Cys-coupled Au NP clusters showed a negative CD peak at 700 nm and a positive chiral signal at 790 nm. When the total concentration of Cys was controlled to $9 \times 10^{-6}$ M, the sign and the magnitude of the chiral peaks changed according to the different ratios of D-/L-Cys (100 : 0, 80 : 20, 60 : 40, 40 : 60, 20 : 80, and 0 : 100). A linear relationship was established between the different ratios of D-/L-Cys and the CD peak intensity at 700 nm, and another linear relationship was also obtained between the different ratios of D-/L-Cys and the chiral signal at 790 nm. This developed correlation was suitable for the chiral recognition of D-/L-Cys [83].

In comparison to Au NPs, Au NRs showed anisotropic properties and Au NR assemblies displayed giant CD responses for the improvement in sensitivity of chiral recognition. Cysteine enantiomers as linkers were also functionalized on the end of Au NRs and enabled the assembly of Au NRs into the end-to-end chain form (Figure 4.17a) [84]. Particularly, L-Cys-driven Au NR chains exhibited negative chiral peaks and positive CD signals in the region of 500–850 nm, but D-Cys-engineered Au NR chains showed contrary chiroptical responses. With the increasing concentration of L-Cys or D-Cys, the chiroptical signals of as-prepared Au NR chains correspondingly enhanced. There was a good linear correlation between the CD peak intensity at 670 nm and the amount of L-Cys or D-Cys. The LODs were 1.4 and 0.8 μM for L-Cys and D-Cys, respectively. Chiral Au NR chains showed better selectivity for Cys and could not recognize other amino acids except L-GSH. L-GSH enabled the assembly of Au NR chains, which also showed CD responses in the range of 500–800 nm. A good linear relationship was obtained between the CD peak intensity at 670 nm and the concentration of L-GSH. The LOD for L-GSH was 1.93 μM. The absolute configuration of Cys and GSH in the visible light region can be determined reliably. The selectivity of Cys and GSH was good, and other amino acids did not interfere with the determination. Cys-functionalized Au/Ag alloy NTs were also reported for chiral recognition of tyrosine and threonine (Figure 4.17b). In comparison to Cys-modified Au NTs and Ag NTs, Cys-modified Au/Ag alloy NTs possessed higher chiral responses due to the strong local electromagnetic field in the inner hollow and the enlargement of the spiral conformation by chiral NTs. L-Cys-Au/Ag alloy NTs modified electrodes displayed higher oxidation current for D-tyr than L-tyr, owing to the chiral electrochemical active sites and large group space effect [85]. At the same time, D-Cys-Au/Ag alloy NTs modified electrodes displayed higher oxidation current for L-tyr than D-tyr. L-Cys-Au/Ag alloy NTs modified electrodes and D-Cys-Au/Ag alloy NTs modified electrodes achieved −0.27 ee ratios and 0.31 ee ratios, respectively. Based on these principles, enantiomer analysis can identify and screen the enantiomer structure, which has great potential in the application of biological medicine and other fields.

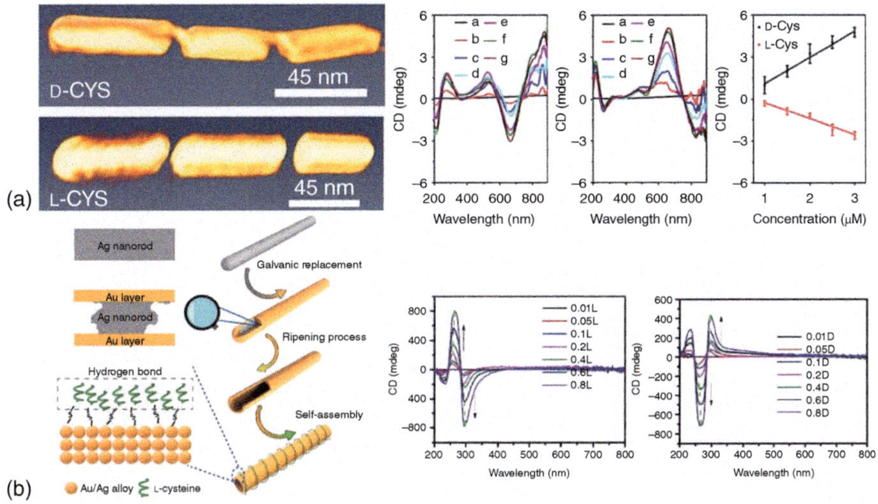


**Figure 4.17** (a) Cysteine enantiomer–driven end-to-end Au NR chains for the chiral recognition of Cys and GSH. (b) Cysteine-modified Au/Ag ANTs for the recognition of the enantiomers of tyrosine and threonine by using CD signal. (a) Reproduced with permission Zhu et al. [84]. Copyright 2015, American Chemical Society; (b) Reproduced with permission Liu et al. [85]. Copyright 2019, Royal Society of Chemistry.

Cystine enantiomers were also determined by using chiral Au substrates as detection probes and Raman signals as detection signals. Liu et al. prepared D-tartaric acid functionalized Au NPs (D-Au NPs) and L-tartaric acid Au NPs modified Au NPs (L-Au NPs) through the reduction of $HAuCl_4$ on the surface of polyethyleneimine-D tartaric acid-$SiO_2$ templates and polyethyleneimine-L tartaric acid-$SiO_2$ templates (Figure 4.18). As-prepared D-Au NPs and L-Au NPs displayed positive and negative plasmonic CD peaks at 530 nm, owing to the distorted Au atoms array in their crystalline structures. The pair of cystine enantiomers showed similar Raman characteristic peaks and intensity, and it was difficult to distinguish each other by using common Raman spectroscopy. However, it was found that D-Au NPs-D-cystine mixtures showed stronger Raman intensity at 497 $cm^{-1}$ (S—S bond) than D-Au NPs-L-cystine mixtures. At the same time, L-Au NPs-L-cystine mixtures showed stronger Raman intensity at 497 $cm^{-1}$ than L-Au NPs-D-cystine mixtures. Chiral Au NPs possessed unique atom arrangement, which was different from the common achiral Au NPs. The unique atom arrangement played a critical part in the enantioselectivity on the Raman spectra via three possible pathways: (i) spatial preferences, (ii) configurational and orientational change, and (iii) electromagnetic effects. The vertical polarized components in the tilted orientation of the S—S bond were more than those in the parallel orientation, resulting in the Raman signal intensity differences between L-cystine and D-cystine on chiral Au NPs. This platform afforded another enantioselective recognition method on ordinary Raman spectroscopy by using chiral plasmonic metallic nanomaterials [86].

Not only Cys enantiomers could be detected by chiral nanostructures, but also other amino acids, such as cystine, phenylalanine, and tryptophan enantiomers,

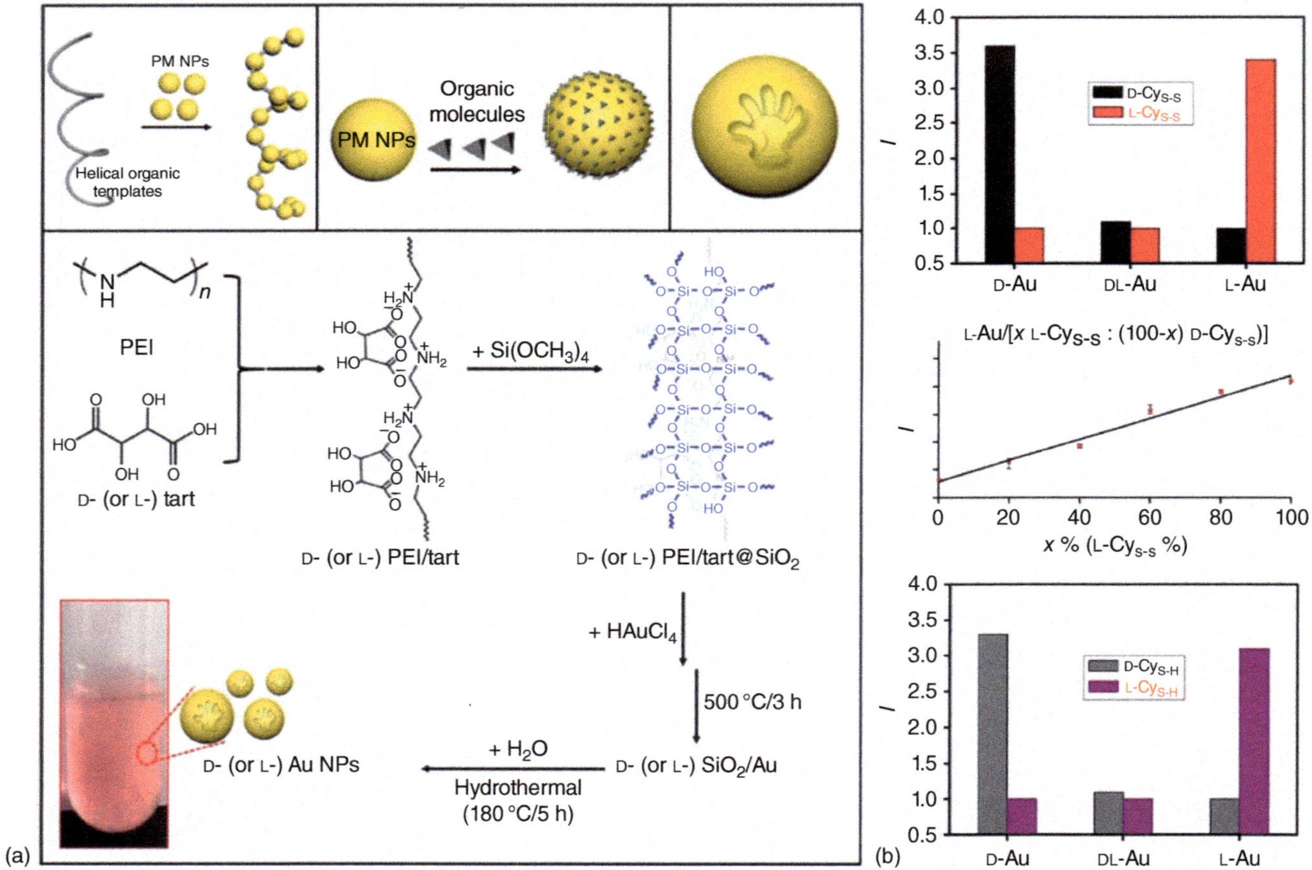


**Figure 4.18** (a) Schematic diagram of determination of cystine enantiomers using chiral Au substrates as detection probes and Raman signal as detection signals. (b) Relative values of *I* for different coupled mixtures. Source: (a, b) Sun et al. [86] / with permission of American Chemical Society.

could be distinguished by chiral $SiO_2$ nanofibers. Chiral $SiO_2$ nanofibers were synthesized by using D-tartaric acid and L-tartaric acid as ligands (Figure 4.19). The weak interaction between $SiO_2$ and Cys could not discriminate D-Cys and L-Cys. Dopamine was polymerized on the surface of chiral $SiO_2$ nanofibers to form chiral $SiO_2$/polydopamine (PDA) nanocomposites. $SiO_2$/PDA nanocomposites still showed well chiroptical responses and were used for chiral recognition of enantiomers. D-$SiO_2$/PDA showed stronger Raman intensity in the presence of D-Cys than L-Cys, and L-$SiO_2$/PDA showed stronger Raman intensity in the presence of L-Cys than D-Dys. This result achieved the accurate chiral recognition of L-Cys and D-Dys by using chiral $SiO_2$/PDA nanocomposites as detection probes and Raman signals as detection signals. PDA contains many groups, such as quinones, hydroxyl and amine groups, and may interact with amino acids through hydrogen bonds. Chiroptically active PDA will spatially match with enantiomeric analytes, which will affect the orientations of enantiomers and the numbers and sites of hydrogen bonds. Chiral $SiO_2$/PDA nanocomposites also achieved the accurate detection of cystine, phenylalanine, and tryptophan enantiomers. These $SiO_2$-based chiral functional nanomaterials are novel inorganic nanomaterials for chiral recognition applications by using Raman signals [87].

In comparison to single-metal NPs, bimetallic NPs show amplified CD responses owing to the intensive plasmonic coupling between different metallic NPs in the presence of chiral molecules. For example, Au-Ag alloy NPs generally display stronger CD responses in the visible region than Au NPs and Ag NPs. Chiral Au-Ag alloy NPs were utilized for the chiral recognition of D/L-penicillamine (D/L-Pen) (Figure 4.20). Pen is a naturally occurring amino acid and a vital medically active thiol compound, but only D-Pen is pharmaceutically and clinically useful and frequently used as a medicinal agent against a number of diseases. However, L-Pen form is harmful due to its excessive toxicity. It is critical to distinguish D-Pen and L-Pen. Liu et al. synthesized chiral Au-Ag alloy NPs by using D-Pen or L-Pen as ligands and $NaBH_4$ as the reducing agent. As-prepared D-Pen-Au-Ag alloy NPs and L-Pen-Au-Ag alloy NPs showed two pairs of Cotton effect at 450 and 548 nm, which were stronger than D-Pen/L-Pen-Au NPs and D-Pen/L-Pen-Ag NPs. L-Pen-Au-Ag alloy NPs showed bright yellow after the addition of L-Pen and displayed red when D-Pen was present. Similarly, D-Pen-Au-Ag alloy NPs displayed the same phenomenon after the addition of D-Pen and L-Pen. The heterochiral recognitions may result in a preferred molecular orientation around the alloy nanoparticles, which have a strong positive effect on the enhancement of the induced plasmonic CD signals. D-Pen-Au-Ag alloy NPs exhibited increased chiroptical responses with the increasing concentration of L-Pen, and L-Pen-Au-Ag alloy NPs exhibited increased chiroptical responses with the increasing concentration of D-Pen. There was good linearity between the amplified plasmonic CD intensity and the concentration of Pen. The detection range was 5.0–80.0 μM and the LOD was as low as 1.7 μM for L-Pen and 1.5 μM for D-Pen, respectively. Furthermore, this proposed chiral recognition platform also possessed better selectivity toward D- and L-Pen in the presence of other enantiomers. This sensing system will open new opportunities

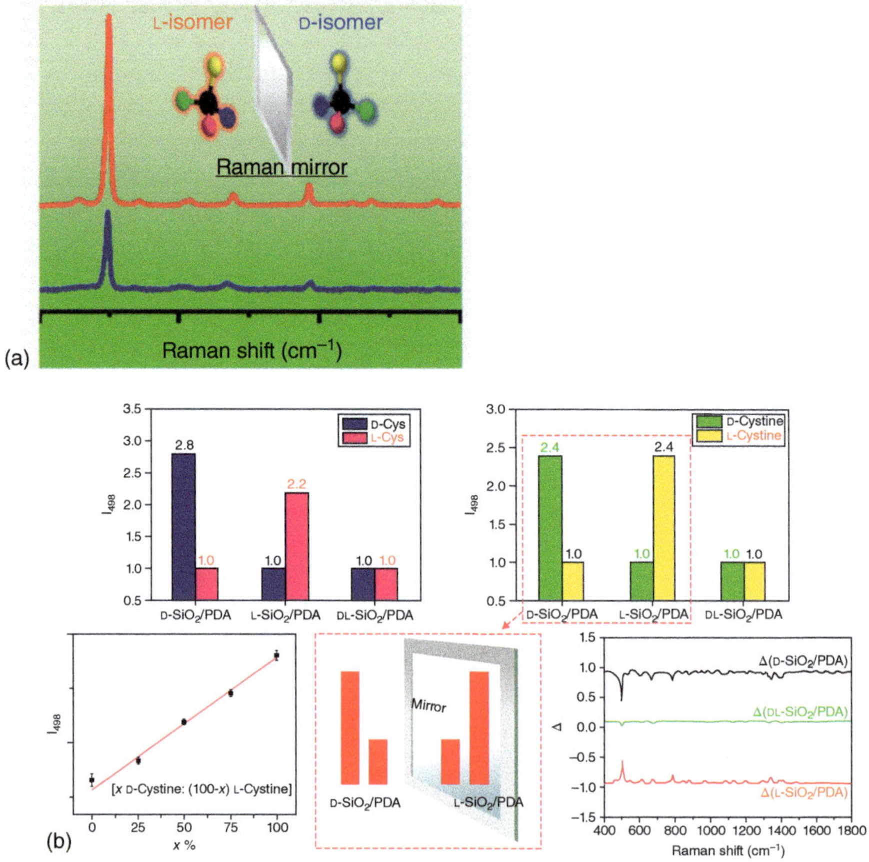


**Figure 4.19** (a) Schematic diagram of chiral $SiO_2$/PDA nanocomposites for detection of amino acid enantiomers. (b) Relative values of I for different coupled mixtures. Source: (a, b) Reproduced with permission Kong et al. [87]. Copyright 2020, American Chemical Society.

for the design of more novel enantioselective analyses of important species related to biological processes [88].

Besides chiral plasmonic NPs and assemblies, chiral carbon dots (CDs) were prepared and utilized for enantiomer recognition. In particular, L-lysine (L-Lys), as an essential amino acid, provides the necessary energy source for repairing nerve cells and normal physiological activities. However, the abnormal metabolism of L-Lys leads to severe diseases. It is of paramount importance to develop efficient strategies for enantioselective recognition of Lys enantiomers. Zheng et al. reported a method for recognition of Lys enantiomers by using chiral CDs (Figure 4.21a). Chiral CDs were prepared through a facile hydrothermal method by using citric acid and L-aspartic acid (L-Asp) as precursors. It was found that the presence of $Sn^{2+}$ induced the quench of fluorescence signals of CDs, due to the coordination of $Sn^{2+}$ on the surface of CDs and the formation of CD–Sn complex. Benefiting from the competition between L-Lys and CDs, $Sn^{2+}$ could preferentially bind with L-Lys, which would

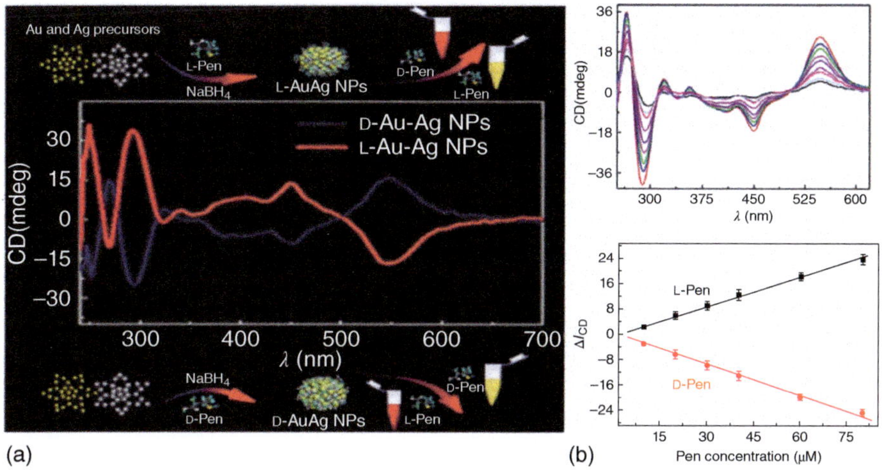


**Figure 4.20** (a) Schematic diagram of chiral identification of D/L-Pen using chiral Au-Ag alloy NPs. (b) CD responses and the linear relationship between CD signal amplification ($\Delta I_{CD}$) at 548 nm and the added concentration of L-Pen or D-Pen [88]. Source: Copyright 2017, American Chemical Society.

induce the separation of $Sn^{2+}$ from the surface of CDs. This resulted in a significant fluorescence recovery of CDs. However, due to the steric hindrance between CD-$Sn^{2+}$ and D-Lys, the presence of D-Lys would not induce the recovery of fluorescence signals of CDs. This chiral CD-based sensor achieved the selective detection of L-Lys in the range of 0–1.0 mM with an LOD of 3.34 μM. This sensor also achieved the detection of $Sn^{2+}$ in the concentration range from 0 to 0.44 mM with an LOD of 0.057 μM. This chiral CD-based sensor was successfully applied for intracellular $Sn^{2+}$ and L-Lys imaging, which presents great potential in clinical diagnosis and drug screening [89]. In comparison to D-Lys, $Sn^{2+}$ possessed better selectivity toward L-Lys. Metal ions as a recognition unit showed different affinity toward enantiomers, enabling accurate enantiomer recognition. Furthermore, $Cu^{2+}$ was reported to show stronger binding constant toward D-Trp than L-Trp. Trp is regarded as a precursor of the neurotransmitter serotonin, and the content of L-Trp is closely related to the hepatic disease. D-Trp was an intermediate for the preparation of peptide antibiotics and immunosuppressant agents. Accurate recognition of D-Trp from L-Trp was challenging due to the similarity of physical and chemical properties of the isomers. Ultrasensitive chiral recognition of D-/L-Trp was achieved by using $Cu^{2+}$ as recognition unit and electroactive Au@Ag NPs as detection probes (Figure 4.21b). The strong interaction between $Cu^{2+}$ and D-Trp enabled the modification of D-Trp-functionalized Au@Ag NPs on the surface of the electrodes. A large number of Au@Ag NPs on electrode would generate a high electrochemical oxidization peak at 0.2 V, owing to the electron transfer from $Ag^0$ to $Ag^+$ in Au@Ag NPs. In sharp contrast to D-Trp, the weak affinity between L-Trp and $Cu^{2+}$ just induced a little number of Au@Ag NPs modified on the surface of the electrodes, resulting in a weak electrochemical oxidization peak at 0.2 V. The different affinity between $Cu^{2+}$ and D-/L-Trp achieved sensitive and accurate quantification of D-/L-Trp. The LOD was 1.21 pM for D-Trp. This

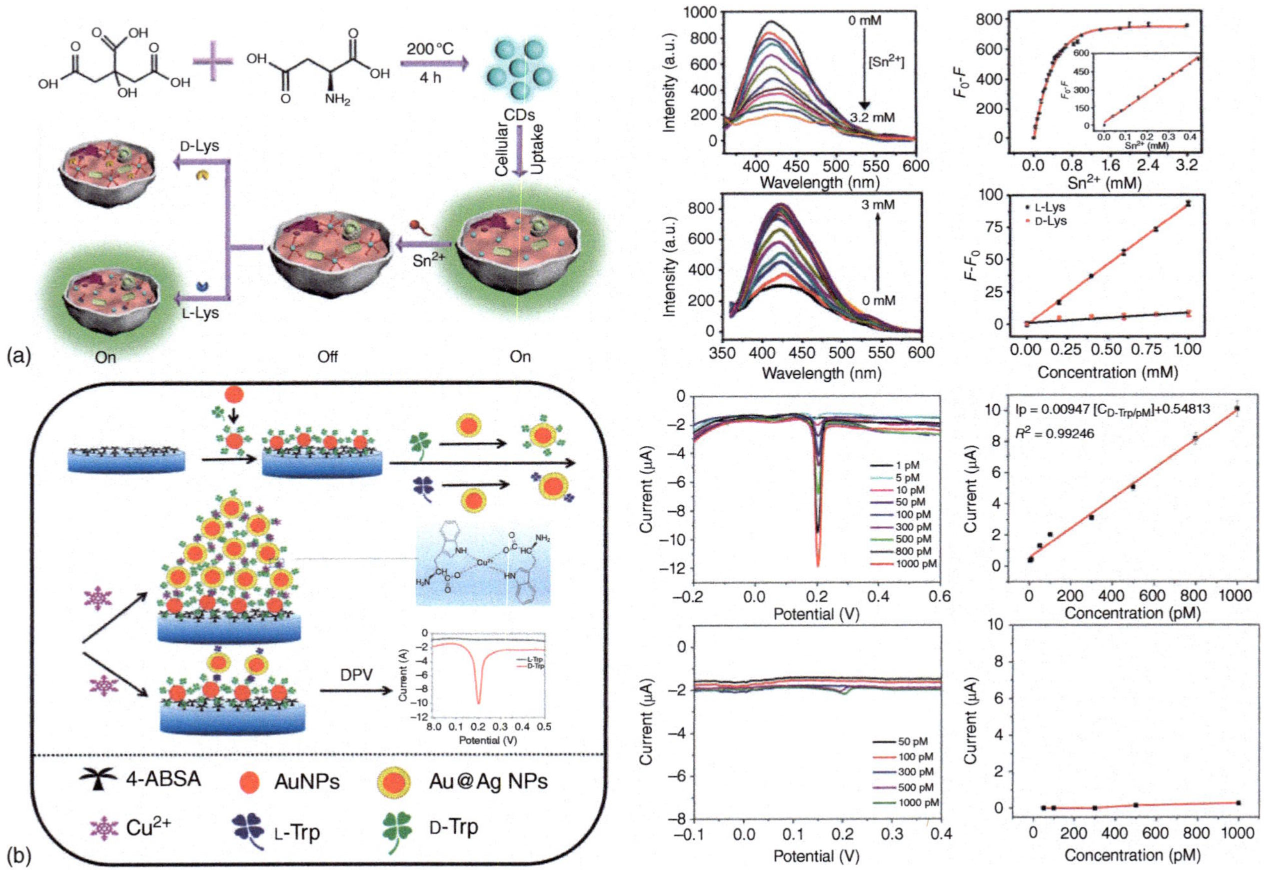


**Figure 4.21** (a) Chiral CD sensor for intracellular $Sn^{2+}$ and L-Lys detection. (b) Ultrasensitive chiral recognition of D-/L-Trp was achieved by using $Cu^{2+}$ as the recognition unit and electroactive Au@Ag NPs as the detection probes. Source: (a) Reproduced with permission Gao et al. [89]. Copyright 2020, Elsevier; (b) Reproduced with permission Zhao et al. [80]. Copyright 2019, American Chemical Society.

electrochemical chiral sensor also achieved specific determination of enantiomeric excess. In comparison to other reported approaches, this proposed electrochemical chiral sensor excels by its sensitivity, simplicity, and good availability of electroactive Au@Ag NP assemblies [80].

## 4.8 Future Prospects and Outlook

Currently, chiral sensors have been widely established and applied for the bioanalysis of DNA, biomarkers, enzymes, metal ions, and small molecules, as well as for bioapplications of chiral recognition. There are still some shortcomings that can be improved and developed. First, to improve the detection sensitivity and accuracy in intracellular and living systems, the chiral configuration of nanomaterials should be further optimized to obtain strong chiroptical responses. It is critical to improve the assembly yields of biomolecule-engineered chiral assemblies, in order to increase the stability of chiral probes and improve detection accuracy. Second, the repeatability and reproducibility of chiral biosensors need to be explored for the bioanalysis in real complex matrices. Third, sensitive detection of multiple markers is conducive to comprehensive bioanalysis and diagnosis of some diseases. This can save detection times and is helpful to the practical application of chiral sensors. But chiral signal–dependent simultaneous detection of multiple targets is still a challenge. Fourth, the biological safety of chiral nanostructures in living systems is also very important. If they can safely coexist with organisms, real-time detection of living organisms can be carried out, which is very helpful for human disease detection. This will enable chiral biosensors to achieve more sensitive, more specific, and more extensive applications in extracellular and intracellular detection and chiral recognition.

## References


1 Sun, M., Xu, L., Fu, P. et al. (2016). Scissor-like chiral metamolecules for probing intracellular telomerase activity. *Advanced Functional Materials* 26 (40): 7352–7358.

2 Xia, Y., Zhou, Y., and Tang, Z. (2011). Chiral inorganic nanoparticles: origin, optical properties and bioapplications. *Nanoscale* 3 (4): 1374–1382.

3 Sun, M., Li, S., Hao, C. et al. (2020). Chiral nanoprobes and their biological effects. *Chinese Journal of Chemistry* 39 (1): 25–31.

4 Ma, W., Xu, L., Wang, L. et al. (2018). Chirality-based biosensors. *Advanced Functional Materials* 29 (1): 1805512.

5 Sun, M., Ma, W., Xu, L. et al. (2014). Chirality of self-assembled metal–semiconductor nanostructures. *Journal of Materials Chemistry C* 2 (15): 2702–2706.

6 Zhao, Y., Sun, M., Ma, W. et al. (2017). Biological molecules-governed plasmonic nanoparticle dimers with tailored optical behaviors. *The Journal of Physical Chemistry Letters* 8 (22): 5633–5642.

**7** Ma, W., Kuang, H., Wang, L. et al. (2013). Chiral plasmonics of self-assembled nanorod dimers. *Scientific Reports* 3: 1934.

**8** Wu, X., Xu, L., Ma, W. et al. (2015). Gold, Core-DNA-silver shell nanoparticles with intense plasmonic chiroptical activities. *Advanced Functional Materials* 25 (6): 850–854.

**9** Hao, C., Xu, L., Ma, W. et al. (2015). Unusual circularly polarized photocatalytic activity in nanogapped gold-silver chiroplasmonic nanostructures. *Advanced Functional Materials* 25 (36): 5816–5822.

**10** Hao, C., Xu, L., Kuang, H., and Xu, C. (2019). Artificial chiral probes and bioapplications. *Advanced Materials* 32 (41): 1802075.

**11** Ma, W., Kuang, H., Xu, L. et al. (2013). Attomolar DNA detection with chiral nanorod assemblies. *Nature Communications* 4: 2689.

**12** Ma, W., Sun, M., Fu, P. et al. (2017). A chiral-nanoassemblies-enabled strategy for simultaneously profiling surface glycoprotein and MicroRNA in living cells. *Advanced Materials* 29 (42): 1703410.

**13** Freeley, M., Attanzio, A., Cecconello, A. et al. (2018). Tuning the coupling in single-molecule heterostructures: DNA-programmed and reconfigurable carbon nanotube-based nanohybrids. *Advanced Science* 5 (10): 1800596.

**14** Xu, Z., Xu, L., Zhu, Y. et al. (2012). Chirality based sensor for Bisphenol A detection. *Chemical Communications* 48 (46): 5760–5762.

**15** Chen, S., Sun, Y., Li, X., and Song, H. (2019). *N*-Acetyl-cysteine modified $CuFe_2O_4$@$SiO_2$ core–shell nanoparticles as chiral probes for recognition of chiral tyrosine. *Journal of Solid State Chemistry* 280: 120994.

**16** Liu, Z. and Su, X. (2017). A novel fluorescent DNA sensor for ultrasensitive detection of helicobacter pylori. *Biosensors and Bioelectronics* 87: 66–72.

**17** Jie, J., Hu, S., Liu, W. et al. (2020). Portable and battery-powered PCR device for DNA amplification and fluorescence detection. *Sensors* 20 (9): 2627.

**18** Zhao, Y., Xu, L., Kuang, H. et al. (2012). Asymmetric and symmetric PCR of gold nanoparticles: a pathway to scaled-up self-assembly with tunable chirality. *Journal of Materials Chemistry* 22 (12): 5574.

**19** Zhao, Y., Shi, L., Kuang, H., and Xu, C. (2020). DNA-driven nanoparticle assemblies for biosensing and bioimaging. *Topics in Current Chemistry* 378 (1): 18.

**20** Li, Z., Zhu, Z., Liu, W. et al. (2012). Reversible plasmonic circular dichroism of Au nanorod and DNA assemblies. *Journal of the American Chemical Society* 134 (7): 3322–3325.

**21** Zhao, Y., Xu, L., Ma, W. et al. (2014). Shell-engineered chiroplasmonic assemblies of nanoparticles for zeptomolar DNA detection. *Nano Letters* 14 (7): 3908–3913.

**22** Zhao, Y., Xu, L., Liz-Marzán, L.M. et al. (2013). Alternating plasmonic nanoparticle heterochains made by polymerase chain reaction and their optical properties. *The Journal of Physical Chemistry Letters* 4 (4): 641–647.

**23** Parvin, N., Jin, Q., Wei, Y. et al. (2017). Few-layer graphdiyne nanosheets applied for multiplexed real-time DNA detection. *Advanced Materials* 29 (18): 1606755.

**24** Yan, W., Xu, L., Ma, W. et al. (2014). Pyramidal sensor platform with reversible chiroptical signals for DNA detection. *Small* 10 (21): 4293–4297.

**25** Wu, X., Xu, L., Ma, W. et al. (2016). Propeller-like nanorod-upconversion nanoparticle assemblies with intense chiroptical activity and luminescence enhancement in aqueous phase. *Advanced Materials* 28 (28): 5907–5915.

**26** Shen, C., Lan, X., Zhu, C. et al. (2017). Spiral patterning of Au nanoparticles on Au nanorod surface to form chiral AuNR@AuNP helical superstructures templated by DNA origami. *Advanced Materials* 29 (16): 1606533.

**27** Zhao, Y. and Xu, C. (2020). DNA-based plasmonic heterogeneous nanostructures: building, optical responses, and bioapplications. *Advanced Materials* 32 (41): 1907880.

**28** Huang, R., He, N., and Li, Z. (2018). Recent progresses in DNA nanostructure-based biosensors for detection of tumor markers. *Biosensors and Bioelectronics* 109: 27–34.

**29** Press, M.F. and Lenz, H.-J. (2007). EGFR, HER2 and VEGF pathways validated targets for cancer treatment. *Drugs* 67 (14): 2045–2075.

**30** Khedri, M., Ramezani, M., Rafatpanah, H.S., and Abnous, K. (2018). Detection of food-born allergens with aptamer-based biosensors. *Trends in Analytical Chemistry* 103: 126–136.

**31** Zhao, Y., Yang, Y., Sun, Y. et al. (2018). Shell-encoded Au nanoparticles with tunable electroactivity for specific dual disease biomarkers detection. *Biosensors and Bioelectronics* 99: 193–200.

**32** Zhao, H., Bian, S., Yang, Y., and Wu, X. (2017). Chiroplasmonic assemblies of gold nanoparticles as a novel method for sensitive detection of alpha-fetoprotein. *Microchimica Acta* 184 (6): 1855–1862.

**33** Wu, X., Xu, L., Liu, L. et al. (2013). Unexpected chirality of nanoparticle dimers and ultrasensitive chiroplasmonic bioanalysis. *Journal of the American Chemical Society* 135 (49): 18629–18636.

**34** Tang, L., Li, S., Xu, L. et al. (2015). Chirality-based Au@Ag nanorod dimers sensor for ultrasensitive PSA detection. *ACS Applied Materials & Interfaces* 7 (23): 12708–12712.

**35** Liu, Y., Wei, M., Zhang, L. et al. (2016). Chiroplasmonic assemblies of gold nanoparticles for ultrasensitive detection of 8-hydroxy-2′-deoxyguanosine in human serum sample. *Analytical Chemistry* 88 (12): 6509–6514.

**36** Wang, X., Xu, L., Hao, C. et al. (2018). Circular dichroism-active interactions between fipronil and neuronal cells. *Environmental Science & Technology Letters* 5 (8): 500–507.

**37** Zhao, Y., Cui, L., Sun, Y. et al. (2018). Ag/CdO NPs engineered magnetic electrochemical aptasensor for prostatic specific antigen detection. *ACS Applied Materials & Interfaces* 11 (3): 3474–3478.

**38** Wang, Y., Rao, Z., Zhou, J. et al. (2019). A chiral assembly of gold nanoparticle trimer-based biosensors for ultrasensitive detection of the major allergen tropomyosin in shellfish. *Biosensors and Bioelectronics* 132: 84–89.

**39** Zhao, Y., Yang, Y., Zhao, J. et al. (2016). Dynamic chiral nanoparticle assemblies and specific chiroplasmonic analysis of cancer cells. *Advanced Materials* 28 (24): 4877–4883.

**40** Li, S., Xu, L., Ma, W. et al. (2016). Dual-mode ultrasensitive quantification of MicroRNA in living cells by chiroplasmonic nanopyramids self-assembled from gold and upconversion nanoparticles. *Journal of the American Chemical Society* 138 (1): 306–312.

**41** Meng, D., Ma, W., Wu, X. et al. (2020). DNA-driven two-layer core-satellite gold nanostructures for ultrasensitive MicroRNA detection in living cells. *Small* 16 (23): 2000003.

**42** Sang, F., Li, G., Li, J. et al. (2019). A hairpin-type DNA probe for direct colorimetric detection of endonuclease activity and inhibition based on the deaggregation of gold nanoparticles. *Microchimica Acta* 186 (2): 100.

**43** Zhou, X., Zhao, M., Duan, X. et al. (2017). Collapse of DNA tetrahedron nanostructure for "Off-On" fluorescence detection of DNA methyltransferase activity. *ACS Applied Materials & Interfaces* 9 (46): 40087–40093.

**44** Yu, Z., Jiang, F., Hu, C., and Tang, B. (2021). Functionalized nanoprobes for in situ detection of telomerase. *Chemical Communications* 57 (31): 3736–3748.

**45** Hao, C., Kuang, H., Xu, L. et al. (2013). Chiral supernanostructures for ultrasensitive endonuclease analysis. *Journal of Materials Chemistry B* 1 (41): 5539–5542.

**46** Liu, Y., Wei, M., Zhang, L. et al. (2015). Evaluation of DNA methyltransferase activity and inhibition via chiroplasmonic assemblies of gold nanoparticles. *Chemical Communications* 51 (76): 14350–14353.

**47** Xu, L., Hao, C., Yin, H. et al. (2013). Plasmonic core–satellites nanostructures with high chirality and bioproperty. *The Journal of Physical Chemistry Letters* 4 (14): 2379–2384.

**48** Ma, F., Zhang, Q., and Zhang, C. (2020). Nanomaterial-based biosensors for DNA methyltransferase assay. *Journal of Materials Chemistry B* 8 (16): 3488–3501.

**49** Song, G., Chen, C., Ren, J., and Qu, X. (2009). A simple, universal colorimetric assay for endonuclease/methyltransferase activity and inhibition based on an enzyme-responsive nanoparticle system. *ACS Nano* 3 (5): 1183–1189.

**50** Ma, W., Fu, P., Sun, M. et al. (2017). Dual quantification of MicroRNAs and telomerase in living cells. *Journal of the American Chemical Society* 139 (34): 11752–11759.

**51** Sun, Y., Shao, J., Li, H. et al. (2019). A portable ratiometric fluorescent strip for sensitive determination of mercuric ions. *Journal of Photochemistry and Photobiology A: Chemistry* 374: 68–74.

**52** Zhao, Y., Yang, X., Cui, L. et al. (2018). PVP-capped Pt NPs-depended catalytic nanoprobe for the simultaneous detection of $Hg^{2+}$ and $Ag^{+}$. *Dyes and Pigments* 150: 21–26.

**53** Zheng, F., Ke, W., Zhao, Y., and Xu, C. (2019). Pt NPs catalyzed chemiluminescence method for $Hg^{2+}$ detection based on a flow injection system. *Electrophoresis* 40 (16–17): 2218–2226.

**54** Salman, M.S., Znad, H., Hasan, M.N., and Hasan, M.M. (2021). Optimization of innovative composite sensor for Pb(II) detection and capturing from water samples. *Microchemical Journal* 160: 105765.

**55** Wu, P., He, Y., Lu, S. et al. (2021). A regenerable ion-imprinted magnetic biocomposite for selective adsorption and detection of $Pb^{2+}$ in aqueous solution. *Journal of Hazardous Materials* 408: 124410.

**56** Mahajan, P.G., Dige, N.C., Vanjare, B.D. et al. (2019). Gallotannin functionalized gold nanoparticles for rapid, selective and sensitive colorimetric detection of $Ag^{+}$

in aqueous medium: approach to eco-friendly water analysis. *Sensors and Actuators B: Chemical* 281: 720–729.
**57** Peng, B., Tang, L., Zeng, G. et al. (2018). Current progress in aptasensors for heavy metal ions based on photoelectrochemical method: a review. *Current Analytical Chemistry* 14 (1): 4–12.
**58** Wang, X., Shen, C., Zhou, C. et al. (2021). Methods, principles and applications of optical detection of metal Ios. *Chemical Engineering Journal* 417: 129145.
**59** Xu, Z., Xu, L., Liz-Marzán, L.M. et al. (2013). Sensitive detection of silver ions based on chiroplasmonic assemblies of nanoparticles. *Advanced Optical Materials* 1 (9): 626–630.
**60** Zhu, Y., Xu, L., Ma, W. et al. (2012). A one-step homogeneous plasmonic circular dichroism detection of aqueous mercury ions using nucleic acid functionalized gold nanorods. *Chemical Communications* 48 (97): 11889–11891.
**61** Wang, X., Hao, J., Cheng, J. et al. (2019). Chiral CdSe nanoplatelets as an ultrasensitive probe for lead ion sensing. *Nanoscale* 11 (19): 9327–9334.
**62** Kuang, H., Yin, H., Xing, C., and Xu, C. (2013). A sensitive DNAzyme-based chiral sensor for lead detection. *Materials* 6 (11): 5038–5046.
**63** Hao, C., Xu, L., Sun, M. et al. (2018). Chirality on hierarchical self-assembly of Au@AuAg yolk-shell nanorods into core-satellite superstructures for biosensing in human cells. *Advanced Functional Materials* 28 (33): 1802372.
**64** Gao, R., Xu, L., Hao, C. et al. (2019). Circular polarized light activated chiral satellite nanoprobes for the imaging and analysis of multiple metal ions in living cells. *Angewandte Chemie International Edition* 58 (12): 3913–3917.
**65** Hao, C., Xu, L., Sun, M. et al. (2019). Circularly polarized light triggers biosensing based on chiral assemblies. *Chemistry – A European Journal* 25 (53): 12235–12240.
**66** Zhao, Y., Luo, Y., Li, T., and Song, Q. (2014). Au NPs driven electrochemiluminescence aptasensors for sensitive detection of Fumonisin B1. *RSC Advances* 4 (101): 57709–57714.
**67** Feng, J., Xu, L., Cui, G. et al. (2016). Building SERS-active heteroassemblies for ultrasensitive Bisphenol A detection. *Biosensors and Bioelectronics* 81: 138–142.
**68** Kuang, H., Yin, H., Liu, L. et al. (2014). Asymmetric plasmonic aptasensor for sensitive detection of Bisphenol A. *ACS Applied Materials & Interfaces* 6 (1): 364–369.
**69** Zhao, Y., Yang, X., Li, H. et al. (2015). Au, Nanoflower-Ag nanoparticle assembled SERS-active substrates for sensitive MC-LR detection. *Chemical Communications* 51 (95): 16908–16911.
**70** Zhao, Y., Zheng, F., Ke, W. et al. (2019). Gap-tethered Au@AgAu raman tags for the ratiometric detection of MC-LR. *Analytical Chemistry* 91 (11): 7162–7172.
**71** Zheng, F., Ke, W., Shi, L. et al. (2019). Plasmonic Au-Ag Janus nanoparticle engineered ratiometric surface-enhanced raman scattering aptasensor for ochratoxin A detection. *Analytical Chemistry* 91 (18): 11812–11820.
**72** Zhao, X., Wu, X., Xu, L. et al. (2015). Building heterogeneous core-satellite chiral assemblies for ultrasensitive toxin detection. *Biosensors and Bioelectronics* 66: 554–558.
**73** Cai, J., Hao, C., Sun, M. et al. (2018). Chiral Shell Core-Satellite Nanostructures for Ultrasensitive Detection of Mycotoxin. *Small* 14 (13): 1703931.

**74** Li, A., Tang, L., Song, D. et al. (2016). A, SERS-active sensor based on heterogeneous gold nanostar core-silver nanoparticle satellite assemblies for ultrasensitive detection of AflatoxinB1. *Nanoscale* 8 (4): 1873–1878.
**75** Zhao, Y., Yang, Y., Luo, Y. et al. (2015). Double detection of mycotoxins based on SERS labels embedded Ag@Au core–shell nanoparticles. *ACS Applied Materials & Interfaces* 7 (39): 21780–21786.
**76** Sun, M., Qu, A., Hao, C. et al. (2018). chiral upconversion heterodimers for quantitative analysis and bioimaging of antibiotic-resistant bacteria in vivo. *Advanced Materials* 30 (50): 1804241.
**77** Ngamdee, K. and Ngeontae, W. (2018). Circular dichroism glucose biosensor based on chiral cadmium sulfide quantum dots. *Sensors and Actuators B: Chemical* 274: 402–411.
**78** Wu, Z. and Cui, B. (2019). Simultaneous fluorometric and chirality based aptasensing of sulfamethazine by using upconversion nanoparticles and Au@Ag@Au core–shell nanoparticles. *Microchimica Acta* 186 (8): 555.
**79** Zor, E., Bingol, H., and Ersoz, M. (2019). Chiral sensors. *Trends in Analytical Chemistry* 121: 115662.
**80** Zhao, Y., Cui, L., Ke, W. et al. (2019). Electroactive Au@Ag nanoparticle assembly driven signal amplification for ultrasensitive chiral recognition of LD-/L-Trp. *ACS Sustainable Chemistry & Engineering* 7 (5): 5157–5166.
**81** Liu, H., Shao, J., Shi, L. et al. (2020). Electroactive NPs and D-amino acids oxidase engineered electrochemical chiral sensor for D-alanine detection. *Sensors and Actuators B: Chemical* 304: 127333.
**82** Xu, L., Xu, Z., Ma, W. et al. (2013). Highly selective recognition and ultrasensitive quantification of enantiomers. *Journal of Materials Chemistry B* 1 (35): 4478–4483.
**83** Wang, R.-Y., Wang, P., Liu, Y. et al. (2014). Experimental observation of giant chiroptical amplification of small chiral molecules by gold nanosphere clusters. *The Journal of Physical Chemistry C* 118 (18): 9690–9695.
**84** Zhu, F., Li, X., Li, Y. et al. (2015). Enantioselective circular dichroism sensing of cysteine and glutathione with gold nanorods. *Analytical Chemistry* 87 (1): 357–361.
**85** Liu, H., Li, Z., Yan, Y. et al. (2019). Chiroptical study of the bimetal-cysteine hybrid composite: interaction between cysteine and Au/Ag Alloyed Nanotubes. *Nanoscale* 11 (45): 21990–21998.
**86** Sun, X., Kong, H., Zhou, Q. et al. (2020). Chiral plasmonic nanoparticle assisted raman enantioselective recognition. *Analytical Chemistry* 92 (12): 8015–8020.
**87** Kong, H., Sun, X., Yang, L. et al. (2020). Chirality detection by raman spectroscopy: the case of enantioselective interactions between amino acids and polymer-modified chiral silica. *Analytical Chemistry* 92 (21): 14292–14296.
**88** Wei, J., Guo, Y., Li, J. et al. (2017). Optically active ultrafine Au-Ag alloy nanoparticles used for colorimetric chiral recognition and circular dichroism sensing of enantiomers. *Analytical Chemistry* 89 (18): 9781–9787.
**89** Gao, P., Xie, Z., and Zheng, M. (2020). Chiral carbon dots-based nanosensors for Sn(II) detection and lysine enantiomers recognition. *Sensors and Actuators B: Chemical* 319: 128265.

# 5

# Chiral Nanomaterials for Emerging Biological Effects

*Chuanlai Xu[1], Maozhong Sun[1], and Yuan Zhao[2]*

[1] *International Joint Research Laboratory for Biointerface and Biodetection, State Key Lab of Food Science and Technology, School of Food Science and Technology, Jiangnan University, Wuxi, Jiangsu Province, 214122, China*
[2] *School of Chemical and Material Engineering, Jiangnan University, Wuxi, Jiangsu Province, 214122, China*

## 5.1 Photothermal Therapy

Chiral nanomaterials featured with intensive localized surface plasmon resonance (LSPR) characteristics effectively absorb photons and possess excellent photothermal conversion ability [1]. It has been realized that the strongly absorbed radiation is converted efficiently into heat on a picosecond time domain [2–4]. This will result in an increase in the kinetic energy and lead to the overheating of the local environment, which selectively causes irreversible damage to cancer cells, protein denaturation, and cell membrane destruction. Chiral nanomaterials have been widely explored for applications in photothermal therapy (PTT) [5]. Chiral nanomaterials with well-defined sizes, shapes, composition, and surface functionalities possess highly efficient light-to-heat conversion for local cell or tissue destruction, thereby providing therapeutic applications.

Plasmonic nanoparticle (NPs) can absorb photons from light due to their unique LSPR properties and have been considered good phototherapeutic agents in the visible and near-infrared (NIR) wavelength range for realizing precise cancer therapy [6]. Although several kinds of plasmonic metals were reported, only a few plasmonic metal NPs could support plasmon resonances in the visible range of the electromagnetic spectrum [7, 8]. In particular, Au and Ag nanostructures with different morphologies possessed tunable LSPR over a broad spectral range from the visible to NIR regions and exhibited better plasmonic activities than Cu and Al nanostructures. In comparison to Ag NPs with low stability and easy oxidation characteristics, Au NPs featured the advantages beyond their plasmonic properties and high chemical stability. Currently, Au-based nanostructures are the most common nanomaterials for applications in PTT. For example, the temperature increased to 47.6 °C by using Au NPs under 808 nm laser irradiation for five minutes, whereas the temperature was increased to 67.2 °C in the presence of Au cups under the same laser irradiation. It was because that Au cups displayed intensive and broad LSPR peaks

*Chiral Nanoprobes for Biological Applications*, First Edition. Edited by Chuanlai Xu.

ranging from 500 to 1000 nm, which could effectively absorb the energy from 808 nm laser irradiation to increase temperatures. After PPT treatment, the cell death was 30.6% ± 1.1% for Au NPs and 51.2% ± 1.7% for Au cups. Au cups achieved as low as 48.2% ± 1.8% cell viability [9]. In comparison to spherical Au NPs, spiky Au NPs generated stronger electromagnetic (EM) fields inside the particles due to the multiple and coupled sharp tips and demonstrated higher LSPR properties for the utilization of light irradiation. Spiky Au shell was deposited on the surface of $Fe_3O_4$ NPs. As-prepared spiky $Fe_3O_4$@Au NPs displayed strong LSPR peaks ranging from 600 to 850 nm and possessed a maximum absorption wavelength at 700 nm. Spiky $Fe_3O_4$@Au NPs also exhibited PTT ability under 808 nm laser irradiation. The temperature was obviously enhanced to 65.7 °C, and the photothermal conversion efficiency reached up to 31% for spiky $Fe_3O_4$@Au NPs [10]. At the same time, core–petal Au NPs were synthesized and exhibited wide and strong LSPR absorption from the visible to NIR region, which showed a deep penetration depth under NIR excitation and possessed the potential for biological applications. The temperature increased from 23 to 53.7 °C under 785 nm irradiation. The photothermal conversion efficiency of core–petal Au NPs reached up to 32%, which was higher than that of Au nanorods (NRs) (21%) (Figure 5.1a,b). With the increasing density of petals on the surface of

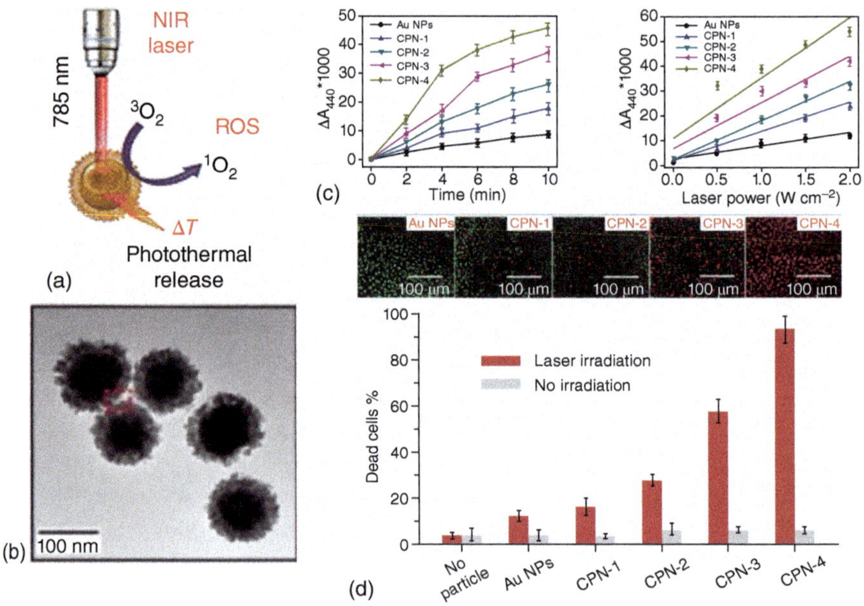


**Figure 5.1** (a) Schematic diagram of the production of ROS and photothermal effects by plasmonic core–petal Au NPs under 785 nm laser irradiation. (b) The TEM image of CPNs. (c) For CPNs with different surface petal densities, the relationship between temperature and laser irradiation time and the relationship between temperature and laser power with six minutes irradiation times. (d) Confocal fluorescence microscopy images of living/dead HeLa cells (green and red dots represent living and dead cells, respectively) and analysis diagram of the causes of cell death. Source: (a–d) Kumar et al. [11] / with permission of American Chemical Society.

core–petal Au NPs, the LSPR properties became stronger and the absorption spectrum became wider, which resulted in core–petal Au NPs with good performance for PTT application (Figure 5.1c,d) [11].

Compared with isotropic spherical NPs, anisotropic nanostructures display superior light-harvesting capacities. Au NRs exhibited strong plasmonic coupling and could be an ideal model for customizing the wavelength range and the optical activity in the visible and NIR regions [12]. Au NRs and Au NR–dependent assemblies were widely studied and applied for PTT applications. Au NRs were modified on the surface of metal–organic framework (MOF), and the temperature achieved fast responses under 808 nm irradiation (Figure 5.2a,b) [13]. Another metal shell was attempted to deposit on the surface of Au NRs to adjust the LSPR peaks at longer wavelengths. For example, Au NRs were covered by a Pt shell, and the LSPR peak shifted from 756 to 870 nm. Au NRs@Pt-evolved nanostructures exhibited good PTT performances, in which the temperature of solution was easily increased to 75 °C under 980 nm irradiation. For the application in mice, the temperature was increased from 28.2 to 59.1 °C by using Au NRs@Pt, which was higher than that of Au NRs (52.8 °C). This indicated better photothermal conversion efficiency of Au NRs@Pt in vivo [15]. Similarly, Au NPs after the growth of $Cu_9S_5$ also displayed broad and strong absorption in the regions of 450–1050 nm. The integration of plasmonic metal and $Cu_9S_5$ semiconductor into a single unit could be advantageous not only for a fundamental understanding of different LSPR couplings but also for efficient PTT applications (Figure 5.2c,d). The mice injected with Au-$Cu_9S_5$ NPs or Au-$Cu_9S_5$ NP-evolved nanostructures demonstrated a high temperature of 55 °C under 980 nm irradiation for four minutes (Figure 5.2e). Au-$Cu_9S_5$ NP-evolved nanostructures were the potential PTT agents for photothermal treatments [14]. Alternatively, NP assemblies with intensive plasmonic coupling showed an amplified EM field and were explored to possess better photothermal conversion efficiency than single NPs. For example, anisotropic Au NR dimers were assembled and displayed 42.3% heat transfer efficiency, which was two times better than single Au NRs. The plasmonic absorption peak of Au NR dimers (ranging from 750 to 900 nm) was well matched with the laser wavelength (808 nm), facilitating the utilization of light energy and accelerating the heat conversion. Different from the temperature increase from a single NR, the temperature increase of Au NR dimers was not only from the absorbed photon energy under LSPR excitation but also from the superposition of the EM fields. The intensive plasmonic coupling in the hot spot improved the EM fields [16]. There are many other reports on PTT of Au, such as D/L-Glu-Au nanobipyramids (NBPs) [17], Au-PGEA nanohybrids [18], GO/Au complex [19], PEA NRs (Pt end-deposited Au nanorods) [20].

The intensive plasmonic properties of NPs could largely utilize the energy for photothermal conversion. The adjustment and improvement of the LSPR responses of nanostructures enabled the enhancement of PTT performances. Alternatively, polarization-sensitive photochemistry phenomenon related to hot electron–mediated phototherapy effect was interesting and showed great potential for biological applications. Chiral NPs were widely reported and could efficiently utilize circularly polarized light (CPL) illumination to produce better PTT effects

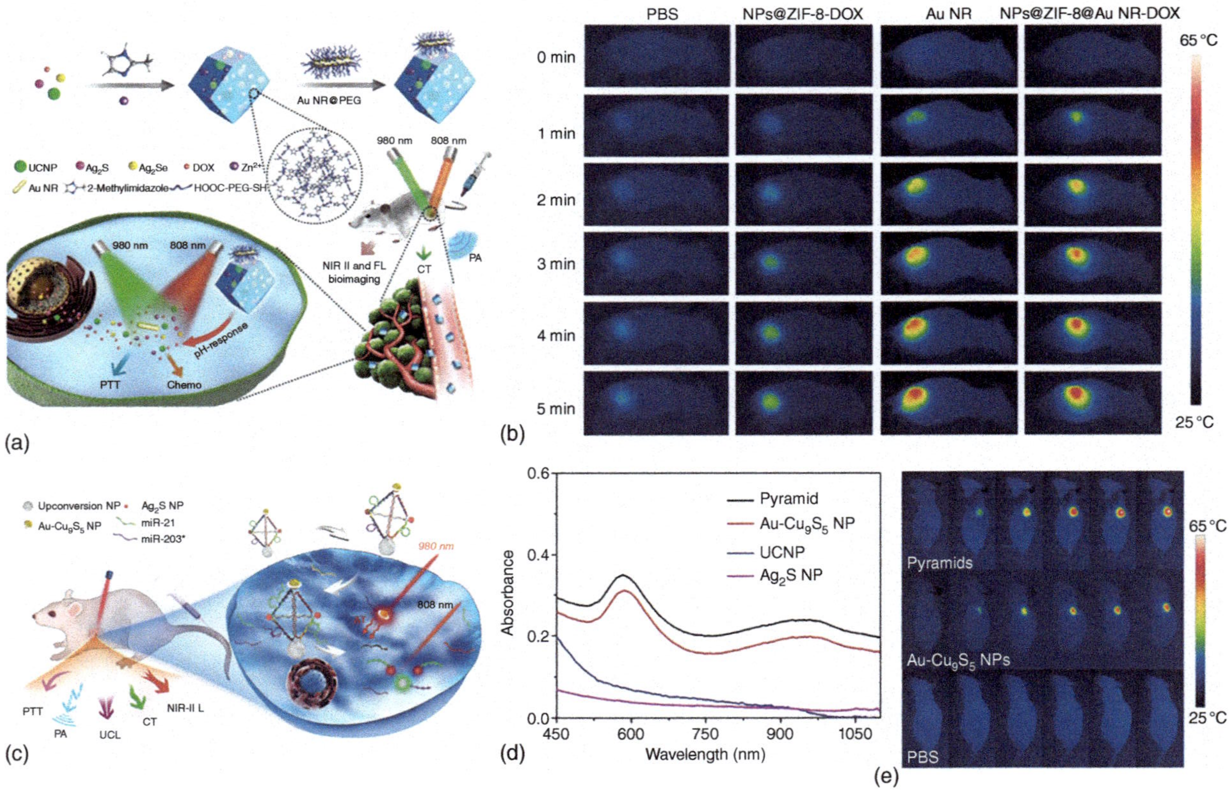


**Figure 5.2** (a) Schematic illustration of photothermal therapy with NPs@ZIF-8@Au NR-doxorubicin (DOX) heterodimer. (b) Thermal images of HeLa tumor-bearing mice with injection of different chiral nanoassemblies under irradiation of 808 nm laser taken at different time intervals. (c) Biological applications of nanopyramids in vivo. (d) UV–vis absorption spectra of different materials in the regions of 450–1050 nm. (e) Thermal images of HeLa tumor-bearing mice with injection of pyramids, Au-$Cu_9S_5$ NPs under irradiation of 980 nm NIR laser taken at different time intervals. Source: (a, b) Zhang et al. [13] / with permission of Wiley-VCH; (c, d) Li et al. [14] / with permission of Wiley-VCH.

on in vivo tumors. For example, AuCuAu NRs were synthesized and showed bisignate circular dichroism (CD) signal in the region of 400–1000 nm after the modification of L- and D-dipeptide cys-phe ligand. The double-sign CD spectrum at the plasma band of AuCuAu NRs was symmetrical and opposite (Figure 5.3a). Chiral molecule–modified AuCuAu NRs showed strong rotational optical activity response with the increasing aspect ratios, due to the formation of rough surface and cavity structure. In particular, L-cys-phe-modified AuCuAu NRs showed better PTT performances under LCP, followed by linearly polarized (LP) and RCP, owing to the strong absorption ability toward LCP (Figure 5.3b,c) [21]. Chiral L/D-cys-$MoO_{3-x}$ NPs were fabricated and exhibited CD activities in the visible

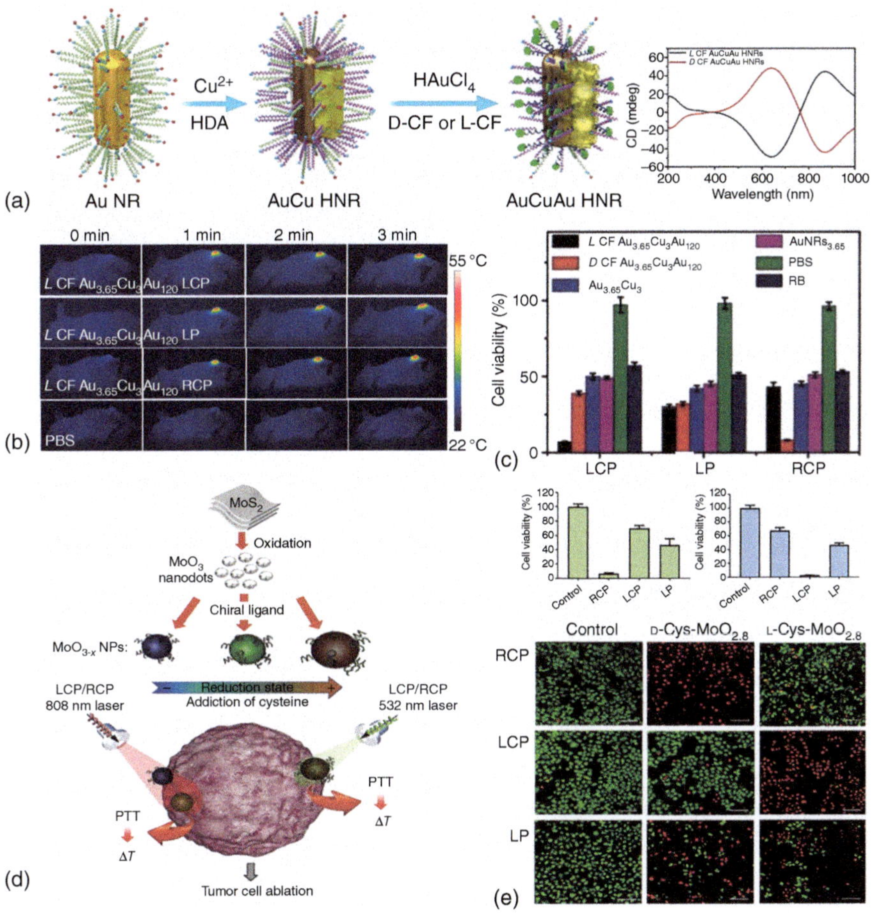


**Figure 5.3** (a) Synthesis and CD signal of AuCuAu HNRs. (b) Thermal images of HeLa tumor-bearing mice with injection of AuCuAu HNRs under different circularly polarized light at 808 nm taken at different time intervals. (c) Cell viability analysis of different materials under different circularly polarized light. (d) Schematic illustration of photothermal therapy with $MoO_{3-x}$ NPs. (e) Cell viability analysis and corresponding fluorescence images (green and red dots represent living and dead cells, respectively) of HeLa cells in the presence of D-/L- $MoO_{2.8}$ NPs under different circularly polarized light irradiation. Source: (a–e) Wang et al. [21] / with permission of Wiley-VCH.

and NIR regions. In particular, L-cys-$MoO_{2.8}$ NPs could increase the temperature above 28 °C under LCP illumination and displayed 41.24% photothermal conversion efficiency, and L-cys-$MoO_{2.8}$ NPs induced 96.95% Hela cellular fatality rate. Meanwhile, D-cys-$MoO_{2.8}$ NPs exhibited better PTT activity under 808 nm RCP illumination, and the Hela cellular fatality rate reached up to 93.01% (Figure 5.3d,e) [7]. The generated chirality participated in inter-band transition of noble metals and thereby accelerated the electronic transfer or oscillations for the production of photothermal effect. Although single chiral NPs exhibit chiroptical activity under illumination, chiral NP assemblies display stronger plasma effect and chirality after the organization of individual NPs into complex nanostructures and may be further explored for PPT applications. Even though much process had been obtained for plasmonic NP-driven PTT applications, current studies mainly focus on Au nanostructures, especially for Au NPs and Au NRs. Future research should explore more anisotropic Au nanostructures with different morphologies and good stability in PTT applications. Chiral Au NBPs were prepared and reported to exhibit good PTT activity in the treatment of bacterial and biofilm infections. In comparison to other Au structures, the nanostructure of plasmonic Au NBPs was special, and the unique structure with sharp ends could make them more quickly penetrate the bacterial cell wall and enter the interior of the biofilm, which could improve the efficiency of PTT treatment. Chiral Au NBPs were synthesized and firstly modified by thiol-PEG-carboxyl (SH-PEG-COOH) and thiol-PEG (SH-PEG) and then were functionalized with chiral D-glutamate (D-Glu) and L-glutamate (L-Glu), respectively. It was found that under 808 nm NIR radiation for five minutes, the temperature of the solution rapidly reached over 50 °C when the concentration of Au NBPs was higher than 37.5 $\mu g\,ml^{-1}$, and the photothermal conversion efficiency was calculated to be 43.16%. The high photothermal conversion capability and strong NIR absorption make Au NBPs highly promising for photothermal antibacterial therapy. In comparison to PEG-Au NBPs, the interaction between D/L-Glu-Au NBPs and bacteria was enhanced, owing to the surface targeting modification of chiral D/L-Glu molecules. The bacteria showed a significant increase in temperature, and the bactericidal effect of D/L-Glu-Au NBPs was significantly stronger than that of PEG-Au NBPs. Moreover, D-Glu-Au NBPs had the ability to kill bacteria in the absence of NIR radiation, reducing bacterial activity by up to 70%. In contrast, under the same conditions, L-Glu-Au NBPs and PEG-Au NBPs showed little bactericidal activity. This chiral-functionalized Au NBPs enhanced targeting, killing bacteria while protecting other normal cells from damage [17]. Alternatively, Ag nanostructures with the strongest plasmonic properties in the whole visible region should be further explored, and even Au-Ag alloy nanostructures with better stability and biocompatibility need to be applied as PTT agents.

In addition to plasmonic metals, pure semiconductor materials also have PTT effects, for example $Cu_9S_8$ NPs. $Cu_2O$ NPs showed very weak absorption in the NIR region (500–1000 nm), and then were transformed to $Cu_9S_8$ NPs in the presence of NaHS solution. In comparison to $Cu_2O$ NPs, $Cu_9S_8$ NPs showed enhanced absorption intensity in the wavelength of 500–1000 nm and exhibited hook-shaped absorption in the NIR region (Figure 5.4a). Under irradiation of 808 nm laser

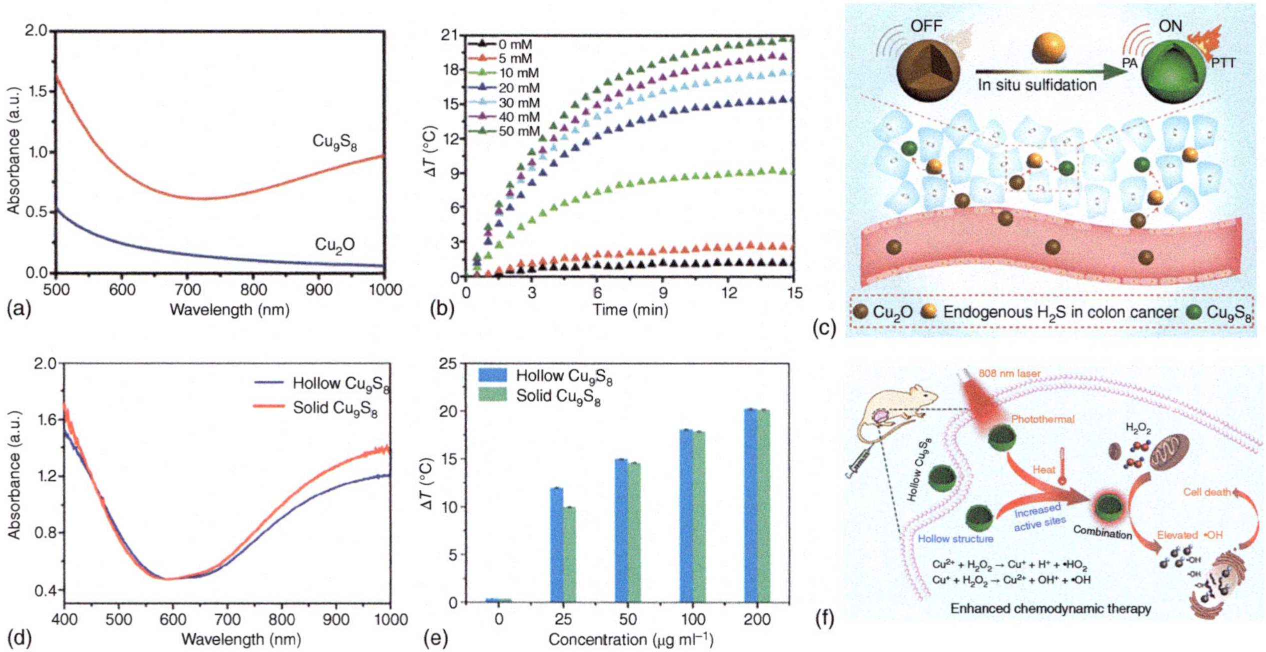


**Figure 5.4** (a) Vis–NIR absorption spectrum of $Cu_2O$ and $Cu_9S_8$. (b) Curve of temperature change with time of $Cu_2O$ solution under 808 nm laser irradiation in the presence of different concentrations of NaHS. (c) In situ reaction of $Cu_2O$ with endogenous $H_2S$ for the treatment of colon cancer. (d) UV–vis absorbance spectrum of the hollow and solid hexagonal $Cu_9S_8$ NPs. (e) The temperature changes of hollow and solid $Cu_9S_8$ solution at different concentrations under 808 nm laser irradiation for 10 minutes. (f) Schematic illustration of photothermal therapy with hollow $Cu_9S_8$ NPs. Source: (a–c) Reproduced with permission An et al. [22]. Copyright 2018, Wiley-VCH; (d–f) Reproduced with permission Wang et al. [23]. Copyright 2020, Elsevier.

for 15 minutes, the thermal imaging color of pure $Cu_2O$ NPs demonstrated little change, while the color of the aqueous $Cu_2O$ solution after the addition of NaHS turned brighter upon an increase in the concentration, in which the temperature was increased by 2.6–20.78 °C after the addition of NaHS with concentrations ranging from 5 to 30 mM. The temperature of $Cu_2O$ aqueous solution increased more obviously under laser irradiation with the increase in NaHS concentration (Figure 5.4b). The photothermal conversion efficiency increased to be 15.6%. Therefore, the $Cu_2O$ in situ vulcanization could be used in the clinical diagnosis and treatment of colon cancer. This diagnostic treatment was minimally invasive and effective, avoiding the risks that might be associated with conventional surgery. Since mice with colon cancer contained a large amount of $H_2S$, in vivo experiments were performed on the mice. The in situ reaction between $Cu_2O$ and endogenous $H_2S$ could achieve the purpose of the diagnosis and treatment of colon cancer. $Cu_2O$ was injected intravenously into the mice and reacted with $H_2S$ to produce $Cu_9S_8$ at the colon tumor site in the mice (Figure 5.4c). $Cu_2O$ itself had no absorption in the near-infrared light (NIR) region, while $Cu_9S_8$ could show obvious absorption in the NIR region, so it could use PA imaging and PTT effect to diagnose and treat colon cancer well [22]. Based on the structural study of $Cu_9S_8$, Tian et al. also constructed a multifunctional hollow $Cu_9S_8$ NP-based nanotherapy platform with low cost and simple treatment. Hollow $Cu_9S_8$ NPs were prepared by using $Cu_2O$ NPs as sacrificial templates, and solid $Cu_9S_8$ NPs were synthesized by the reaction of copper salt and sodium sulfide directly. Compared with solid $Cu_9S_8$ NPs, although hollow $Cu_9S_8$ NPs had no changes in the crystal phase and composition, it had a larger specific surface area, which was 1.7 times that of solid $Cu_9S_8$ NPs, and could provide more active sites. Both solid $Cu_9S_8$ NPs and hollow $Cu_9S_8$ NPs dispersed in water had a strong absorbance in the NIR region, which can be attributed to the LSPR effect caused by self-doping of the hexagonal $Cu_9S_8$ NPs. The strong NIR absorption enabled as-prepared solid $Cu_9S_8$ NPs and hollow $Cu_9S_8$ NPs to be alternative agents for photothermal and photoacoustic applications (Figure 5.4d). Under irradiation of 808 nm laser for 10 minutes, the temperature increased by 5–23 °C with the increasing amount of $Cu_9S_8$ NPs ranging 25–200 $Cu_9S_8$ μg per ml, indicating their similar outstanding PTT activity of solid $Cu_9S_8$ NPs and hollow $Cu_9S_8$ NPs (Figure 5.4e). The combination of increased active sites and strong photothermal performance could play a better role in chemodynamic therapy. In particular, the PTT performance of hollow $Cu_9S_8$ NPs could promote the Fenton-like reaction, and the reaction with $H_2O_2$ could generate more hydroxyl radicals, thereby killing more tumor cells and providing hollow $Cu_9S_8$ NPs with enhanced chemodynamic therapy performance (Figure 5.4f). The photoacoustic imaging properties of hollow $Cu_9S_8$ NPs provided convenience for tumor treatment. Various experimental studies had also confirmed that $Cu_9S_8$ NPs had good hydrophilicity, stability, and biocompatibility and did not damage other normal tissues and had low toxicity [23].

Plasmonic metal–semiconductor heterostructures with intense plasmonic coupling may strongly contribute to the EM intensity for heat transfer efficiency, and these should be studied in future research. The assembly of semiconductors and plasmonic metal NPs into different chiral models needs further discussion. In addition to the generation of distinct optical properties, incorporated semiconductors can be used in fluorescence imaging in living cells to better detect the state of cells assisting PTT. Importantly, multicomponent heterostructures can give rise to targeted LSPR properties and intense EM coupling for optical enhancement. The combination of multiple metal alloy NPs and the composites of plasmonic metal NPs and semiconductors not only adjust the LSPR properties and the chiroptical responses but also may possess excellent photothermal conversion efficiency. The facile fabrication of multicomponent heterostructures has great prospects and may become potential PTT agents for in vivo applications. For example, copper selenide ($Cu_{2-x}Se$)-Au semiconductor-plasmonic metal Janus heterostructures were fabricated, and they had the sizes of $30.4 \pm 2.5$ nm with the Au NP core sizes of $14.4 \pm 1.4$ nm. $Cu_{2-x}Se$-Au Janus NPs showed broad and elevated absorption in the NIR region with a maximum absorption peak of 635 nm (Figure 5.5a), owing to the strong coupling between Au NP core and $Cu_{2-x}Se$ semienclosed shell. It had a good PTT conversion effect because both $Cu_{2-x}Se$ NPs and Au NPs themselves could promote PTT conversion. The Janus NPs formed by $Cu_{2-x}Se$-Au coupling showed obvious absorption in the NIR region and enhanced the photothermal conversion ability. Under irradiation of 808 nm laser for eight minutes, the temperature increased from 29.2 to 76.2 °C in the presence of $Cu_{2-x}Se$-Au Janus NPs (Figure 5.5b), and the temperature of deionized water barely varied under the same condition. The photothermal conversion efficiency reached up to 67.2%, which was much higher than for $Cu_{2-x}Se$ NPs (22%), Au NPs (21%), and Au@$Cu_{2-x}S$ NPs (52.1%). According to the experimental analysis, the photocurrent response of pure $Cu_{2-x}Se$ NPs was very weak, while the photocurrent response of Janus NPs with the addition of Au core was enhanced, indicating that Janus NPs had a good separation effect of photogenerated electrons and holes. The Fermi energy of Au was much lower than the valence band (VB) of $Cu_{2-x}Se$, and the photogenerated electrons would transfer from the VB of $Cu_{2-x}Se$ shell to the Au core, and the Au core with unique LSPR effect could improve the photothermal conversion efficiency. The photo-generated holes located on the conduction band (CB) of $Cu_{2-x}Se$ shell would reach with $OH^-$ for the formation of ·OH, and $Cu^+$ would react with $H_2O_2$ for the generation of ·OH. The generation of reactive oxygen species (ROS) promoted the application of photocatalytic therapy (Figure 5.5c) [24]. $MnO_2$@Au NPs were synthesized through the reduction of $KMnO_4$ on the surface of 40 nm Au NPs. The color changed from pink to dark green, and $MnO_2$@Au NPs showed two characteristic peaks at 380 and 600 nm (Figure 5.5d), which was due to the absorption of $MnO_2$ layer and the redshifted absorbance of Au NPs. As-prepared $MnO_2$@Au NPs also showed inherent Raman signatures at 569 and 651 $cm^{-1}$. In comparison to bare $MnO_2$ NSs, as-prepared $MnO_2$@Au NPs showed a twofold enhancement of the Raman intensity at 569 $cm^{-1}$, which could be used as Raman signals to real-time track the changes of $MnO_2$@Au NPs structures. The PTT ability of $MnO_2$@Au NPs was explored upon irradiation of 808 nm laser, due to the enhanced tissue penetration through the biological window. It was found that the temperature was increased to 70 °C in the presence of $MnO_2$@Au NPs under light irradiation for 10 minutes, whereas the temperature was just increased to 35 °C in the presence of Au NPs, which indicated the enhanced PTT performances of $MnO_2$@Au NPs under 808 nm laser (Figure 5.5e). The photothermal conversion efficiencies were calculated to be 64.8% for $MnO_2$@Au NPs and 54.3% for Au NPs. The high concentration of $H_2O_2$ in cancer cells would trigger the degradation of $MnO_2$ layer in the intracellular environment. At the same time, the characteristic peaks at 380 nm completely disappeared, and the characteristic peak at 569 $cm^{-1}$ underwent a drastic decline, in comparison to untreated $MnO_2$@Au NPs. After PTT treatment, $MnO_2$ layer was degraded, and the Raman signature at 832 and 1210 $cm^{-1}$ underwent obvious

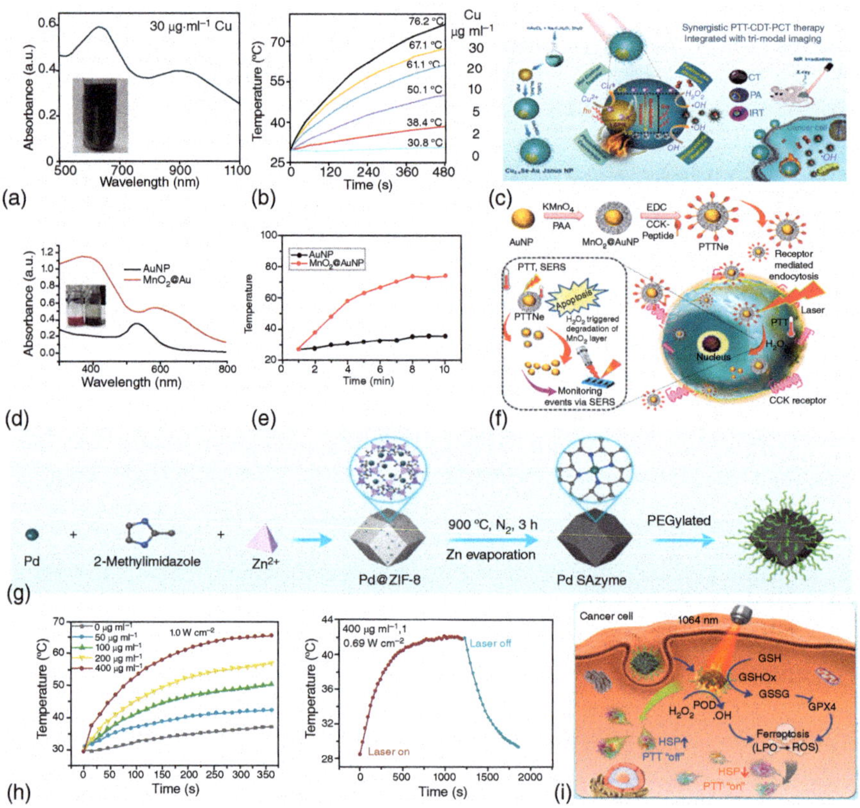

**Figure 5.5** (a) UV–vis–NIR absorption spectrum of $Cu_{2-x}$Se-Au Janus NPs. (b) Photothermal activity of $Cu_{2-x}$Se-Au Janus NPs at different concentrations under 808 nm laser irradiation. (c) Schematic illustration of $Cu_{2-x}$Se-Au Janus NPs promoting Fenton-like and photocatalytic reactions for tumor therapy. (d) UV–vis absorbance spectrum of Au NPs and $MnO_2$@Au NPs. (e) The temperature change of Au NPs and $MnO_2$@Au NPs with time under 808 nm laser irradiation. (f) Schematic illustration of photothermal therapy with $MnO_2$@Au NPs. (g) Schematic diagram of the formation of PEGylated Pd SAzyme. (h) Exploring the photothermal activity of Pd SAzyme under 1064 nm laser irradiation. (i) Schematic illustration of Pd SAzyme-mediated mild PTT promoting ferroptosis. Source: (a–c) Wang et al. [24] / with permission of Elsevier; (d–f) Reproduced with permission Sujai et al. [25]. Copyright 2021, American Chemical Society; (h, i) Reproduced with permission Chang et al. [26]. Copyright 2021, Wiley-VCH.

changes due to O–P–O backbone breakage, and the intensity at 1523 $cm^{-1}$ increased due to the $N_7$-H vibration from nucleic bases. These both results substantiated the apoptotic cell death mechanism (Figure 5.5f) [25]. A new type of cell death (Ferroptosis) was a new non-apoptotic cell death caused by iron-dependent lipid peroxidation, which was first reported by Brent Stockwell research group of Columbia University in 2012. A series of studies have shown that ferroptosis have played a very important role in pathophysiological processes such as cancer, neurodegenerative diseases, and organ damage. It is now known that ferroptosis is caused by lipid peroxidation of cell membrane or organelle membrane to kill the cells. Li et al. proposed a PTT model of ferroptosis-boosted mild PTT based on monatomic Pd nanozymes. PTT converted the light energy into heat, which was used to kill cancer cells. However, if the temperature was too high (above 50 °C), it can damage the normal tissue surrounding the cancer cells. Therefore, a mild temperature (38–43 °C) was needed to kill the cancer cells and protect the normal tissue from damage. The ultrasmall metal Pd nanoparticles were encapsulated into the zeolite imidazolate framework-8 (ZIF-8) to form Pd@ZIF-8 NPs and then treated at 900 °C under the protection of $N_2$ for three hours to obtain Pd nanozymes (Figure 5.5g). The Pd concentration in Pd@ZIF-8 NPs was quantified to be 0.18 wt% via inductively coupled plasma optical emission spectrometry, and the surface area of Pd@ZIF-8 NPs was measured to be 1141.22 $m^2\,g^{-1}$. Pd@ZIF-8 NPs had strong light absorption ability in the NIR II region of 1000–1400 nm and could achieve better PTT conversion activity. The photothermal conversion efficiency of Pd@ZIF-8 NPs was calculated to be 33.98% (Figure 5.5h). In a in vivo therapy, a complete tumor regression could be achieved by the ferroptosis-promoted mild PTT (Figure 5.5i). These findings innovatively introduced and validated the concept of ferroptosis-boosted mild PTT based on Pd@ZIF-8 NPs, which provides a hopeful direction for future cancer therapy [26].

## 5.2 Photodynamic Therapy

Photodynamic therapy (PDT) is an emerging cancer treatment method in comparison to radiation therapy and chemotherapy. Through a complex chain of photochemical reactions, ROS are generated, and the strong oxidizability produces selective cytotoxicity to tumors, leading to cell death and tissue destruction [27]. The characteristics of PDT are its specific site photoactivation of targeted photosensitizers (PSs) by the visible or NIR light, which have minimal damages to normal tissue. There are mainly two kinds of photochemical reactions for PDT applications. Type I mechanism is that PSs directly interact with surrounding substrates (such as cell membranes and other biological molecules), forming free radicals with the reaction with oxygen. Type II mechanism involves the energy transfer from the excited PSs to triplet oxygen ($^3O_2$) to promote the formation of reactive oxygen intermediates, such as singlet oxygen ($^1O_2$) [28].

The hydrophobic PS chlorine e6 (Ce6) has commonly been used in PDT applications because of its highly sensitive efficiency and high fluorescence emission at long wavelengths of 660–670 nm, which can provide a clear window and ideal penetration for optical image and PDT in vivo. Yu et al. used a polyphosphoester-based nanocarrier (NP-PPE) as the carrier of Ce6, which could promote the internalization of Ce6 cells and promote the production of ROS, which was of great help in the treatment of pancreatic cancer (Figure 5.6a). Ce6-encapsulated nanocarrier NP-PPE/Ce6 was stable in the blood environment, so the drug could be delivered to tumor cells through the drug delivery system. NP-PPE/Ce6 could keep the Ce6 in the hydrophobic core during circulation but rapidly release the encapsulated Ce6 in the acid endosome or lysosome after internalization into cancer cells. Furthermore, experiments had shown that under acidic conditions at pH 5.5, the hydrophobicity of polyphosphoester core was reduced, which would increase the release of the encapsulated Ce6 (Figure 5.6b) [29]. Ce6 was also modified on the surface of Au cups and promoted the generation of $^1O_2$ under 660 nm laser irradiation (Figure 5.6c). About 67.5% ± 1.3% cell viability was obtained for Au NC-Ce6, and 100% ± 1.5% for phosphate buffer saline (PBS)-treated control cells (Figure 5.6d) [9]. Ce6 was also attached on the surface of upconversion nanoparticles (UCNPs) and was excited by the UCL peak at 650 nm from UCNPs under 980 nm laser irradiation. The energy was transferred from UCNPs to Ce6 for the production of $^1O_2$ in PDT applications [16]. Use of UCNPs enabled the utilization of NIR laser for the generation of UCL signals for the excitation of the PS, which would produce little damage to cells and tissues [31]. These studies mainly focused on PS-attached nanostructures for PDT applications. Furthermore, UCNPs could also transfer the visible energy to methylene blue (MB) for the production of $^1O_2$ in NIR-triggered antimicrobial PDT applications, by utilizing the advantages of high tissue penetration, broad antibacterial spectrum, and less acquired resistance of UCNPs. For example, UCNPs with the size of 100 nm and a thin inert shell of 3.5 nm exhibited strong UCL intensity (Figure 5.6e). UCNPs were encapsulated inside dendritic mesoporous silica, and MB as PS and lysozyme (LYZ) were loaded on the surface of dendritic mesoporous silica. In particular, MB showed maxima absorption peaks at 590 and 654 nm, which were due to the presence of dimers and monomers, respectively. When UCNPs existed under 980 nm laser irradiation with a relatively low power density of 0.5 W cm$^{-2}$, the emission peak at around 660 nm would be absorbed by MB monomer, which further resulted in the generation of $^1O_2$ for PDT applications. Besides, most lysozyme would be released from the surface of dendritic mesoporous silica for treatment. About 37% wound healing rate was achieved for LYZ+PDT treatment on Day 2, which was better than that of single treatment (25%) and the control (18%). After 20 days, the wound healing rate reached up to 95% for the LYZ+PDT treatment group (Figure 5.6f). The synergistic LYZ-PDT effect achieved effective bactericidal capacity and conspicuous bacteriostasis on methicillin-resistant *Staphylococcus aureus*. This work proposed a novel bioinorganic nanohybrid and for the first time combined lysozyme with the UCNP-PDT system to enhance the efficiency against resistant bacteria (Figure 5.6g). These findings have paved the way for designing efficiently antibacterial nanomaterials and provided a new synergistic strategy for combating deep-tissue bacterial infection [30].

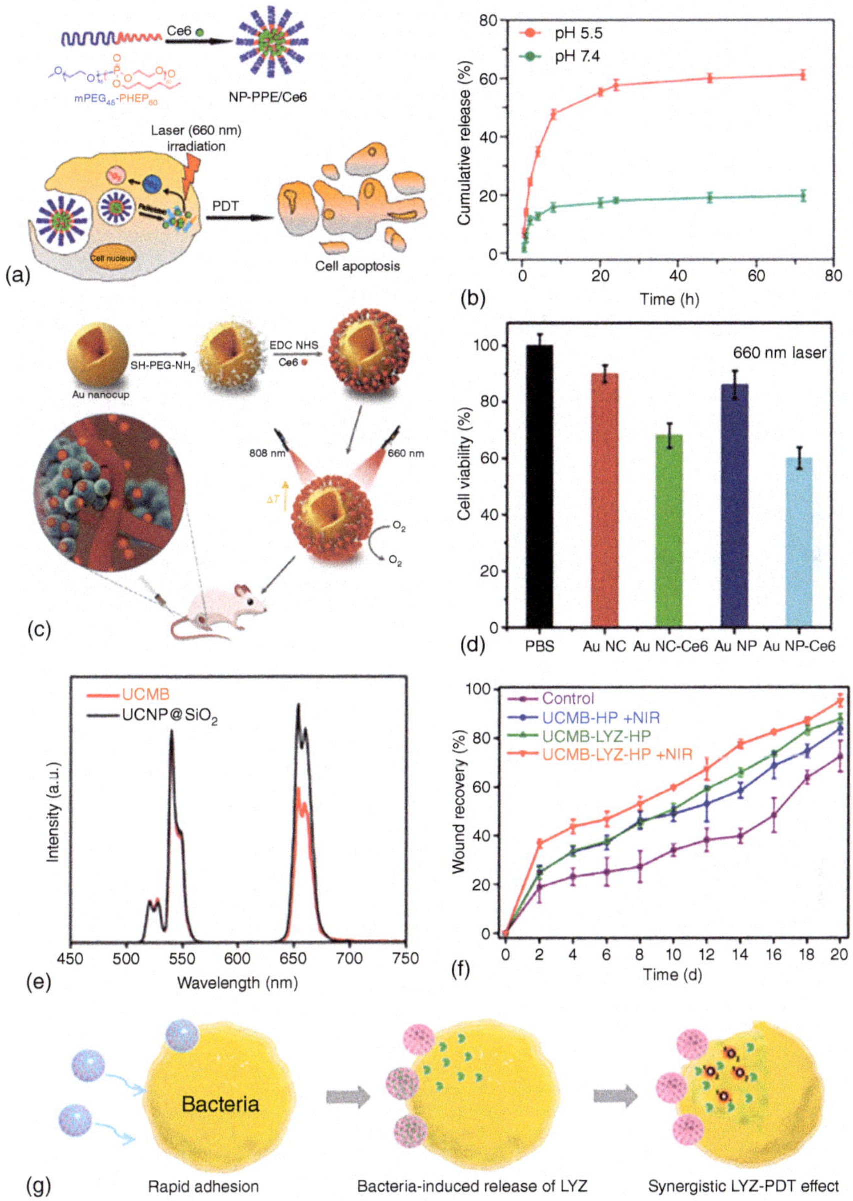


**Figure 5.6** (a) Schematic illustration of phototherapy with NP-PPE/Ce6. (b) Release levels of Ce6 under different pH conditions. (c) Schematic illustration of phototherapy with Au NC-Ce6. (d) Cell viability analysis of cells incubated with different materials under 660 nm laser irradiation. (e) up-conversion luminescence (UCL) spectra of UCNP@$SiO_2$ and UCMB. (f) Wound recovery in different treatment groups. (g) Antibacterial mechanism of lysozyme combined with the UCNP-PDT system. Source: (a, b) Reproduced with permission Ding et al. [29]. Copyright 2015, American Chemical Society; (c, d) Reproduced with permission Gao et al. [9]. Copyright 2017, Wiley-VCH; (e–g) Reproduced with permission Li et al. [30]. Copyright 2021, Wiley-VCH.

For PDT technique, $O_2$ was inherently required to generate ROS for theranostic systems. However, the hypoxia environment in solid tumors often limited the therapeutic efficacy of PDT. In comparison to low hydrophilicity and poor photostability of conventional organic PSs, semiconducting polymer NPs possessed high photothermal conversion ability, resulting in the generation of $O_2$ in the tumor microenvironment for enhanced PDT applications. For example, a semiconducting polymer NP core and an inorganic shell consisting of poly(cyclopentadithiophene-alt-benzothiadiazole) and $MnO_2$ NSs were prepared. In particular, poly(cyclopentadithiophene-alt-benzothiadiazole) was used as PS, and $MnO_2$ NSs were able to react with $H_2O_2$ to produce supply $O_2$ in the acidic tumor microenvironment for PDT applications (Figure 5.7a). Under NIR laser irradiation (808 nm), the photothermal conversion efficiency was calculated to be 22.3%, and $^1O_2$ was generated with the aid of poly(cyclopentadithiophene-alt-benzothiadiazole) in the presence of enough $O_2$.

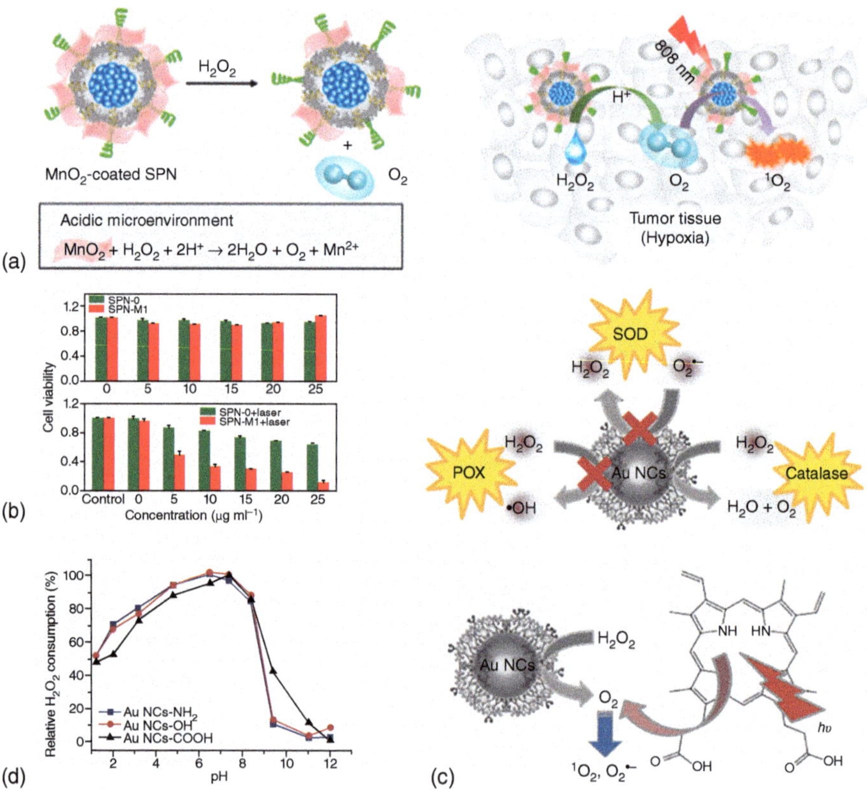


**Figure 5.7** (a) Schematic illustration of phototherapy with semiconducting polymer nanoparticles-$MnO_2$ NS core–shell NPs. (b) Cell viability analysis of SPN-0 and SPN-M1. (c) Schematic illustration of the enzyme-like activities of Au NCs-$NH_2$. (d) Comparison of relative $H_2O_2$ consumption of different materials under different pH solutions. Source: (a, b) Reproduced with permission Zhu et al. [32]. Copyright 2017, American Chemical Society; (c, d) Reproduced with permission Liu et al. [33]. Copyright 2017, Wiley-VCH.

In comparison to poly(cyclopentadithiophene-alt-benzothiadiazole), as-prepared core–shell NPs produced 2.68-fold more $^1O_2$ at the hypoxic and acidic conditions under NIR laser irradiation at 808 nm, which was attributed to the oxygen-evolution performances in the presence of $MnO_2$ NSs. With the increasing concentration of poly(cyclopentadithiophene-alt-benzothiadiazole)-$MnO_2$ NS core–shell NPs, the cell viabilities decreased (Figure 5.7b). This provided a new way to prepare oxygen-generating PDT agents for efficient PDT applications in a tumor-microenvironment-sensitive system [32]. Similarly, Au NCs also showed catalase-like activity for the production of enough $O_2$ in PDT applications. Amine-terminated, PAMAM dendrimer-encapsulated Au NCs (Au NCs-$NH_2$) were prepared and exhibited catalase-like activity, which could catalytic $H_2O_2$ for the production of $O_2$ over the physiological pH range (i.e. pH 4.8–7.4), but also could extend such activity to acidic conditions, in comparison to other nanomaterial-based catalase mimics (Figure 5.7c,d). The mechanism may be because the enriched tertiary amines of dendrimers were easily protonated in acidic solutions to facilitate the preadsorption of OH on the metal surface, thereby favorably triggering the catalase-like reaction. This proof-of-concept study provides a simple way to combine current $O_2$-dependent cancer therapy of PDT to overcome cancer cell hypoxia, thus achieving more effective anticancer treatments [33].

Similarly, to increase the concentration of $O_2$ in tumor cells, a Fenton reaction was introduced to enhance the PDT effect. Zhou et al. prepared the $Fe_3O_4$@Dex-TPP NPs through the coprecipitation of triphenylphosphine (TPP)-grafted dextran (Dex-TPP) and $Fe^{2+}/Fe^{3+}$. $Fe_3O_4$@Dex-TPP NPs showed the sizes of around $11.13 \pm 1.23$ nm and also possessed the magnetic resonance imaging (MRI) effect. Protoporphyrin IX (PpIX) as PS and mPEG-ss-COOH were both attached on the surface of $Fe_3O_4$@Dex-TPP NPs, resulting in the formation of $Fe_3O_4$@Dex/TPP/PpIX/ss-mPEG NPs (Figure 5.8a). The sizes of $Fe_3O_4$@Dex/TPP/PpIX/ss-mPEG NPs reached up to $64.34 \pm 2.63$ nm, and the core sizes were approximately $37.33 \pm 1.46$ nm. In the acidic lysosome, $Fe_3O_4$@Dex/TPP/PpIX/ss-mPEG NPs were internalized, and the $Fe_3O_4$ NPs would be decomposed into $Fe^{2+}/Fe^{3+}$, which would diffuse into the cytoplasm. The $Fe^{2+}$ would react with the overproduced $H_2O_2$, resulting in the direct generation of $O_2$ and ·OH. In particular, ·OH was highly oxidative, and the produced $O_2$ by the Fenton reaction could serve as a raw material for PDT to overcome the tumor oxygen consumption (Figure 5.8b). This had a great capacity of overcoming the drawback of the short half-life of ROS through mitochondrial targeting, more effectively prompting mitochondrial apoptosis, and finally significantly improving the PDT effect. At the same time, the presence of superparamagnetic $Fe_3O_4$ NPs endowed $Fe_3O_4$@Dex/TPP/PpIX/ss-mPEG NPs with well MRI effects [34].

The efficient PDT application required maximization of phototoxic damage on tumor and minimal side effects on normal tissue. It was essential to improve the selectivity toward malignant tissues for highly efficient PDT of cancer. Alternatively, aptamers as biological recognition molecules were screened from a large library through SELEX (Systematic Evolution of Ligands by Exponential Enrichment), and

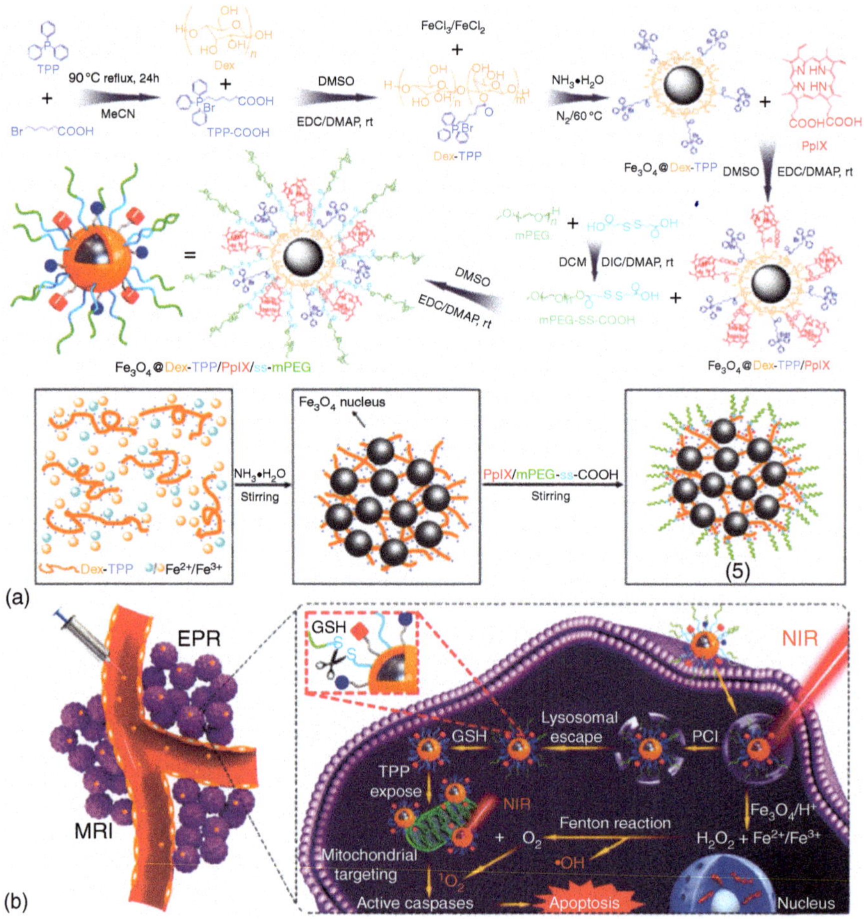


**Figure 5.8** (a) Schematic diagram of the formation of $Fe_3O_4$@Dex/TPP/PpIX/ss-mPEG NPs. (b) Schematic illustration of antitumor therapy with $Fe_3O_4$@Dex/TPP/PpIX/ss-mPEG NPs. Source: (a, b) Reproduced with permission Hou et al. [34]. Copyright 2019, American Chemical Society.

showed unique affinity towards clinic application. Aptamer-modified Zr-based porphyrinic MOF (ZrMOF) NPs were designed and used for target-induced bioimaging and PDT applications (Figure 5.9a). ZrMOF NPs were synthesized and showed a uniform size of 110 nm. A phosphate-terminal aptamer was anchored on the surface of ZrMOF NPs through strong coordination between phosphate and zirconium. Aptamer-conjugated ZrMOF NPs possessed much better stability after 24 hours, in comparison to ZrMOF NPs without any surface modification. TAMRA-modified Sgc8 aptamers showed quenched fluorescence signals after the conjugation on the surface of ZrMOF NPs, owing to the $\pi$–$\pi$ stacking effect (Figure 5.9b). However, the affinity between the aptamers and target membrane protein PTK7 expressed on HeLa cells induced the structural change of the aptamers and the detachment of TAMRA from the surface of ZrMOF NPs. This not only could specifically

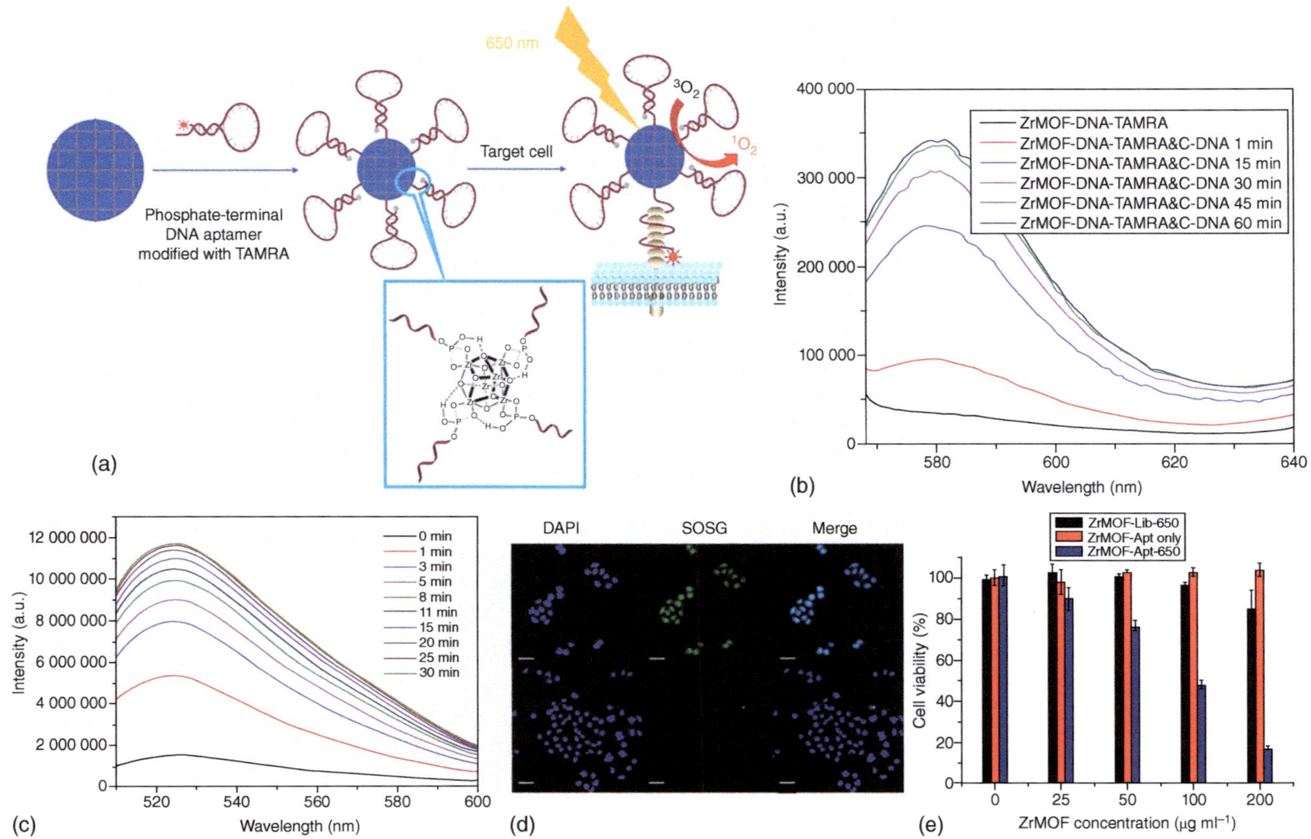


**Figure 5.9** (a) Schematic diagram of aptamer-modified ZrMOF used in photodynamic therapy. (b) Fluorescence spectra of TAMRA-modified Sgc8 aptamers conjugated with ZrMOF NPs. (c) Singlet oxygen generation efficiency of aptamer-modified ZrMOF under 650 nm laser irradiation. (d) Confocal imaging to monitor singlet oxygen after treatment. (e) Cell viability analysis of ZrMOF-aptamer and ZrMOF-library. Source: (a–e) Reproduced with permission Liu et al. [35] / with permission of Royal Society of Chemistry.

recognize the target membrane protein PTK7 expressed on HeLa cells but also showed the recovered fluorescence signals for bioimaging when incubated with HeLa cells. Furthermore, porphyrinic ZrMOF NPs were prepared and modified by aptamers for the selective combination with HeLa cells. Under laser irradiation (650 nm, 200 mW cm$^{-2}$) for five minutes, singlet oxygen was generated from 1 to 30 minutes in the presence of porphyrinic ZrMOF-aptamer NPs, which was confirmed by confocal imaging (Figure 5.9c,d). The cell viabilities were 48% and 17% for HeLa cells when treated with ZrMOF-aptamer NPs at a concentration of 100 and 200 μg ml$^{-1}$. A significant reduction in cell viability was observed for HeLa cells when incubated with ZrMOF-aptamer NPs (Figure 5.9e). Aptamers provided opportunities to develop MOF-based target-directed biosensors, through the rational design of engineering DNA aptamers and MOF nanomaterials for biomedical studies [35].

Highly cancer-cell-specific PS synthesis in vitro and in vivo was reported. Recently, Liu et al. proposed a cancer-cell-active PS synthesis route for highly specific cancer PDT. A simple and efficient mitochondrial targeting PS with aggregation-induced emission (AIE) (TPATrzPy-3+) was designed as the image-guided PDT agent. Two photochemically inert precursors, TPA-alkyne-2+ and MePy-$N_3$, were carried by a $Cu^{II}$-based metal–organic framework (MOF-199) NPs. MOF-199 NPs were cubic and showed an average size of about 80 nm. The presence of GSH in cancer cells induced the reduction of MOF-199, and divalent Cu ions in MOF-199 were reduced to monovalent Cu ions, resulting in the formation of $Cu^{I}$. The generated $Cu^{I}$ further catalyzed the reaction between TPA-alkyne-2+ and MePy-$N_3$, inducing the generation of TPATrzPy-3+ in cancer cells (Figure 5.10a). The yield of intracellular TPATrzPy-3+ synthesis was calculated to be 21.4%, 61.2%, 70.4%, and 70.5% at 8, 12, 16, and 24 hours post incubation with MOF-199 NPs, respectively (Figure 5.10b). Under light irradiation, the mitochondria-targeting TPATrzPy-3+ not only emitted bright red fluorescence with the wavelength ranging from 570 to 650 nm but also produced efficient $^1O_2$ for image-guided cancer-cell ablation. This proposed novel tumor-specific in situ photosensitizer synthesis and molecular design strategy offered a new approach for the rational design of targeted-position PSs inside targeted lesions for efficient and advanced PDT applications [36].

AIE NPs combined NPs and organic dyes or PSs together and had showed potential promising applications for fluorescence bioimaging and PDT applications. Although AIE NPs were stable in water, their stability in saline solution still needed to be improved for further applications in vivo and in vitro. Recently, He et al. prepared nanographene oxide (NGO)-capsulated AIE PSs and further modified NGO-capsulated AIE NPs with polyethylene glycol (PEG), to improve the stability of NGO-capsulated AIE NPs in PBS solution. Fluorescence bioimaging and PDT applications were performed in vitro and in vivo. NGO was synthesized using a modified Hummers method and had a height of about 2 nm and a size of around 30 nm. NGO was modified by PEG, inducing the formation of NGP. And then TPEred as PS was used to combine with NGP, and NGP-TPEred was formed and had the sizes of 50–100 nm (Figure 5.11a). NGP-TPEred exhibited a peak absorption at around 450 nm and displayed peak fluorescence signals at 630 nm with a quantum yield

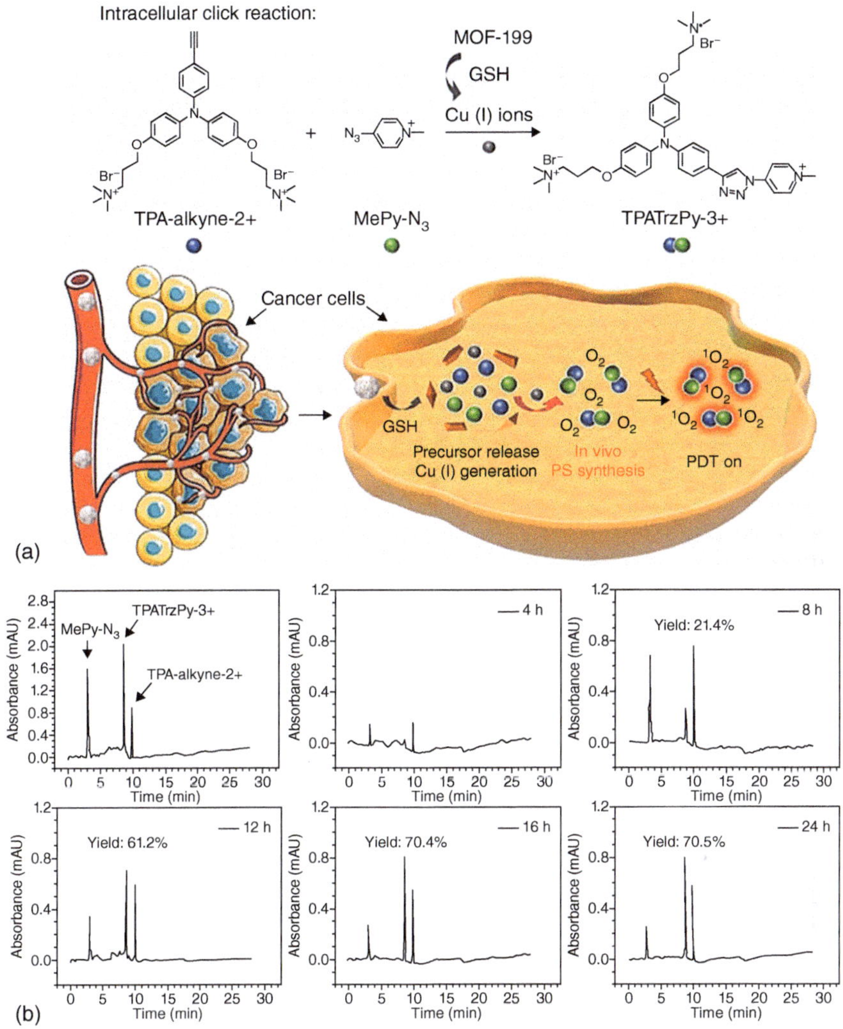


**Figure 5.10** (a) Synthesis and PDT application of TPATrzPy-3+. (b) Yield of intracellular TPATrzPy-3+ synthesis. Source: (a, b) Reproduced with permission Wang et al. [36]. Copyright 2021, Wiley-VCH.

of 3.55% (Figure 5.11b). Under 450 nm laser irradiation, NGP-TPEred produced a large amount of ROS, which largely induced a decrease in absorption at 377 nm for ABDA (ROS probe) (Figure 5.11c). The growth of tumors had been inhibited for several days by NGP-TPEred, in comparison to PEG-capsulated TPEred NPs (Figure 5.11d,e). As-prepared NGP-TPEred also had good stability in PBS buffer for 24 hours. This demonstrated the capability of NGP-AIE NPs as a powerful tool in bioimaging and PDT applications [37].

Currently, PS-free nanostructures were also reported to possess better PDT effects. For example, $Fe_3O_4$@Au NPs possessed unique plasmonic electron transfer

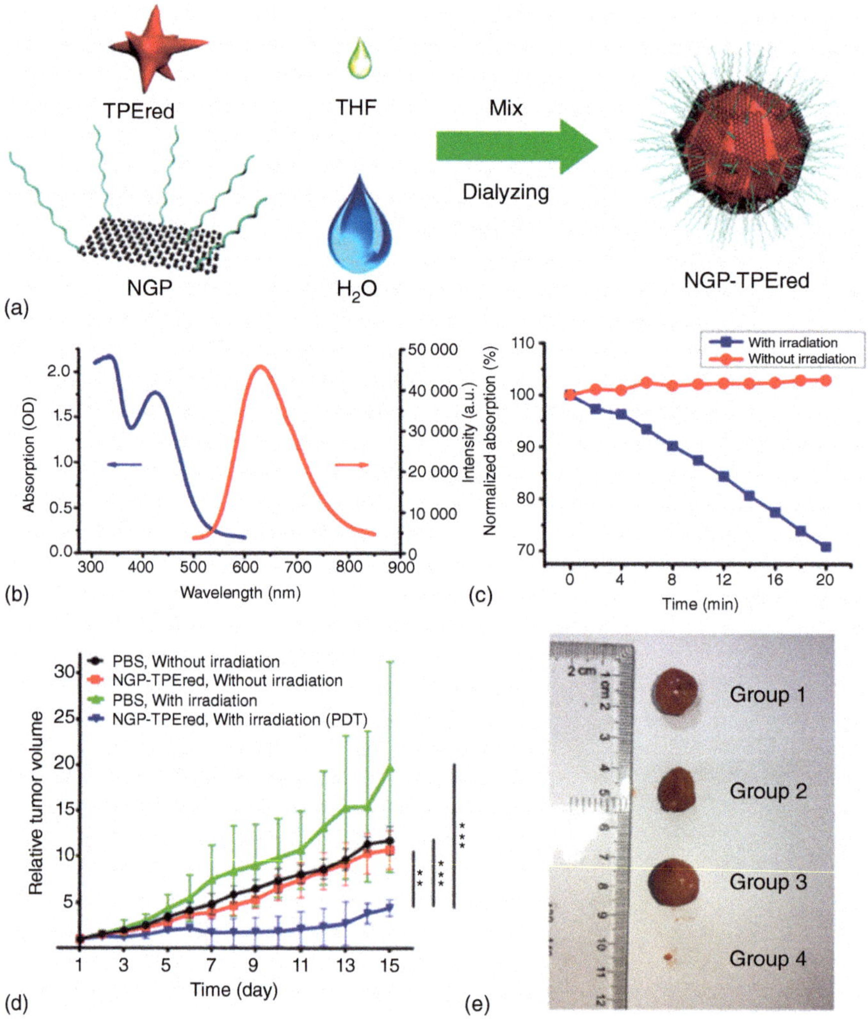


**Figure 5.11** (a) Synthesis of NGP-TPEred. (b) Absorption and fluorescence spectrum (450 nm excited) of NGP-TPEred. (c) Absorption of NGP-TPEred and ABDA mixture under 450 nm laser irradiation. Relative tumor volume (d) and tumor photos (e) under different treatment conditions during the 15-day observation period. Source: (a–e) Sun et al. [37] / with permission of American Chemical Society.

ability, which could transfer electrons to $O_2$ for the efficient promotion of $^1O_2$ [10]. Plasmonic core–petal Au NPs transferred the electrons to $^3O_2$ and further photosensitized $^3O_2$ to convert into $^1O_2$ under 785 nm laser. High concentration of $^1O_2$ was obtained with the increasing irradiation time and laser power. The amount of $^1O_2$ produced by core–petal Au NPs was almost twice than that of Au NRs, which was also higher than that of Au NPs. A strong EM field was obtained from core–petal Au NPs with multiple and sharp tips for plasmonic coupling [11]. The introduction of semiconductors with plasmonic metal NPs could promote the separation of photoexcited electrons and holes for the efficient production of ROS. Au NP-CdSe/CdS QD-composite NRs were synthesized and served as PS for oxygen-independent PDT applications. Under 450 nm LED excitation, the photogenerated electrons would transfer to Au tips, and the holes were confined to CdSe/CdS QDs, which provided hydrogen and oxygen evolution sites for water splitting to produce ROS for the PDT application (Figure 5.12a,b). The composite of semiconductors and plasmonic metal NPs provided another kind of PS for

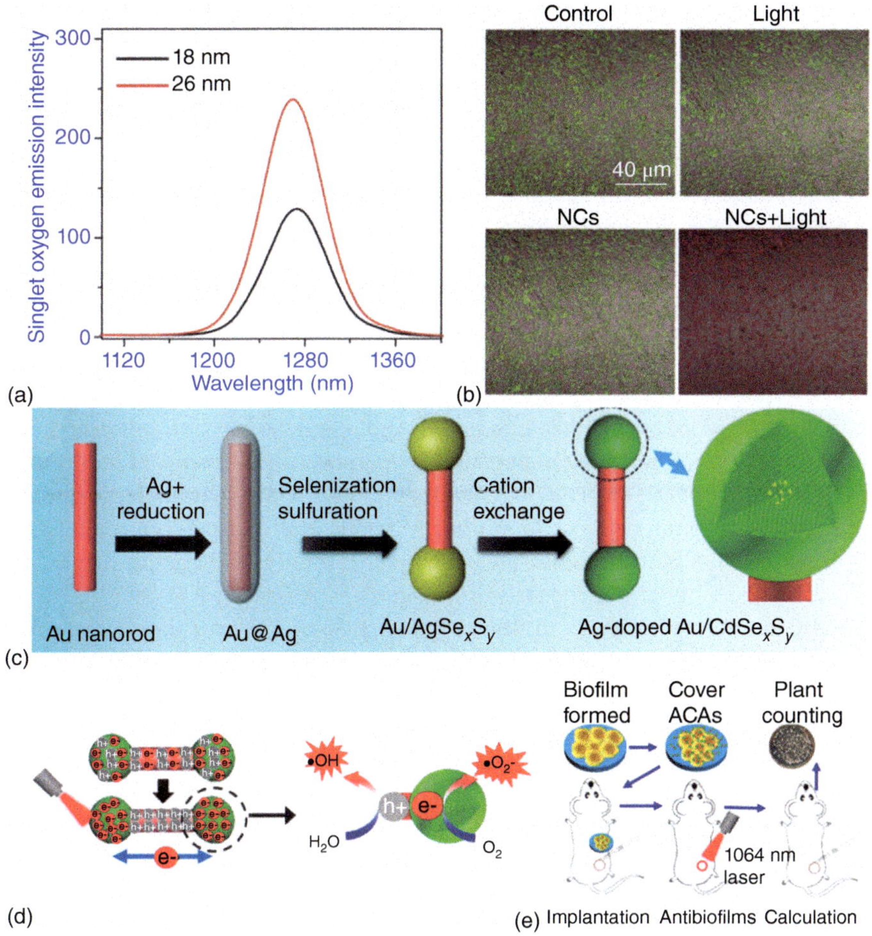


**Figure 5.12** (a) 18- and 26-nm-long CdSe/CdS dot/rod NCs produce singlet oxygen. (b) The therapeutic effect of L-cys-CdSe/CdS dot/rod NCs on cancer cells with or without two-photon irradiation (green and red dots represent living and dead cells, respectively). (c) The preparation process of the Au/CdSe$_x$S$_y$/Ag nanodumbbells. (d) Mechanism diagram of Au/CdSe$_x$S$_y$/Ag nanodumbbells enhancing PDT under NIR-II laser irradiation. (e) Schematic diagram of Au/CdSe$_x$S$_y$/Ag nanodumbbells used for antibacterial applications under 1064 nm laser irradiation. Source: (a, b) He et al. [38] / with permission of Royal Society of Chemistry; (c–e) Reproduced with permission Wang et al. [39]. Copyright 2021, American Chemical Society.

oxygen-free PDT applications [38]. Plasmonic Au/Cd$Se_xS_y$/Ag nanodumbbells were studied to exhibit PDT performances under a second NIR-II region light irradiation (1000–1700 nm). The NIR-II region light irradiation possessed deeper tissue penetration than visible and NIR-I light for effective PDT applications. Au@Ag core–shell NRs were prepared through the coating of Ag layer on the surface of Au NRs with an average length of 50–70 nm and a width of 10 nm. The chalcogenide reaction between selenium/sulfur precursor mixture and Ag induced the formation of Au/Ag$Se_xS_y$ nanodumbbells, and the cation exchange reaction between $Ag^+$ and $Cd^{2+}$ resulted in the generation of Ag-doped Au/Cd$Se_xS_y$ nanodumbbells (named as Au/Cd$Se_xS_y$/Ag nanodumbbells) (Figure 5.12c). Under irradiation of 1064 nm light, Au/Cd$Se_xS_y$/Ag nanodumbbells generated hot electrons and holes, achieved the separation of hot electrons located at the ends of nanodumbbells and holes located in the middle of nanodumbbells. And the electrons would react with $O_2$ to form $\cdot O_2^-$, and the holes would react with water to form ·OH (Figure 5.12d). The generation of more ROS enabled Au/Cd$Se_xS_y$/Ag nanodumbbells to be served as PDT agents for the antibacterial applications under NIR-II light irradiation (Figure 5.12e) [39].

## 5.3 Multimodal Therapy Methods

In phototherapy, of particular interest is the engineering of multifunctional nanoplatforms to allow simultaneous tumor diagnosis and therapy. Phototherapy, including PTT and PDT, has drawn tremendous attention due to its high therapeutic efficiency for targeting cells while minimizing collateral damage to normal tissue and organs [40]. Clinical studies are still fundamentally challenging due to the individual variability and metastasis of cancer cells. Although PTT and PDT both showed potential applications in phototheranostics, one therapy alone could not overcome the tumor metastasis, and usually suffered from incomplete tumor killing, and was not enough for effective cancer treatments. With the discovery of a variety of new antitumor therapies, multipronged approach of antitumor therapies with different mechanisms can realize the ultimate objective of curing tumors [41]. Combined therapeutic platforms integrated different treatment modalities and overcame the limitations of single therapy and had been gradually developed for efficient cancer treatments [42]. Synergistic combination of phototherapy and other therapeutic approaches with multimechanisms can optimize cancer treatment.

Plasmonic nanomaterials have emerged as ideal light-driven therapeutic agents due to their LSPR properties. In particular, plasmonic nanomaterials not only could convert NIR light into localized heat energy in the application of cancer cell apoptosis or necrosis based on cellular hyperthermia but also display excellent PDT effects to selectively kill cells and destroy tissues through the generation of cytotoxic ROS under illumination at an appropriate wavelength. Plasmonic nanomaterials featured with PTT and PDT effects were of great importance and urgently needed for cancer treatment. For example, core–petal Au NPs possessed PTT ability, which only killed 17% cells. However, nearly 100% cell death rate was obtained by core–petal

Au NPs under illumination of 785 nm laser for six minutes, benefiting from the synergistic effects of PTT and PDT. NIR-activated core–petal Au NPs efficiently killed cancer cells through apoptotic pathways by dual phototherapeutic effects and ROS-mediated oxidative intracellular damage with a relatively mild increase in temperature, low laser power, and short laser exposure time [11]. Chiral AuCuAu NPs were also regarded as phototherapy agents with the combination of PTT and PDT. The intense CD signals of chiral AuCuAu NPs allowed the use of CPL to induce PTT and PDT effects in cell lines and in vivo tumor. In particular, L-CF-AuCuAu NPs almost killed all cancer cells under LCP illumination and R-CF-AuCuAu NPs almost killed all cancer cells under RCP illumination. Under the synergistic effects of PTT and PDT, the illumination time was reduced from 20 to 3 minutes. Based on the combination of PTT and PDT, we can further optimize the structures of chiral materials for effective cancer treatments. Using multimodal imaging, the sizes and location of tumors and the efficiency of treatment greatly affected the diagnosis of cancer treatment. Imaging techniques, such as photoacoustic (PA) imaging, computed tomographic (CT) imaging, fluorescence imaging, and MRI should be employed for multimodal imaging and are also a critical auxiliary technique for multimodal therapy. Chiral AuCuAu NPs as photoacoustic agents were applied for PA imaging. It was shown that chiral AuCuAu NPs were mainly located in the tumor area than heart, liver, spleen, lung, and kidney through determining Au amounts by ICP-MS technique. The development of plasmonic NPs with the LSPR in NIR-II biowindow received increasing attention in phototheranostics. NIR-II Au nanoframeworks (Au NFs) with large mesopores (≈40 nm) were synthesized through liposome template–guided route and were applied for multimodal imaging along with therapeutic robustness. The synthesized Au NFs exhibit strong absorbance in the NIR-II region ranging from 750 to 1150 nm with the increasing concentration of $HAuCl_4$ from $1 \times 10^{-3}$ to $8 \times 10^{-3}$ M. Under continuous irradiation with a 1064 nm laser for six minutes, the temperature increased from 25.0 to 52.5 °C, and the photothermal conversion efficiency of Au NFs under 1064 nm laser irradiation was calculated to be 23.9%. These results indicated that NIR-II Au NFs could serve PTT agents to kill cancer cells. Furthermore, Au NFs also exhibited a high PA signal in the NIR-II region (about 1200 nm), and also displayed intense optoacoustic signals at 1200 nm wavelength. Therefore, as-prepared Au NFs could also be used as NIR-II PA contrast agents for imaging-guided cancer theranostics. The unique mesopores and framework structures as well as the formed sharp tips enabled Au NFs to exhibit strong SERS activity, benefiting from the strong hot spots and the intensive electromagnetic field. 4-ATP as Raman molecules were modified on the surface of Au NFs and showed amplified SERS intensity. Due to the reaction between the amino groups of 4-ATP and the carboxylic groups of hyaluronic acid (HA), HA-4-ATP-Au NFs were formed, and they were endowed with the targeting capacity for CD44-overexpressed tumor cells and were used to carry drugs in the large mesopores with the protection of HA. The PTT activity of Au NFs loosened the structures of the HA layer, which induced the release of cargo molecules. In vivo evaluation of HA-4-ATP-AuNFs-DOX on tumor-bearing xenografts demonstrated its high efficacy in eradication of solid tumors in NIR-II

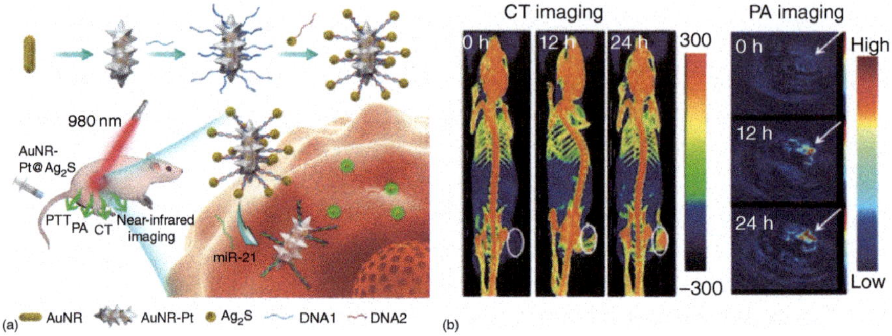


**Figure 5.13** (a) Schematic illustration of phototherapy with Au NR@Pt-$Ag_2S$ NP core–satellite assemblies. (b) Injecting Au NR@Pt-$Ag_2S$ NP core–satellite assemblies can exhibit strong CT and PA signals. Source: (a, b) Qu et al. [15] / with permission of Wiley-VCH.

under PA-Raman dual-image-guided photo-chemotherapy [43]. Besides PA imaging, CT imaging possessed high resolution and deep penetration and was also performed for clinical diagnosis [21]. Great efforts had been devoted to combining PTT and PDT with the aim of achieving synergistic therapeutic performance with attenuated light power. Plasmonic NP-evolved nanostructures were also used as CT agents. For example, as-prepared Au NR@Pt-$Ag_2S$ NP core–satellite assemblies [15], Au@$Cu_9S_5$ NP-UCNP-$Ag_2S$ pyramids [14], and NPs@ZIF-8@Au NRs [13] not only possessed PTT effects but also were used as promising CT and PA agents for imaging applications. Au NR@Pt-$Ag_2S$ NP core–satellite assemblies were largely accumulated at the tumor sites and exhibited strong CT and PA signals after injection for 24 hours (Figure 5.13a,b). CT and PA imaging also clearly demonstrated the accumulation of Au@$Cu_9S_5$ NP-UCNP-$Ag_2S$ NP pyramids in the tumor area. Significant CT and PA signals were observed at the tumor sites after injection of NPs@ZIF-8@Au NRs for 24 hours.

Some plasmonic nanostructures all displayed PTT and PDT effects for multimodal therapy and also served as CT and PA agents for accurate clinical diagnosis. It was reported that Au cups generated $67.5\% \pm 1.3\%$ cell viability after illumination of 660 nm laser for PDT applications and yielded $48.2\% \pm 1.8\%$ cell viability after exposure to 808 nm laser for PTT applications. Benefiting from the increased phototherapeutic capacities of Au cups by using single-laser irradiation, the combination of PDT and PTT effects would induce high rates under laser excitation at 660 and 808 nm. The cell viability by the combination of PDT and PTT was as low as $10.3\% \pm 1.4\%$, which was lower than that of after PDT alone ($67.5\% \pm 1.3\%$) and PTT alone ($48.2\% \pm 1.8\%$). Combination therapy was better than single treatments. Furthermore, Au cups also served as excellent CT and PA contrast agents for CT and PA imaging, which clearly demonstrated the location of the tumor section for further treatments [9]. In comparison to Au NPs and Au cups, Au NRs showed tunable LSPR ranging from the visible to NIR regions and became excellent NIR light activatable PDT and PTT reagents. Prof. Kuo Chu Hwang's group proposed the use of Au NR-mediated PDT without the addition of organic PSs. Lipid-bilayer-coated Au NRs with an average length and diameters of $37.3 \pm 2.4$

and 11 ± 1.2 nm were prepared, and they showed a strong NIR absorption band with an absorption maximum at ~808 nm. It was found that Au NRs emitted visible light luminescence upon UV light excitation. In particular, green, yellow, red, and NIR (700–800 nm) fluorescence emission was observed under light excitation at 488, 533, and 600 nm. The longer the excitation wavelengths, the longer the luminescent wavelengths Au NRs emitted. Au NRs could sensitize the formation of singlet $O_2$ upon photoexcitation, as evidenced by the singlet $O_2$ phosphorescence emission at ~1263 nm. Under light irradiation of 550 nm (28 mW $cm^{-2}$, 80 minutes) and 940 nm (54 mW $cm^{-2}$, 40 minutes), the amount of cellular death was 6% and 62%, respectively. The number of photoinduced cell death at 940 nm was tenfold higher than that at 550 nm when treated with LED light of different wavelengths. This may be due to the PDT effect that played a major role at 940 nm, while the PTT effect was caused by 550 nm. The study of tumor cells in mice also proved that Au NRs could produce singlet oxygen ($^1O_2$) under 915 nm single-photon NIR irradiation, and its PDT effect was significantly stronger than that of the PTT effect under 780 nm NIR irradiation. Therefore, Au NRs had both PTT and PDT effects, and the PTT and PDT effects could be exchanged by changing the excitation wavelength of NIR [12]. Besides, Pt dots were attempted to grow at the end of Au NRs, to form Pt-end-deposited Au NRs (PEA NRs). As-prepared anisotropic PEA NRs exhibited a maximum LSPR peak at 808 nm, which could be excited under 808 nm laser irradiation to serve as PTT and PDT agents. For the controllable growth of Pt metal at the end of Au NRs, 5-bromosalicylic acid was applied as ligands to selectively block the sides of Au NRs. This would inhibit the deposition of Pt metal on the sides of Au NRs and allowed the preferential reduction of $H_2PtCl_6$ solution at the end of Au NRs. The deposition of Pt at the end of Au NRs enabled the redshift of the LSPR peak of Au NRs from 740 to 808 nm. It was found that PEA NRs exhibited a significant electric field enhancement at the end of PEA NRs, in comparison to PBA NRs (Pt was not deposited at the end of Au NRs but distributed on the surface of entire Au NRs) and single Au NRs. PBA NRs generated a strong electric field and exhibited more effective electron–hole spatial separation efficiency for the generation of abundant hot electrons. The generated hot electrons induced the production of ROS through both chemical and energy transformation processes and also could release the heat through the electron–phonon relaxation process. This significantly enhanced hot electron density of PEA NRs could promote the enhanced PDT and PTT performances. The experimental results showed that the photothermal conversion efficiency of PEA NRs could reach 55.4% after 10 minutes of 808 nm NIR irradiation, which was slightly higher than that of PBA NRs (52.9%) and Au NRs (49.3%), and PEA NRs could produce more ROS, which was checked by the electron spin resonance signals of $^1O_2$, $O_2{\cdot}^-$, and ·OH. This was because Pt grew anisotropically at the ends of Au NRs, forming a strong electric field density at both ends, which could generate more hot electrons, which in turn could promote the chemical energy conversion to produce more ROS. This heterostructure could effectively promote the separation of electron–hole pairs, which was significantly better than PBA NRs [20]. $Fe_3O_4$@Au NPs both displayed PDT and PTT ability after exposure to 808 nm laser, and the cell viability was down to 14.2% ± 1.3% treated by $Fe_3O_4$@Au NPs, which was lower than 39.8% ± 7.5% treated by $Fe_3O_4$ NPs (Figure 5.14a,b). At the same time, $Fe_3O_4$@Au NPs displayed plasmonic properties and superparamagnetic characteristics and showed potential applications in CT and PA imaging as well as MRI (Figure 5.14c). It was shown that $Fe_3O_4$@Au NPs were good contrast agents at the tumor sites for CT imaging after injection for 24 hours and were perfect PA agents for PA imaging due to the distinct absorbance in the visible range. All this clearly illustrated the accumulation area in the tumor section. Furthermore, $T_2$-weighted MRI also visualized the location of $Fe_3O_4$@Au NPs, in which the $T_2$ MR signal decreased in the tumor area after injection of $Fe_3O_4$@Au NPs from 12 to 24 hours. Therefore, $Fe_3O_4$@Au NPs served as multifunctional

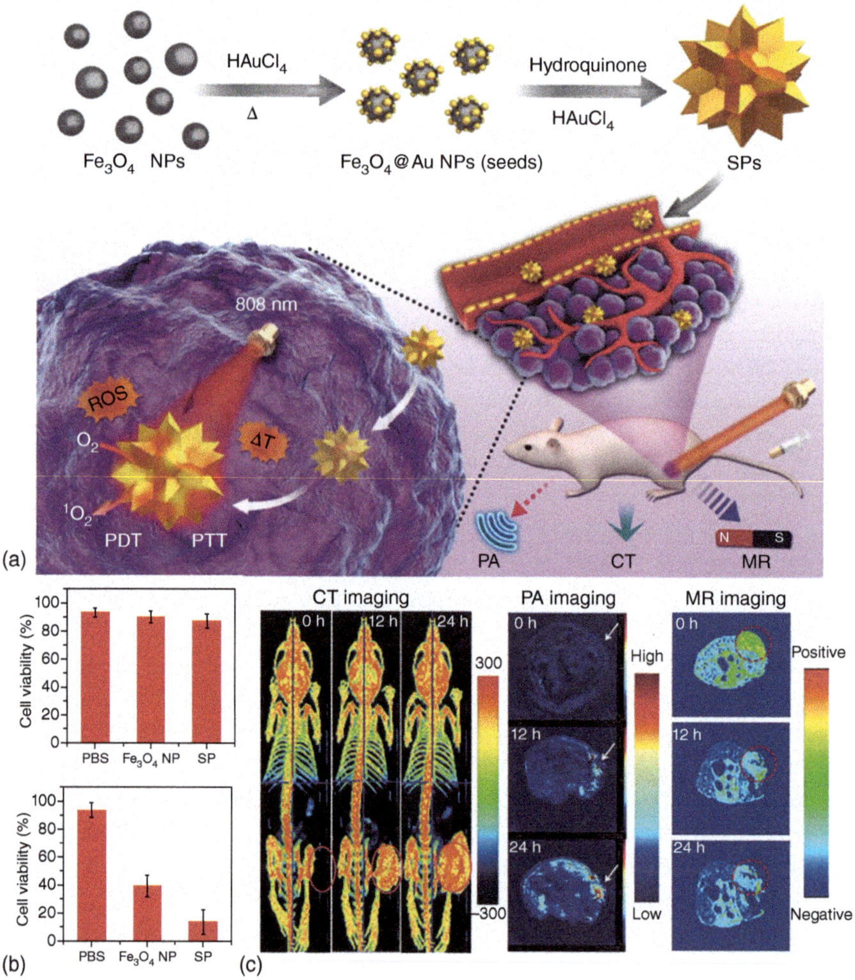


**Figure 5.14** (a) The synthetic spiky $Fe_3O_4$@Au SPs can be used for phototherapy under 808 nm laser. (b) Cell viability analysis of cells incubated with different materials before and after laser irradiation. (c) $Fe_3O_4$@Au SPs process multifunctional CT, PA, and MRI imaging capabilities. Source: (a–c) Wang et al. [10] / with permission of Wiley-VCH.

imaging CT, PA, and MRI agents could accurately illustrate the accumulation position, which gave a critical guide for tumor treatments [10].

Compared to single plasmonic NPs, NP assemblies displayed intensive LSPR properties for effective and multimodal therapy. Au NR dimers exhibited better PTT ability than single Au NRs, which displayed high rates of apoptosis under laser excitation at 808 nm. The intensive plasmonic coupling between Au NR dimers in Au NR dimer@UCNP structures induced an enhanced EM field and produced more thermal energy to the cells. Under 980 nm excitation, Ce6-modified Au NR dimer@UCNP structures exhibited PDT ability. With laser excitation at 808 and 980 nm, Au NR dimer@UCNP structures would display PTT and PDT effects for efficient cancer treatment (Figure 5.15a). A large number of cells were killed in the presence of Au NR dimer@UCNP structures by the combination of PTT and PDT, which was higher than PTT alone and PDT alone. This synergistic interaction induced better therapeutic effect, and this could reduce the doses of Au NR dimer@UCNP structures, in comparison to Au NR@UCNP structures (Figure 5.15b). Au NR dimers served as powerful PTT and the PDT agents were indirectly triggered through UCNPs, and UCNPs showed high UCL intensity in the visible range for UCL imaging. Therefore, Au NR dimer@UCNP structures were also used as multifunctional agents for CT, PA, UCL, and MR imaging (Figure 5.15c). Bright CT contrast imaging occurred in the tumor area after incubation of Au NR dimer@UCNP structures for 24 hours. The intensive NIR absorption enabled Au NR dimer@UCNP structures to serve as an outstanding agent for PA imaging. UCL signals were observed at 500–700 nm under laser excitation at 980 nm, originating from the enhanced UCL signals from UCNPs. Magnetic Gd elements in UCNPs provided Au NR dimer@UCNP structures with unique magnetic resonance properties for T1-weighted MRI application [16]. Common imaging techniques such as CT, PA, and MRI, as well as fluorescence imaging, had been incorporated into a single therapeutic agent. Plasmonic-nanostructure-dependent multimodal imaging probes with high spatial resolution, excellent depths, and sensitivity are the prerequisites for phototherapy. Under imaging guidance, the size and location of tumors as well as the treatment efficiency can be visualized directly, which shows great promise for early warning and diagnosis in cancer therapy. Plasmonic nanostructures with NIR region absorption need to be further explored for multimodal therapy, owing to low and safe power doses in the NIR tissue optical window (700–1100 nm). The biotoxicity of plasmonic nanostructures is still a potential challenge for the usage of high dose in vivo. Plasmonic nanostructures with strong phototherapic ability need to be further explored in multimodal treatments and imaging. This will reduce the usage dose of nanostructures and eliminate the damage toward tissues in vivo.

NIR-laser-induced phototherapy provided a promising strategy for removal of deep tumors and had been considered as a noninvasive option for cancer therapy. In comparison to plasmonic NPs, plasmonic semiconductor nanostructures (such as $Cu_{2-x}S$, $MoO_x$, $Co_3S_4$, $W_{18}O_{49}$) could effectively utilize the NIR light energy for the generation of electrons and holes in PTT and PDT applications. In particular, copper-based NPs were widely used in the biomedical field because of their low cost, easy synthesis, good biocompatibility, and antimicrobial activity under NIR

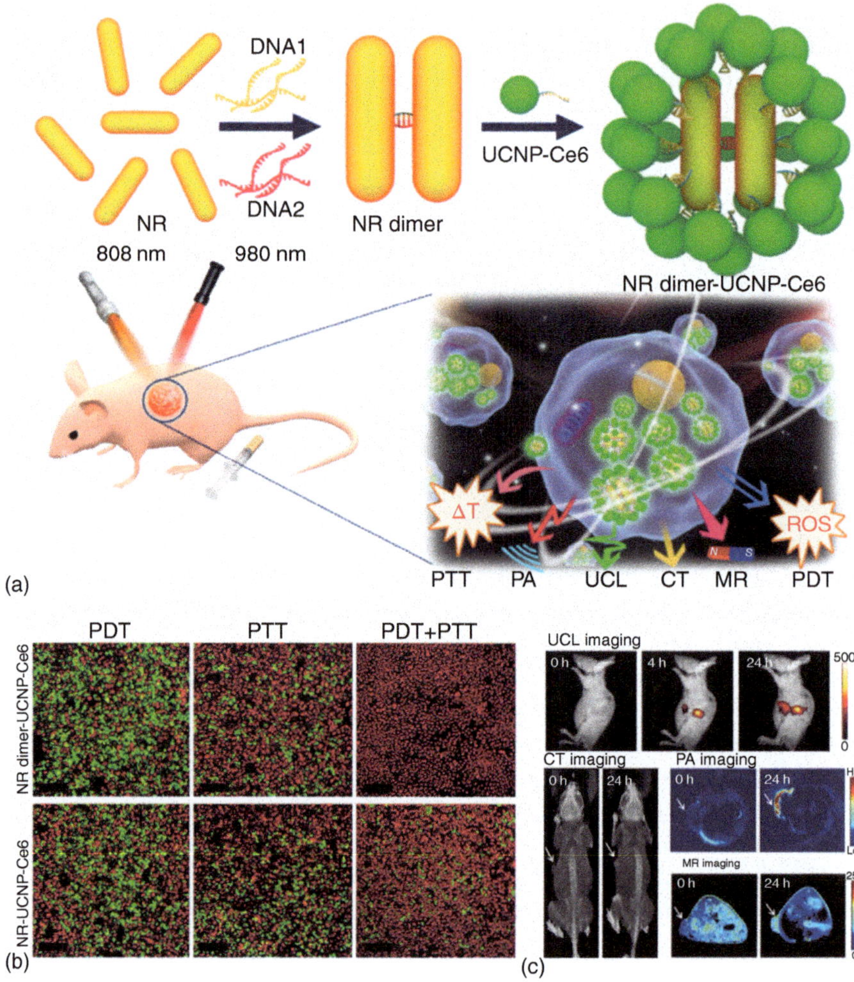


**Figure 5.15** (a) Schematic illustration of synthetic NR dimer-UCNP-Ce6 for phototherapy. (b) HeLa cell death after PDT and PTT with NR dimer-UCNP-Ce6 and NR-UCNP-Ce6 assembly (green and red dots represent living and dead cells, respectively). (c) Au NR dimer@UCNP structures served as multifunctional agents for CT, PA, UCL, and MR imaging. Source: (a–c) Sun et al. [16] / with permission of Wiley-VCH.

irradiation. For example, Lu et al. also reported PEG-modified $Cu_9S_8$ NPs that not only had PTT performance but also had PDT effects, the synergistic effect of which can be used to eliminate bacterial infection and biofilm destruction. As we all know, bacterial infection is very common in clinical medicine. In general, antibiotics are used for treatment, but once the bacterial biofilm is formed, the entry of antibiotics will be blocked, and therapeutic effect cannot be achieved. Therefore, the design of such PEG-modified spherical $Cu_9S_8$ NPs, under NIR irradiation, synergistically exerted PTT and PDT effects, produced a large amount of ROS, destroyed the biofilm of *S. aureus* and the bacterial structure, and had good therapeutic effects

(Figure 5.16a). $Cu_9S_8$ NPs with the average sizes of 120 nm were prepared, and they showed a broad absorption band ranging from 700 to 1000 nm (Figure 5.16b), which could efficiently utilize NIR light. Under 808 nm laser irradiation for 10 minutes, the temperature increased from 32 to 58 °C in the presence of 125 $\mu g\,ml^{-1}$ $Cu_9S_8$ NPs, and further increased to 75.6 °C in the presence of 500 $\mu g\,ml^{-1}$ $Cu_9S_8$ NPs. With the increasing amounts of $Cu_9S_8$ NPs, the PTT effects became strong. The photothermal conversion efficiency of the $Cu_9S_8$ NPs was calculated to be 41.55% (Figure 5.16c). Singlet oxygen and $H_2O_2$ were also generated in the presence of $Cu_9S_8$ NPs (Figure 5.16d). Therefore, $Cu_9S_8$ NPs could provide an effective thermal treatment and produce high levels of ROS under NIR light irradiation, and proved to be the promising PTT and PDT agents, which showed synergistic effects than thermal therapy alone. $Cu_9S_8$ NPs showed excellent antibacterial activity and antibiofilm ability against the clinical pathogen *S. aureus* on Ti plates [44]. Besides $Cu_9S_8$ NPs, $Cu_{7.2}S_4$ NPs were prepared and served as PTT and PDT agents for killing the clinical pathogens *S. aureus* and *Escherichia coli* to prevent bacterial infection (Figure 5.16e). $Cu_{7.2}S_4$ NPs were synthesized with the aid of L-cys and showed flower-like morphology and good dispersibility. $Cu_{7.2}S_4$ NPs exhibited a broad absorption band ranging from 700 to 900 nm (Figure 5.16f). Under 808 nm light irradiation for 10 minutes, the temperature increased to 41 °C in the presence of 125 $\mu g\,ml^{-1}$ $Cu_{7.2}S_4$ NPs and reached 55 °C in the presence of 500 $\mu g\,ml^{-1}$ $Cu_{7.2}S_4$ NPs, which were higher than the 32 °C in the presence of $H_2O$ (Figure 5.16g). This indicated that $Cu_{7.2}S_4$ NPs could efficiently convert NIR light into heat. The ROS generation of $Cu_{7.2}S_4$ NPs were also explored under 808 nm (Figure 5.16h). It was found that a continuous production of singlet oxygen occurred for $Cu_{7.2}S_4$ NPs, compared to the control group of only NIR or $Cu_{7.2}S_4$ NPs. Furthermore, a large amount of intracellular ROS was also generated for $Cu_{7.2}S_4$ NPs, while much less ROS was generated for NIR without $Cu_{7.2}S_4$ NPs. Therefore, $Cu_{7.2}S_4$ NPs were alternative PTT and PDT agents. As-prepared $Cu_{7.2}S_4$ NPs also possessed good biocompatibility and could be applied for clinical antibacterial phototherapy effects with NIR irradiation. The simultaneous PTT and PDT using $Cu_{7.2}S_4$ NPs under NIR laser could efficiently improve rheumatoid arthritis, which has opened a new door for treating rheumatoid arthritis [45].

Transition metal sulfides show narrow bandgap and excellent photoelectric properties. Besides copper sulfides, cobalt sulfides were widely reported to exhibit great performance under full spectrum even under NIR illumination, including $Co_{2.67}S_4$ NPs, CoS NSs, $Co_3S_4$ NPs, etc. Lin et al. prepared hollow $Co_3S_4$ NPs by using polyacrylonitrile NPs as the template, and the chitosan as a carbon source was added to obtain N-doped carbon-coated hollow $Co_3S_4$ NPs (hollow $Co_3S_4$@N-doped carbon) (Figure 5.17a). The average sizes of hollow $Co_3S_4$ NPs were about 80 nm and the carbon shell was around 3–4 nm. As-prepared hollow $Co_3S_4$@N-doped carbon showed high absorption in full spectrum as well as in the NIR region. Compared to $Co_3S_4$ and N-doped carbon, hollow $Co_3S_4$@N-doped carbon possessed enhanced absorption beyond 700 nm, which was beneficial to the production of well PTT performances upon 808 nm laser irradiation. The temperature increased by 38 °C in the presence of hollow $Co_3S_4$@N-doped carbon, which was much higher

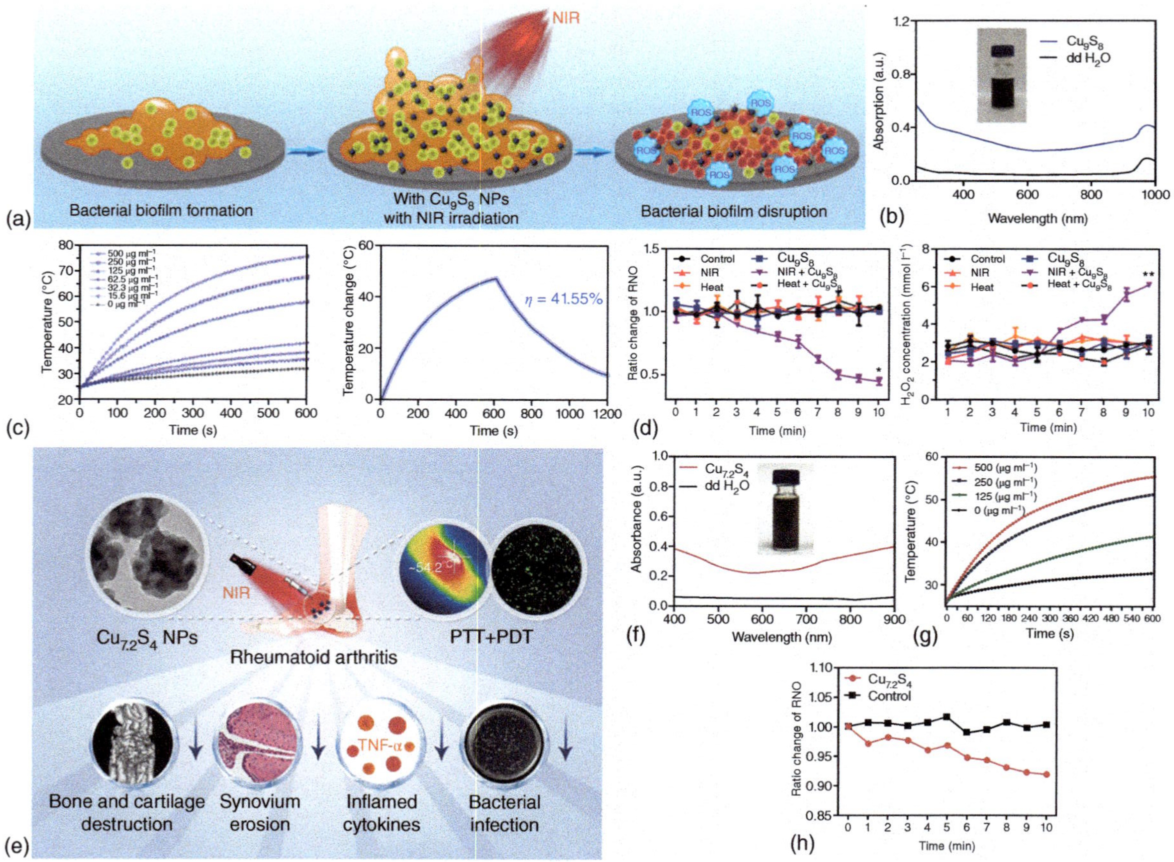


**Figure 5.16** (a) Diagram of $Cu_9S_8$ NPs destroying the staphylococcus aureus biofilm. (b) UV–vis absorbance spectrum of $Cu_9S_8$ and dd $H_2O$. (c) Exploring the photothermal activity of $Cu_9S_8$ NP solution under 1064 nm laser irradiation. (d) Exploring the photodynamic effect of $Cu_9S_8$ NP solution under NIR irradiation. (e) Schematic illustration of phototherapy with $Cu_{7.2}S_4$ NPs. (f) UV–vis absorbance spectrum of $Cu_{7.2}S_4$ and dd $H_2O$. (g) Exploring the photothermal activity of $Cu_{7.2}S_4$ NPs solution under 808 nm laser irradiation. (h) Exploring the photodynamic effect of $Cu_{7.2}S_4$ NPs solution under 808 nm laser irradiation. Source: (a–d) Wang et al. [44] / with permission of American Chemical Society; (e–h) Lu et al. [45] / with permission of Wiley-VCH.

than the 3 °C for pure water. The photothermal conversion efficiency was 57.5% for hollow $Co_3S_4$@N-doped carbon. The narrow bandgap (1.38 eV) enabled hollow $Co_3S_4$@N-doped carbon to exhibit great performance for the direct adsorption of NIR light. It was significant for practical applications of PTT agents [46].

$MoO_x$ NPs are another plasmonic semiconductor NPs showing tunable blueshifted and redshifted LSPR peaks ranging from the NIR-I to NIR-II region by adjusting the morphology of $MoO_x$ NPs. $MoO_x$ NPs were also reported as promising PTT and PDT agents for synergistic cancer therapy effects. $MoO_x$ NPs with the sizes ranging from 15 to 30 nm were synthesized and showed extended NIR absorption covering from 600 to 1200 nm. Therefore, two kinds of lasers (808 and 1064 nm) were utilized for exploring the PTT and PDT performances of $MoO_x$ NPs. The temperature increased from 25 to 54.5 °C in the presence of 200 μg ml$^{-1}$ $MoO_x$ NPs under 1064 NIR nm irradiation for 600 seconds and changed to 43.1 °C in the presence of 200 μg ml$^{-1}$ $MoO_x$ NPs under 808 nm NIR irradiation for 600 seconds (Figure 5.17b). The photothermal conversion efficiency of $MoO_x$ NPs was quantified to be 27.3% (808 nm irradiation) and 37.4% (1064 nm irradiation). The photothermal conversion efficiency under 1064 nm irradiation was much higher than that at the wavelength of 808 nm. Such an increased temperature was largely dependent on the concentration of $MoO_x$ NPs, indicating that $MoO_x$ NPs could efficiently convert NIR light energy into heat for PTT applications. The capability of $MoO_x$ NPs for the production of ROS was explored by using 1,3-diphenylisobenzofuran (DPBF) as a probe for the capture of $^1O_2$. It was found that $^1O_2$ could be produced in the presence of $MoO_x$ NPs under 1064 nm NIR nm irradiation, while there was no obvious $^1O_2$ generated under 808 nm NIR nm irradiation (Figure 5.17c). The remarkable distinction in $^1O_2$ production under two different NIR-I and NIR-II windows could be justified on the grounds of heat generation based on the different photothermal conversion abilities. The remarkable synergetic PTT and PDT functionalities enabled $MoO_x$ NPs to be emerging nanostructures for phototherapy of cancer upon 1064 nm light irradiation [47].

Alternatively, plasmonic metal NP–semiconductor heterojunction nanostructures possessed collective performances deriving from the materials combination and had raised a great social concern together with the impetus for exploring synergistic multimodal therapy effects. In particular, CuS NPs as p-type semiconductors showed narrow bandgaps and could generate a large amount of electrons and holes, inducing the occurrence of LSPR properties in the NIR region. However, CuS NPs still suffered from low light conversion efficiency and could not produce satisfactory therapeutic effects for cancer treatment. The integration of CuS NPs with plasmonic NPs would improve the photothermal performances, benefiting from the LSPR coupling between CuS NPs and plasmonic NPs. For example, Au-CuS hybrid nanostructures were widely fabricated for PTT applications. The PDT performances and the synergistic effects of PTT and PDT of Au-CuS hybrid nanostructures were not studied in detail. Recently, Zhang et al. fabricated Au core@void@CuS shell yolk–shell NPs (YSNPs) for potential chemo-, PTT, and PDT combination performances (Figure 5.17d). Au cores (including Au NPs and Au NRs) with the LSPR absorption peaks at 520, 700, 808, 860, and 980 nm were

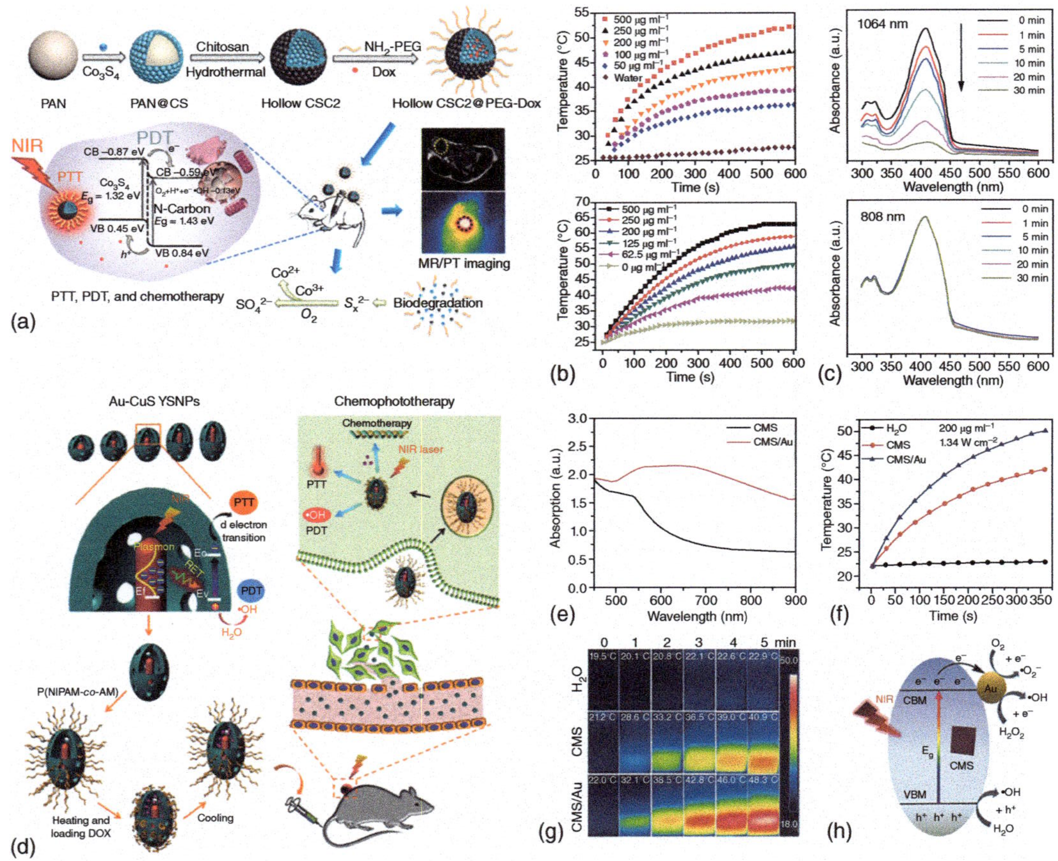


**Figure 5.17** (a) Schematic illustration of synthetic CSC2@PEG-Dox for phototherapy. (b) Exploring the photothermal activity of PEG-$MoO_x$ NPs under NIR irradiation. (c) Exploring the photodynamic effect of PEG-$MoO_x$ NPs under NIR irradiation. (d) Schematic diagram of Au-CuS YSNPs for phototherapy. (e) Vis–NIR absorption spectra of CMS and CMS/Au. (f) Photothermal activity of CMS and CMS/Au under 808 nm laser irradiation. (g) Thermal images of $H_2O$, CMS, and CMS/Au under 808 nm laser irradiation. (h) Mechanism diagram of ROS produced by CMS/Au. Source: (a) Lv et al. [46] / with permission of Elsevier; (b, c) Reproduced with permission Yin et al. [47]. Copyright 2018, Royal Society of Chemistry; (d) Reproduced with permission Chang et al. [48]. Copyright 2018, American Chemical Society; (e–h) Chang et al. [49] / with permission of Wiley-VCH.

embedded in CuS hollow NPs to optimize the efficient resonance energy transfer process for Au-CuS YSNPs. When the LSPR absorption peak of Au cores was matched with the wavelength of incident light, the efficient resonance energy transfer process occurred from the Au core to CuS shell. This would improve the electron–hole pair magnitude in the CuS shell, and d–d energy band transition of Cu(II) ions would be promoted, resulting in enhanced photothermal performance. $Au_{808}$-CuS YSNPs and $Au_{980}$-CuS YSNPs, respectively, showed high photothermal transduction efficiencies of 62.44% and 61.34% upon light irradiation of 808 and 980 nm laser, which were much higher than those of physical mixtures of CuS HNPs and Au cores (25.46% for $Au_{808}$ + CuS under 808 nm laser irradiation; 26.41% for $Au_{980}$ + CuS under 980 nm laser irradiation). At the same time, the increased hole density in the valence band could produce more ·OH radicals, exhibiting a significant photodynamic performance. Au core-CuS NP-dependent plasmonic metal NP–semiconductor heterojunction nanostructures showed synergetic PTT and PDT functionalities for applications in cancer treatment. Furthermore, the void space in Au-CuS YSNPs could be also utilized for carrying small-molecule drugs for potential chemotherapy applications [48]. $Cu_2MoS_4$/Au heterojunction NSs were also prepared for multipronged approach of antitumor therapy. $Cu_2O$ NPs were used as the sacrificial template and turned into hollow NPs with tiny NSs on the surface and were finally disintegrated into tiny $Cu_2MoS_4$ NSs. Au NPs were in situ deposited on the surface of $Cu_2MoS_4$ NSs, resulting in the formation of $Cu_2MoS_4$/Au heterojunction NSs. $Cu_2MoS_4$/Au heterojunction NSs showed broad absorption in the range from 500 to 900 nm (Figure 5.17e), which was higher than that of $Cu_2MoS_4$ NSs, owing to unique LSPR absorption of Au NPs on the surface of $Cu_2MoS_4$ NSs. Due to such enhanced plasmonic absorption, $Cu_2MoS_4$/Au heterojunction NSs was endowed with improved PTT and PDT activities. The temperature increased to 28.2 °C in six minutes for $Cu_2MoS_4$/Au heterojunction NSs under excitation of 808 nm laser irradiation, which was higher than that of $Cu_2MoS_4$ NSs (20.0 °C) and deionized water (0.9 °C) (Figure 5.17f,g). The photothermal conversion efficiency was 34.69% for $Cu_2MoS_4$/Au heterojunction NSs and 28.42% for $Cu_2MoS_4$ NSs. The enhanced photothermal conversion ability for $Cu_2MoS_4$/Au heterojunction NSs was attributed to the encapsulated Au NPs with unique LSPR properties. Therefore, another plasmonic metal NPs, such as Ag NPs and Au-Ag alloy NPs, could also be further considered for the enhancement of PTT ability of plasmonic metal NP–semiconductor heterojunction nanostructures. It was also explored that $Cu_2MoS_4$/Au heterojunction NSs displayed stronger PDT ability than $Cu_2MoS_4$ NSs. Under excitation of 808 nm laser irradiation, the electrons in the valence band of $Cu_2MoS_4$ NSs would be excited and transitioned to the conduction band of $Cu_2MoS_4$ NSs, which promoted the separation of electron–hole pairs. The electrons in the conduction band of $Cu_2MoS_4$ NSs further transferred to the surface of Au NPs and reacted with $O_2$ for the formation of $\cdot O_2^-$ and were captured by $H_2O_2$ for the generation of ·OH. At the same time, the separated holes in the valence band of $Cu_2MoS_4$ NSs would react with $H_2O$, resulting in the production of ·OH (Figure 5.17h). The production of a large number of ROS provided $Cu_2MoS_4$/Au heterojunction NSs with efficient PDT ability. $Cu_2MoS_4$/Au heterojunction NSs showed better PTT and PDT ability than $Cu_2MoS_4$ heterojunction NSs for multimodal therapy. $Cu_2MoS_4$/Au heterojunction NSs are also the promising PA/CT imaging agents, which is vital for multimodal cancer diagnosis and treatment [49].

The metal–semiconductor composite nanostructures displayed excellent photothermal conversion properties and largely improved the separation efficiency of the electron–hole pairs under irradiation of NIR light, leading to considerable generation of ROS. Another kind of n-type semiconductors were also attempted to integrate with plasmonic Au NPs for exploring the PTT and PDT performances. The urchin-shaped Au@$Bi_2S_3$ core–shell structures were prepared by a rapid ion exchange reaction using the as-obtained Au@ZnS core–shell NRs and $Bi(NO_3)_3$. Au@$Bi_2S_3$ core–shell NRs showed strong and broad absorption peaks ranging from 400 to 900 nm, indicating high photothermal conversion efficiency. Under 808 nm light irradiation for 180 seconds, the temperature was increased by 40.4 °C in the presence of 125 μg $ml^{-1}$ Au@$Bi_2S_3$ core–shell NRs, whereas there was almost no enhancement for the temperature of pure water. The photothermal conversion efficiency was calculated to be 35.30% for Au@$Bi_2S_3$ core–shell NRs, which was much higher than that of $Bi_2S_3$ (25.55%). Au@$Bi_2S_3$ core–shell NRs possessed good PTT ability and chemical stability. The conduction band of $Bi_2S_3$ was 0.108 V (versus the normal hydrogen electrode), which was higher than the Fermi level of Au (0.5 V). This induced the formation of the typical Schottky junction between Au@$Bi_2S_3$ core–shell NRs. The photogenerated electrons were preferentially migrated from the $Bi_2S_3$ to Au NRs under light illumination, enhancing the separation efficiency of the electron–hole pairs. The electrons on the surface of Au NRs reacted with the surrounding $O_2$ for the production of $\cdot O_2^-$, and the holes located on the valence band of $Bi_2S_3$ reacted with OH- for the generation of ·OH. Meanwhile, there was no ROS generated for the control groups (without NIR light or without NPs). Au@$Bi_2S_3$ core–shell NRs both displayed good photothermal conversion ability and high charge separation efficiency for the generation of ROS in PTT and PDT applications. The urchin-shaped Au@$Bi_2S_3$ core–shell NRs exhibited superior antibacterial activities against both *E. coli* and *S. aureus*. They are alterative to fabricate metal–semiconductor composite nanostructures for synthetic treatment [50]. Furthermore, Au/$Bi_2Se_3$ metal/semiconductor heterostructures were also reported through the site reduction of $HAuCl_4$ solution on the surface of hollow mesoporous structured $Bi_2Se_3$ NPs, in order to form Au NPs. As-prepared Au/$Bi_2Se_3$ metal/semiconductor heterostructures showed obvious absorption at around 532 nm owing to the LSPR effect of Au NPs. Compared to the single $Bi_2Se_3$ NPs, Au/$Bi_2Se_3$ metal/semiconductor heterostructures exhibited higher photothermal conversion property and photoinduced electron hole separation efficacy. The PTT performances of Au/$Bi_2Se_3$ metal/semiconductor heterostructures were explored under light irradiation of 808 nm for six minutes. The temperature was increased by 24.2 °C in the presence of 250 μg $ml^{-1}$ Au/$Bi_2Se_3$ metal/semiconductor heterostructures, which was higher than the 16.1 °C in the presence of 250 μg $ml^{-1}$ $Bi_2Se_3$ NPs and 1.0 °C in the presence of deionized water. The photothermal conversion efficiency was calculated to be 37.85% for Au/$Bi_2Se_3$ metal/semiconductor heterostructures and 30.29% for $Bi_2Se_3$ NPs. The better PTT activity of Au/$Bi_2Se_3$

metal/semiconductor heterostructures was attributed to the mutual electron migrations in p-orbital of Bi and Se and effective orbital overlaps between the Se p-orbital and Au-4d, resulting in efficient circuit loops for electron migrations with enhanced spectral absorption and photothermal conversion efficiency. Au/$Bi_2Se_3$ metal/semiconductor heterostructures showed the sizes of hollow mesoporous structured $Bi_2Se_3$ NPs of 100 nm, and the surface area and pore volume were evaluated to be 44.59 $m^2 g^{-1}$ and 0.79 $cm^3 g^{-1}$, which enabled successful loading of DOX on the surface of Au/$Bi_2Se_3$ metal/semiconductor heterostructures. The loading efficiency was 25.56%. Under NIR irradiation, the photothermal capacity of Au/$Bi_2Se_3$ metal/semiconductor heterostructures induced the temperature increment, enhancing the thermal vibration and weakening the interaction between DOX and NP carrier. Photothermal-triggered drug release occurred for chemotherapy. This proposed Au/$Bi_2Se_3$ metal/semiconductor heterostructures not only exhibited good PTT performances but also showed PTT induced enhanced chemotherapy activity for multimodal antitumor therapy. Furthermore, Au/$Bi_2Se_3$ metal/semiconductor heterostructures could also generate and achieve the separation of electrons and holes under 808 nm light irradiation. The presence of $H_2O_2$ could react with electrons to generate ·OH, and $O_2$ would be seized o produce $\cdot O_2^-$ and $^1O_2$ for photocatalytic therapy [51] (Figure 5.18).

## 5.4 Photochemical Effects

Chiral metamaterials show better absorption toward RCP or LCP light. This largely enhances the generation of a huge number of charge carriers and promotes the separation of $e^-$ and $h^+$. The generation of $e^-$ not only accelerates the process of photocatalytic reaction but also the produced $h^+$ engineers the production of ROS for special photochemical applications [52]. Semiconductors can efficiently utilize light energy, which promotes the $e^-$ to transition from VB to CB for the generation of $e^-/h^+$ pairs. Chiral semiconductors promote the production of a large number of ROS for photochemical applications.

However, due to the small dielectric constants of semiconductors, the dipole interaction of semiconductors is relatively weak. Alternatively, chiral plasmonic NPs possess stronger chiroptical properties than chiral semiconductors and can largely and efficiently absorb RCP or LCP light. This will enhance the radiation characteristics of NPs (scattering and absorption) and promote the generation of high-energy hot electrons for photocatalytic applications. For example, chiral Au-gap-Ag NPs were fabricated by using L-/D-cys as ligands and displayed chiroptical peaks in the regions from 350 to 650 nm. The sizes of the interior gap greatly affected the CD responses (Figure 5.19a). Chiral L-/D-cys-modified Au-gap-Ag NPs were applied for the photocatalytic reduction of 4-nitrophenol to 4-aminophenol to evaluate the circularly polarized photocatalytic activity. Under irradiation by LCP light, chiral L-cys-modified Au-gap-Ag NPs exhibited better photocatalytic reduction ability, and chiral D-cys-modified Au-gap-Ag NPs exhibited better photocatalytic reduction ability upon irradiation by RCP light. This was due to the

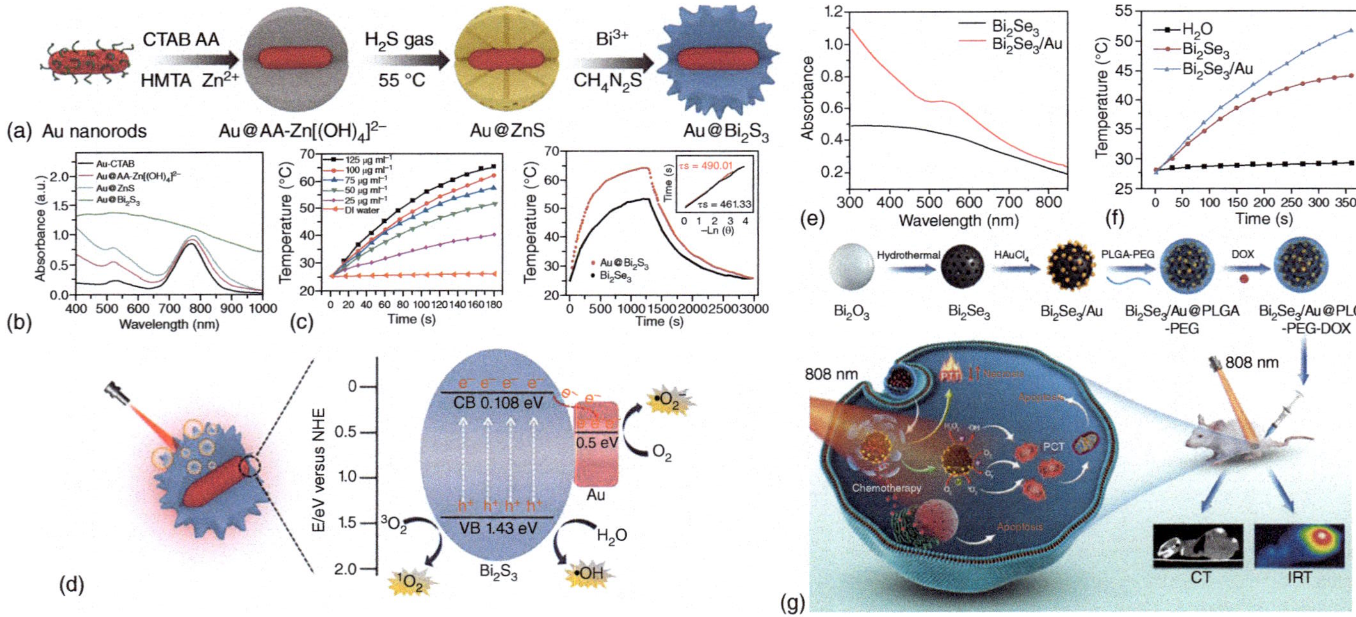


**Figure 5.18** (a) Schematic diagram of the formation of $Au@Bi_2S_3$ core–shell NRs. (b) UV–vis–NIR absorption spectra of different materials. (c) Exploring the photothermal activity of $Au@Bi_2S_3$ solution under 808 nm laser irradiation. (d) Mechanism diagram of ROS produced by $Au@Bi_2S_3$ core–shell NRs. (e) UV–vis–NIR absorption spectra of BS and BS/Au NPs. (f) Photothermal activity of $Bi_2Se_3$ and $Bi_2Se_3$/Au under 808 nm laser irradiation. (g) Schematic illustration of synthetic $Bi_2Se_3$/Au@PLGA-PEG-DOX for phototherapy. Source: (a–d) Reproduced with permission Wang et al. [50]. Copyright 2020, Elsevier; (e–g) Reproduced with permission Yang et al. [51] / with permission of Wiley-VCH.

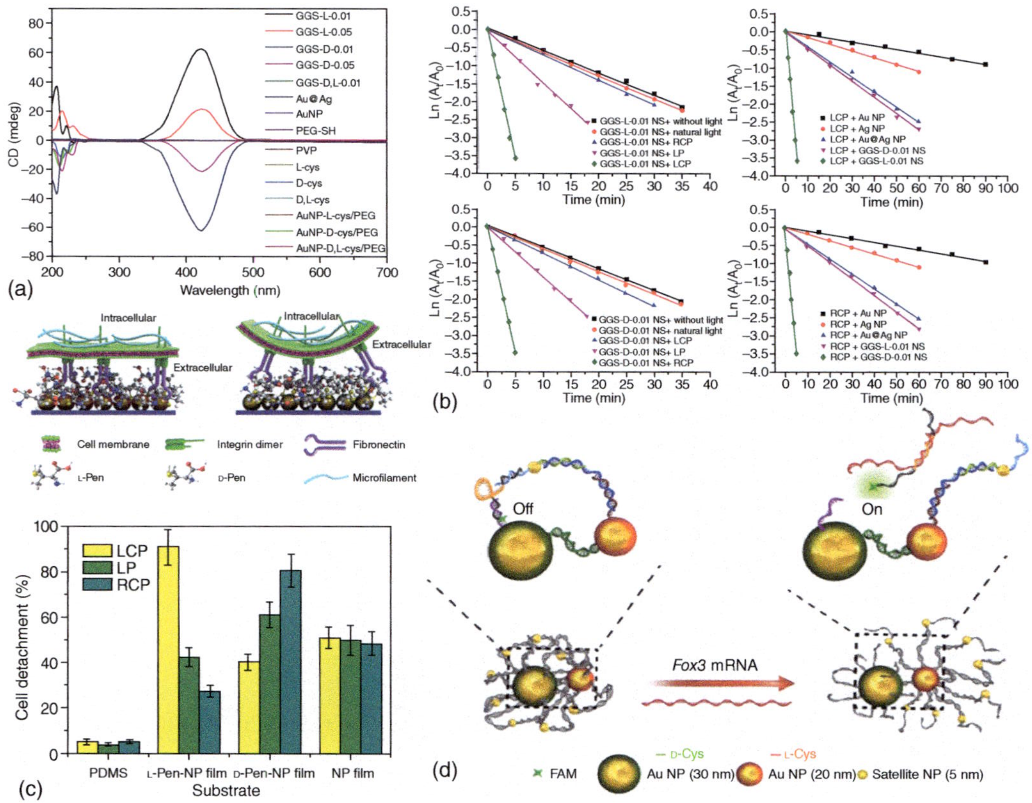


**Figure 5.19** (a) CD spectra of GGS NSs (the number represents the concentration of cysteine). (b) Comparison of photocatalytic activities of GGS NSs for the reduction of 4-nitrophenol. (c) Schematic illustration of Au NP films for cell proliferation application and statistical diagram of cell detachment rates under different circularly polarized light. (d) Schematic illustration of $C_{30(D)}$-$C_{20(L)}S_5$ configurational transformation in the presence of Fox3 mRNA. Source: (a, b) Reproduced with permission Hao et al. [53]. Copyright 2015, Wiley-VCH; (c) Zhao et al. [54] / Springer Nature / CC BY-4.0; (d) Reproduced with permission Qu et al. [55]. Copyright 2021, Springer Nature.

different absorption ability of L-cys-modified Au-gap-Ag NPs and D-cys-modified Au-gap-Ag NPs toward LCP light and RCP light, respectively. The high absorption toward light induced the generation of more hot electrons for the reduction reaction of 4-nitrophenol. Compared to achiral NPs (such as Au NPs, Ag NPs, Au@Ag NPs), chiral L-/D-cys-modified Au-gap-Ag NPs showed the best photocatalytic activity (Figure 5.19b) [53]. Furthermore, chiral plasmonic NP assemblies showed higher CD responses than chiral NPs. Chiral NP assemblies under RCP or LCP illumination were widely used for the introduction of much more hot electrons in the applications of cell proliferation and cell detachment. For example, chiral Au NP film was assembled by Au NPs modified by chiral L- and D-Pen molecules. Particularly, chiral L-Pen Au NP films showed positive CD responses in the regions of 400–900 nm, and chiral D-Pen Au NP films showed negative CD responses at the same position. It was found that chiral L-Pen Au NP film rapidly promoted cell proliferation, and the cell morphology was round on the surface of chiral D-Pen Au NP film. Different chiral Pen NP films induced different proliferation behaviors of the cells, due to the regulation ability of cell function of Pen molecules and the distinct cell adhesion between fibronectin and the films. Fibronectin is a key protein to promote the adhesion of cells on substrates, and fibronectin was found to prefer to absorb on the surface of L-Pen Au NP films, accelerating cell proliferation. Not only chiral L- and D-Pen Au NP films promoted distinct cell proliferation states, but also chiral L-/D-cys Au NP films and chiral L-/D-Phe Au NP films possessed the same functions. This indicated that chiral NP-film–engineered cell proliferation exhibited good universality. Strong CD responses occurred for chiral Au NP films, which was ascribed to the plasmonic hot spot from Au NP films and the dipolar interaction of chiral cells and plasmon. In particular, chiral L-Pen Au NP films exhibited high absorption toward LCP light, and chiral D-Pen Au NP films largely absorbed RCP light. This induced the generation of a large number of photoejected hot electrons; about 91.2% cells were detached from chiral L-Pen Au NP films by LCP illumination, and 80.7% cells were detached from chiral D-Pen Au NP films under RCP irradiation (Figure 5.19c) [54].

Compared to the utilization of one type of CPL (RCP or LCP) light, combined RCP and LCP light would largely increase the usage of light for photochemical effects. D-cys-modified 30 nm Au NPs and L-cys-modified 20 nm Au NPs as well as satellite 5 nm Au NPs were assembled to $C_{30(D)}S_5$-$C_{20(L)}$ and $C_{30(D)}$-$C_{20(L)}S_5$ by using ssDNA as linkers. There existed stem–loop sequences in the DNA skeletons, which was complementary to Fox3 mRNA in the process of neural stem cell(NSC) differentiation into neurons. As-prepared $C_{30(D)}S_5$-$C_{20(L)}$ exhibited negative chiral signal at 524 nm, and the CD response became positive for $C_{30(D)}$-$C_{20(L)}S_5$, due to the configuration transformation in the presence of Fox3 mRNA (Figure 5.19d). The changes in CD responses were an indicator for the monitoring of cell differentiation. In particular, $C_{30(D)}S_5$–$C_{20(L)}$ induced obvious neuronal differentiation by using 532 nm RCP light than LCP light, LP light, and in the dark, and the neurite length reached up to 51 ± 4.4 μm, whereas the neurite length was 53 ± 5.0 μm for $C_{30(D)}$-$C_{20(L)}S_5$ upon LCP irradiation. The different absorption of photons for $C_{30(D)}S_5$-$C_{20(L)}$ and $C_{30(D)}$-$C_{20(L)}S_5$

toward RCP and LCP light could be applied for exploring the cell differentiation process through switching the CPL polarity. CPL irradiation promoted the transfer of a periodic force to cytoskeleton and nucleus 42, which further accelerated the process of cell differentiation. The biological effects of CPL light in the presence of distinct chiral metamaterials need to be further explored for biomedical applications [55].

## 5.5 Future Prospects and Outlook

Chiral nanomaterials have made great progress in their preparation, functionalization, chiral properties studying, and phototheranostic treatments as well as in biocatalysis and photochemical applications. There are still some shortcomings that can be improved and developed. First, currently reported chiral metamaterials mainly focused on chiral plasmonic NPs, chiral semiconductors, and their composites. The long-term biosecurity of chiral nanomaterials in vivo is still highly questionable. Alternatively, carbon nanomaterials and metal oxides possess good biocompatibility and are the potential candidates for bioapplications in living cells. Chiral carbon nanomaterials and metal oxide–included chiral nanocomposites need to be further fabricated and explored, and their chiroptical responses need to be enhanced for sensitive bioapplications. Second, combination treatments possess efficient and better effects than single treatments. Single-component chiral nanomaterials had many limitations and could not satisfy diversified biological applications. Multimodal imaging and treatments based on multicomponent chiral nanomaterials are greatly beneficial to the early warning and diagnosis of cancer treatment and to improve treatment efficiency. Third, biocatalytic nanomaterials with multicomponent synthesis and multifunction will gradually become the mainstream materials in the field of biocatalysis. It is important to precisely and rationally design the structural structures of chiral nanocatalysts of high affinity for corresponding substrates in enantioselective biocatalysis. Fourth, natural amino acids and peptides endowed chiral nanomaterials with well biospecific interaction, but more types and high affinity between chiral metamaterials and complex biological sites in vivo need still to be explored for efficient photochemical effects.

## References

1 Wen, L., Chen, L., Zheng, S. et al. (2016). *Advanced Materials* 28: 5072.
2 Wu, L., Sedgwick, A.C., Sun, X. et al. (2019). *Accounts of Chemical Research* 52: 2582.
3 Langer, J., de Aberasturi, D.J., Aizpurua, J. et al. (2020). *ACS Nano* 14: 28.
4 Lai, W.F., Wong, W.T., and Rogach, A.L. (2020). *Advanced Materials* 32: e1906872.
5 Chen, J., Ning, C., Zhou, Z. et al. (2019). *Progress in Materials Science* 99: 1.
6 Huang, P., Pandoli, O., Wang, X. et al. (2012). *Nano Research* 5: 630.
7 Li, Y., Miao, Z., Shang, Z. et al. (2019). *Advanced Functional Materials* 30.

8 Lu, M., Qu, A., Li, S. et al. (2020). *Angewandte Chemie (International Ed. in English)* 59: 8698–8705.
9 Gao, F., Sun, M., Xu, L. et al. (2017). *Advanced Functional Materials* 27: 1700605.
10 Wang, W., Hao, C., Sun, M. et al. (2018). *Advanced Functional Materials* 28: 1800310.
11 Kumar, A., Kumar, S., Rhim, W.K. et al. (2014). *Journal of the American Chemical Society* 136: 16317.
12 Vankayala, R., Huang, Y.K., Kalluru, P. et al. (2014). *Small* 10: 1612.
13 Zhang, H.Y., Hao, C.L., Qu, A.H. et al. (2018). *Advanced Functional Materials* 28: 1805320.
14 Li, S., Xu, L., Sun, M. et al. (2017). *Advanced Materials* 29: 1606086.
15 Qu, A., Xu, L., Sun, M. et al. (2017). *Advanced Functional Materials* 27: 1703408.
16 Sun, M., Xu, L., Ma, W. et al. (2016). *Advanced Materials* 28: 898.
17 Zhang, M., Zhang, H., Feng, J. et al. (2020). *Chemical Engineering Journal* 393: 124778.
18 Zhao, N., Li, J., Zhou, Y. et al. (2016). *Advanced Functional Materials* 26: 5848.
19 Gao, S., Zhang, L., Wang, G. et al. (2016). *Biomaterials* 79: 36.
20 Chang, Y., Feng, Y., Cheng, Y. et al. (2019). *Advanced Science (Weinh)* 6: 1900158.
21 Wang, J., Wu, X., Ma, W., and Xu, C. (2020). *Advanced Functional Materials* 30: 2000670.
22 An, L., Wang, X., Rui, X. et al. (2018). *Angewandte Chemie (International Ed. in English)* 57: 15782.
23 Wang, Y., An, L., Lin, J. et al. (2020). *Chemical Engineering Journal* 385: 123925.
24 Wang, Y., Li, Z., Hu, Y. et al. (2020). *Biomaterials* 255: 120167.
25 Sujai, P.T., Shamjith, S., Joseph, M.M., and Maiti, K.K. (2021). *ACS Applied Bio Materials* 4: 4962.
26 Chang, M., Hou, Z., Wang, M. et al. (2021). *Angewandte Chemie (International Ed. in English)* 60: 12971.
27 Chen, J., Fan, T., Xie, Z. et al. (2020). *Biomaterials* 237: 119827.
28 Shibu, E.S., Hamada, M., Murase, N., and Biju, V. (2013). *Journal of Photochemistry and Photobiology C: Photochemistry Reviews* 15: 53.
29 Ding, F., Li, H.J., Wang, J.X. et al. (2015). *ACS Applied Materials & Interfaces* 7: 18856.
30 Li, Z., Lu, S., Liu, W. et al. (2021). *Angewandte Chemie (International Ed. in English)* 133: 19350.
31 Liu, Y., Meng, X., and Bu, W. (2019). *Coordination Chemistry Reviews* 379: 82.
32 Zhu, H., Li, J., Qi, X. et al. (2018). *Nano Letters* 18: 586.
33 Liu, C.P., Wu, T.H., Liu, C.Y. et al. (2017). *Small* 13: 1700278.
34 Hou, H., Huang, X., Wei, G. et al. (2019). *ACS Applied Materials & Interfaces* 11: 29579.
35 Liu, Y., Hou, W., Xia, L. et al. (2018). *Chemical Science* 9: 7505.
36 Wang, Y., Xu, S., Shi, L. et al. (2021). *Angewandte Chemie (International Ed. in English)* 60: 14945.

37 Sun, X., Zebibula, A., Dong, X. et al. (2018). *ACS Applied Materials & Interfaces* 10: 25037.

38 He, T., Qiu, X., Li, J. et al. (2019). *Nanoscale* 11: 15245.

39 Wang, D., Wang, H., Ji, L. et al. (2021). *ACS Nano* 15: 8694.

40 Abbas, M., Zou, Q., Li, S., and Yan, X. (2017). *Advanced Materials* 29: 1605021.

41 Li, X., Liu, Y., Fu, F. et al. (2019). *Nano-Micro Letters* 11: 1.

42 Yang, H., Liu, R., Xu, Y. et al. (2021). *Nano-Micro Letters* 13: 1.

43 Wang, J., Sun, J., Wang, Y. et al. (2020). *Advanced Functional Materials* 30: 1908708.

44 Wang, W., Cheng, X., Liao, J. et al. (2019). *ACS Biomaterials Science & Engineering* 5: 6243.

45 Lu, Y., Li, L., Lin, Z. et al. (2018). *Advanced Healthcare Materials* 7: e1800013.

46 Lv, K., Lin, H., and Qu, F. (2020). *Chemical Engineering Journal* 392: 124555.

47 Yin, W., Bao, T., Zhang, X. et al. (2018). *Nanoscale* 10: 1517.

48 Chang, Y., Cheng, Y., Feng, Y. et al. (2018). *Nano Letters* 18: 886.

49 Chang, M., Hou, Z., Wang, M. et al. (2020). *Small* 16: e1907146.

50 Wang, W.-N., Pei, P., Chu, Z.-Y. et al. (2020). *Chemical Engineering Journal* 397: 125488.

51 Yang, C., Chang, M., Yuan, M. et al. (2021). *Small* 17: e2100961.

52 Zhou, R., Zhang, D., Wang, P., and Huang, Y. (2021). *Applied Catalysis B: Environmental* 285: 119820.

53 Hao, C., Xu, L., Ma, W. et al. (2015). *Advanced Functional Materials* 25: 5816.

54 Zhao, X., Xu, L., Sun, M. et al. (2017). *Nature Communications* 8: 2007.

55 Qu, A., Sun, M., Kim, J.Y. et al. (2021). *Nature Biomedical Engineering* 5: 103.

# 6

# Chiral Nanomaterials for Biocatalysis

*Chuanlai Xu, Wei Ma, and Hua Kuang*

*Jiangnan University, International Joint Research Laboratory for Biointerface and Biodetection, State Key Lab of Food Science & Technology, and School of Food Science and Technology, 1800 Lihu Road, Wuxi, Jiangsu Province, 214122 China*

It is of great interest to develop artificial inorganic nanomaterials that mimic natural enzymes for biocatalysis for considerable potential applications in pharmaceutical processes, agrochemical production, biosensing, food industry applications, and so on [1–6]. Sometimes, nanomaterials may possess low catalytic activity and may only work effectively with a narrow temperature range [7–9]. Nowadays, tremendous efforts have been made to design and fabricate biomimetic catalysts. Among the already developed artificial enzymes, the emergence of recent advances in functional nanomaterials with enzyme-like characteristics for are crucial for biocatalysis [10, 11]. The catalytic active nanomaterials could serve as novel and promising nanocatalysts with facile preparation, tunability of catalytic activity, and high stability [12–17]. Due to potential great academic and practical value, the research on nanocatalyst has been extremely hot in recent years [18–24]. Many efforts have been put on the discovery of new nanocatalysts, the optimization of catalytic activity, and the mechanism of catalytic nanocatalysts [25, 26]. Although there is vast literature on improving the catalytic activity of nanocatalysts, more future research is needed on catalytic activity and their substrate selectivity.

## 6.1 Chiral Nanomaterials Based Enantioselectivity in Biocatalysis

Natural enzymes are usually chiral consisting of L-amino acids in which the catalytic activity is dependent on the steric configuration [27–30]. The natural enzymes usually possess high catalytic activity but also have some shortcomings, such as low stability, sensitivity to environmental conditions [31–33]. Nowadays, remarkable progress has been achieved in fabrication of enzyme-like nanocatalysts [12, 34]. Different nanomaterials have been reported to display enzyme mimetic activities, including superoxide oxidase (SOD), catalase, and oxidase; however, these catalytic activities are not related to chiral selectivity, which is the core property of enzymes

*Chiral Nanoprobes for Biological Applications*, First Edition. Edited by Chuanlai Xu.

[35, 36]. The stereoselectivity is a nonnegligible factor in fabrication of high-quality artificial enzymes [37, 38].

The CeNPs were reported to construct as chiral artificial oxidase functionalized by D- or L-amino acids for stereoselective catalysis of 3,4-dihydroxyphenylalanine (DOPA), which is the most commonly used drug for treating Parkinson's disease [39]. Amino acids are deemed as cofactors of natural enzymes that play an important role in stereochemical selection. Generally, the phenylalanine (Phe), alanine (Ala) and tryptophan (Trp), and polar amino acids ([histidine (His), glutamic acid [Glu], arginine [Arg], lysine [Lys] and tyrosine [Tyr]) was used for modifying the Ce NPs. These amino acids were functionalized onto the CeNPs surface by amidation approach. It was found that phenylalanine-modified CeNPs (CeNP@Phe) exhibited optimal enzymatic activity. Importantly, the chiral CeNP@Phe displayed good stereoselectivity for DOPA enantiomers, in which CeNP@L-Phe had higher catalytic activity against D-DOPA, and CeNP@D-Phe was preferred to L-DOPA. The poly(acrylic acid)-coated CeNPs were fabricated in situ with uniform diameter of 4.3 nm as indicated by TEM. X-ray photoelectron spectroscopy (XPS) and Fourier transform (FT) IR spectroscopy revealed the mixed oxidation states and the acylamide bonds, respectively. The powder X-ray diffraction (XRD) results correlated well with CeNPs (Figure 6.1a) with the unspoiled crystal structure.

First, the CeNPs oxidase-like activity toward DOP was tested based on the color change from colorless to brown, which demonstrated the oxidation of DOPA to dopachrome (Figure 6.1b). It was found that modification of CeNPs with amino acids decreased the catalytic activity against DOPA, which may be due to the shielding effect by surface amino acids. Then the optimal amino acids were screened for constructing the stereoselective nanocatalyst. Overall, the nonpolar amino acids (Phe with aromatic side chain, Ala with alkyl side chain) and polar amino acids (His with alkaline side chain, Glu with acid side chain) were chosen as chiral ligands in the catalysis. Among them, phenylalanine-anchored CeNPs(CeNP@Phe) gave rise to the highest catalytic activity toward DOPA, while Ce NP@His had medium catalytic activity and CeNP@Glu displayed the lowest catalytic activity. From the catalytic activity, it was assumed that the aromatic properties (e.g. Phe, His) and positively charged properties (e.g. His) may have contributed to the high catalytic activity. The aromatic amino acid–functionalized CeNPs were found to have higher catalytic activity, in which CeNP@Phe had the fastest catalytic rate. Based on the Arrhenius equation, active energies of DOPA oxidation were calculated, where the CeNP@Phe displayed the lowest activation energy which correlated well with its highest catalytic activity.

The achiral CeNPs did not display any stereoselectivity for oxidation of L- and D-DOPA, while the modification of amino acids could lead to enantioselective oxidation of DOPA (Figure 6.1c). The chiral CeNP@Phe displayed high stereoselectivity. Based on the Michaelis–Menten constant ($K_M$), the smaller value of $K_M$ indicated higher binding affinity of the chiral nanocatalyst. The catalytic number ($K_{cat}$) is also important that demonstrates the ability of an enzyme to catalyze a certain substrate. The parameter $K_{cat}/K_M$ is used for evaluating the catalytic efficiency. $K_{cat}$ for CeNP@D-Phe for L- and D-DOPA was $3.12 \times 10^{-3}$ $s^{-1}$ and $1.94 \times 10^{-3}$ $s^{-1}$,

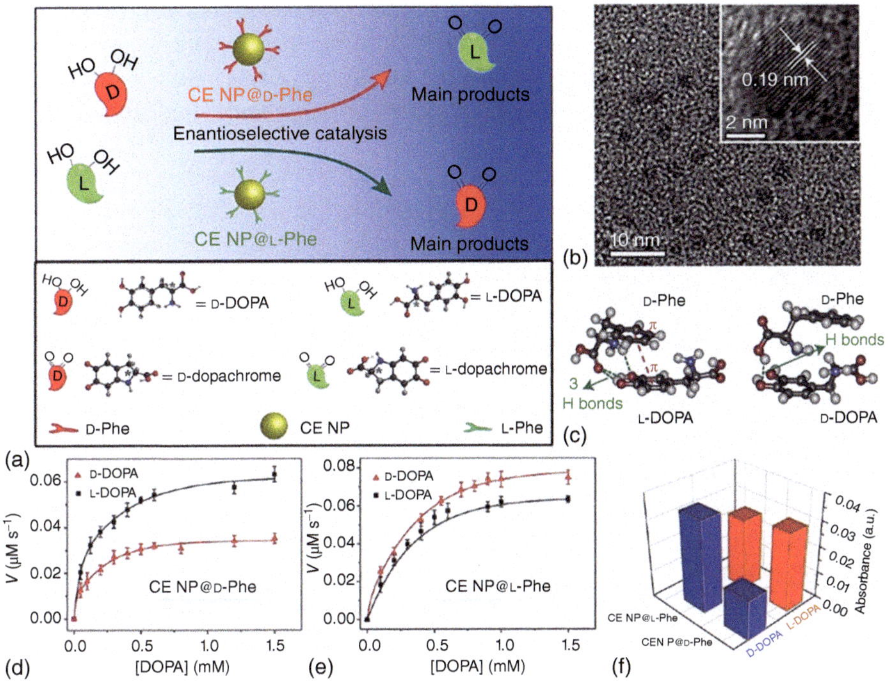


**Figure 6.1** (a) Illustration of stereoselective catalytic oxidation of the DOPA enantiomers by the phenylalanine-modified CeNPs. With CeNP@D-Phe as a catalyst, L-DOPA was oxidized more effectively into L-dopachrome. Conversely, D-DOPA was oxidized more effectively than L-DOPA with CeNP@L-Phe. (b) TEM image of CeNPs. The inset is magnification of one segment. (c) Energy-minimized average interaction models of D-Phe with L-DOPA, D-Phe with D-DOPA. (d) Saturation curves for the enantiomers of DOPA catalytic oxidation by CeNP@D-Phe and (e) CeNP@L-Phe. (f) The absorbance change of DOPA enantiomers with CeNP@L-Phe and CeNP@D-Phe at 100-second intervals. Source: Sun et al. [39] / with permission of Wiley-VCH.

which demonstrated higher catalytic activity of CeNP@D-Phe for oxidizing L-DOPA than D-DOPA (Figure 6.1d). The $K_M$ value also indicated stronger binding affinity for CeNP@D-Phe for L-DOPA. The $K_{cat}/K_M$ value was higher for CeNP@D-Phe against L-DOPA than D-DOPA, suggesting higher catalytic efficiency (1.87-fold) for Ce NP@D-Phe against L-DOPA than D-DOPA. Similarly, the catalytic efficiency of CeNP@L-Phe for DOPA showed 1.13-fold difference (Figure 6.1e), which suggested weaker stereoselectivity for CeNP@L-Phe. At the same time, CeNP@D-His also displayed stereoselectivity in which CeNP@D-His had higher catalytic activity for D-DOPA (Figure 6.1f).

The interaction between L-DOPA and D-Phe was calculated and it was found that three hydrogen bonds were formed, while p–p aromatic packing interaction enhanced the interactions. The result revealed that L-DOPA bind strongly to D-Phe, correlated well with the highest catalytic activity for CeNP@D-Phe against L-DOPA. This result further indicated that chirality of amino acids mostly determined stereoselectivity. The different interactions between D-or L-His and D- or

L-DOPA suggest that stereoselectivity is dependent on different parameters such as chirality of amino acids and reaction conditions. It is urgent to design more nanomaterials as nanocatalysts with stereoselectivity. To this end, peptide, nucleic acids, and saccharides are promising cofactors for developing stereoselective nanocatalysts.

Inspired by natural enzymes, various nanomaterials have been reported for biocatalysis for accelerating chemical reactions with high enantioselectivity. Based on Au NPs as active centers, an enantioselective nanocatalyst was constructed using chiral cysteine as selectors for chiral recognition. The nanocatalyst modified with D-cys displayed preference to L-DOPA for oxidation of chiral 3,4-dihydroxyphenylalanine (DOPA), and the nanocatalyst modified with L-cys preferred to bind to D-DOPA. The calculated steady-state kinetic and activation energies by molecular dynamics simulations showed that the chiral cys and DOPA formed hydrogen bonding, which was the origin of chiral selectivity. L-DOPA is a very important biomolecule in clinics and neurochemistry, while D-DOPA is toxic. To realize the chiral selectivity, the cysteine was chosen to bind DOPA enantiomers, where the cys was the active site of the natural enzymes bind to L-DOPA. The expanded mesoporous silica (EMSN) was designed as the skeleton to support the Au NP catalyst to prevent its aggregation (Figure 6.2a–e). The EMSN and Au NPs-EMSN were synthesized individually and then complexed together. AuNPs-EMSN reacted with D-cys/L-cys by Au—S bond to form D-cys/L-cys@AuNPs-EMSN. A weak band at 2550 $cm^{-1}$ corresponding to the –SH group was characterized by Fourier transform infrared (FTIR) spectroscopy. The TEM image indicated that Au NPs–encoded EMSN were well dispersed and the Au NPs had an average size of 2.3 nm. The average size was not changed after functionalization with cys. The functionalization of Au NPs with cys did not change the crystal structure of Au NPs as revealed by XRD results of Au NPs-EMSN and D-cys/L-cys@AuNPs-EMSN. The D-cys/L-cys@AuNPs-EMSN displayed peroxidase-like activity as shown in the catalysis of substrates 2,2′-azinobis(3-ethylbenzthiazoline-6-sulfonate) (ABTS) and 3,3′,5,5′-tetramethylbenzidine (TMB) in the presence of $H_2O_2$. The study showed that the nanocatalyst was stable in the reaction conditions in which the Au@S bond was not affected by hydrogen peroxide in repeat cycles.

The catalytic chiroselectivity of D-cys/L-cys@AuNPs-EMSN was investigated through kinetic parameters. The $K_M$ value for D-cys/L-cys@AuNPs-EMSN increased due to the reduced affinity between the Au NPs and DOPA after cys functionalization. Based on $K_{cat}$ results, the catalytic activity of D-cys/L-cys@AuNPs-EMSN decreased due to the modification of cys. It was found that D-cys@AuNPs-EMSN had higher $K_M$ and $K_{cat}$ for L-DOPA against D-DOPA, which demonstrated that the interaction between D-cys@AuNPs-EMSN and D-DOPA was stronger. Meanwhile, the catalytic efficiency $K_{cat}/K_M$ was used to identify the chiroselectivity. The Au NPs-EMSN showed similar $K_{cat}/K_M$ values for D-DOPA and L-DOPA, which indicated no chiral selectivity for the oxidation of DOPA. The $K_{cat}/K_M$ for D-cys@AuNPs-EMSN against D-DOPA and L-DOPA was calculated to be 1.52 and 2.24, respectively. Thus, D-cys@AuNPs-EMSN reacted with L-DOPA was more effective with a 1.47-fold difference. The L-cys@Au NPs-EMSN reacted effectively with

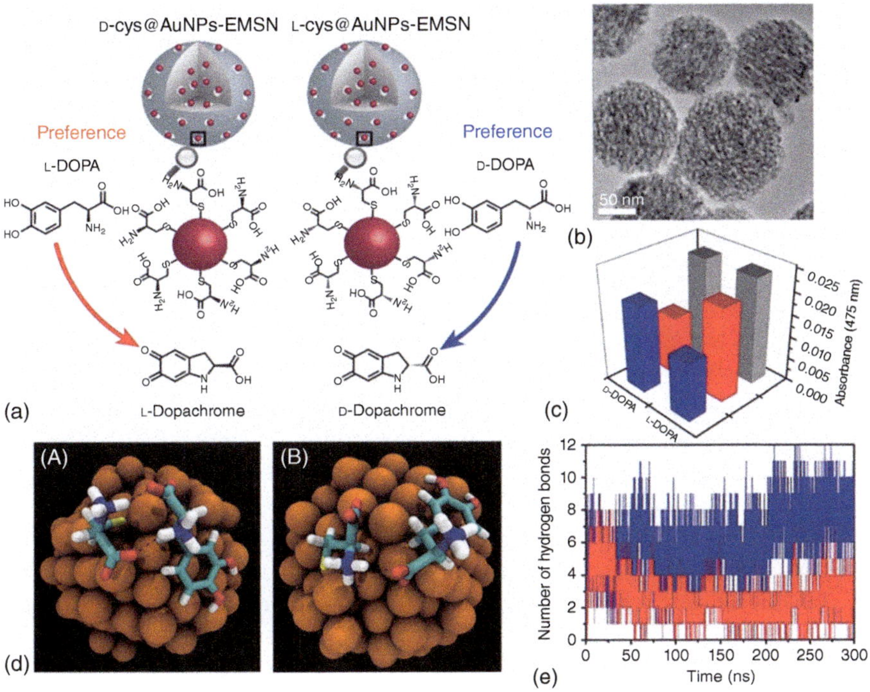


**Figure 6.2** (a) Enantioselective oxidation of chiral DOPA by D-cys/L-cys@AuNPs-EMSN. (b) TEM images of L-cys@AuNPs-EMSN. (c) The absorbance change of DOPA enantiomers with L-cys@AuNPs-EMSN (blue), D-cys@AuNPs-EMSN (red), and AuNPs-EMSN (gray) at 180-second intervals. (d) (A) L-cys and L-DOPA hydrogen bonds (dashed blue lines) formation on a gold nanoparticle (orange), and (B) L-cys and D-DOPA hydrogen bond (dashed blue line) formation on a gold nanoparticle. (e) Time series of average number of hydrogen bonds per molecule between L-cys and L-DOPA (blue), L-cys and D-DOPA (red). Source: Zhou et al. [40] / with permission of Wiley-VCH.

D-DOPA with a selectivity factor of 1.69. The enantioselectivity was attributed to the chiral cys on the surface of Au NPs, which could be due to higher binding affinity. The D-cysteine- or L-cysteine-modified Au NPs (D-cys/L-cys@AuNPs-EMSN) showed higher specificity to DOPA, where L-cys@AuNPs-EMSN was specific to L-DOPA. The enantioselectivity of the nanocatalyst D-cys/L-cys@AuNPs-EMSN played a role in nantioselective catalysis. The hydrogen bond calculation between two molecules revealed that more hydrogen bonds were formed between L-cys and L-DOPA (L–L) than between L-cys and D-DOPA (L–D), which induced the selective catalytic activity. The characteristic hydrogen bond was formed between the carboxyl group and amino group in cys and DOPA molecules, where two hydrogen bonds were formed between L-cys and L-DOPA and one hydrogen bond for mismatching chirality (L-D). The two hydrogen bonds and the strength of bifurcated hydrogen bond resulted in a significant improvement in intermolecular interactions. It was found that the competing effect of hydrogen bonds from the DOPA amino group retarded the catalytic kinetics. This construction strategy has provided exciting opportunities for fabricating stereoselective enzyme mimics.

Natural enzymes are powerful catalysts due to their high activity, selectivity, and reliability that have attracted great interest both in chemical and biological fields, such as biosensing, disease diagnosis and therapy, combating bacteria, detoxification. Nanomaterials modified with chiral ligands form uniform architectures that can display intrinsic enzymatic activity against bio-related substrates. Usually, nanoarchitectures have the advantages of facile production, long storage time, low cost, and greater resistance to biodegradation. Naturally, enzymes are chiral and possess high catalytic activity and enantioselectivity. However, fabrication of these types of novel nanocatalysts has special challenges of achieving high activity and enantioselectivity, and great efforts have been made to improve the activity and selectivity. In the past decade, designing the nanocatalysts with high activity and stereoselectivity was in great demand. It was demonstrated that the right path to address this problem was to mimic the structure and the catalytic mechanism of natural enzymes in the design and fabrication of nanocatalysts including well-defined binding pockets consisting of organized active centers and substrate binding sites, where the chiral amino acid residues at the binding sites determine the stereoselectivity. Various materials have been designed for biocatalysis by incorporating different geometric nanomaterials and ligands.

Crystalline porous materials, such as metal–organic frameworks (MOFs) have already been proved powerful for the construction of active nanozymes [36, 41]. Recently, crystalline porous covalent organic framework (COF) materials have been reported for the construction of active biocatalysts. The COFs usually possess high density, well-defined and designable porous structures, which are superior candidates to achieve high activity and selectivity. The catalytic active sites are believed to be dispersed in the skeleton of COFs and can promote electron transport that shorten the electron transportation pathway that help to improve enzyme-like activity. To achieve stereoselectivity in biocatalysis, it's important to incorporate chiral units into COFs to realize stereocontrol in various enantioselective catalysis in which the chiral amino acids at the binding sites within COF materials can be used for chiral recognition. As inspired by the natural structure of horse radish peroxidase, the COF nanozyme is designed with amino acids for chiral recognition [42]. The Qu's group reported the COF nanocatalyst that possesses 21.7 times higher activity than natural HRP, the activity and selectivity of which were modulated by the amount of amino acids (Figure 6.3a). Based on the backbone of the COF for designing COF nanocatalysts, the iron 5,10,15,20-tetrakis(40-tetraphenylamino) porphyrin unit (Fe-ATPP) covalently bound to the COF backbone served as the active center and L-His was modified as the substrate binding site. In the research, the dispersed iron porphyrin and L-histidine units incorporated in the COF provided the nanocatalyst with high activity. The COF nanocatalyst imbued with the covalent L-His units displayed good selectivity for L-DOPA in the oxidation of DOPA enantiomers with a selectivity factor of up to 1.86, while the D-His induced the oxidation of D-DOPA with a selectivity factor of 1.51 (Figure 6.3b,c). The Fe-COF, L-Ala@Fe-COF, and L-His50@Fe-COF displayed the oxidation catalysis of ABTS in the presence of $H_2O_2$, which indicated that Fe-ATPP was the active site for the catalytic activity of the COF nanocatalyst. The varying content of the L-His modulated the activity and selectivity of the COF nanocatalyst against DOPA, while

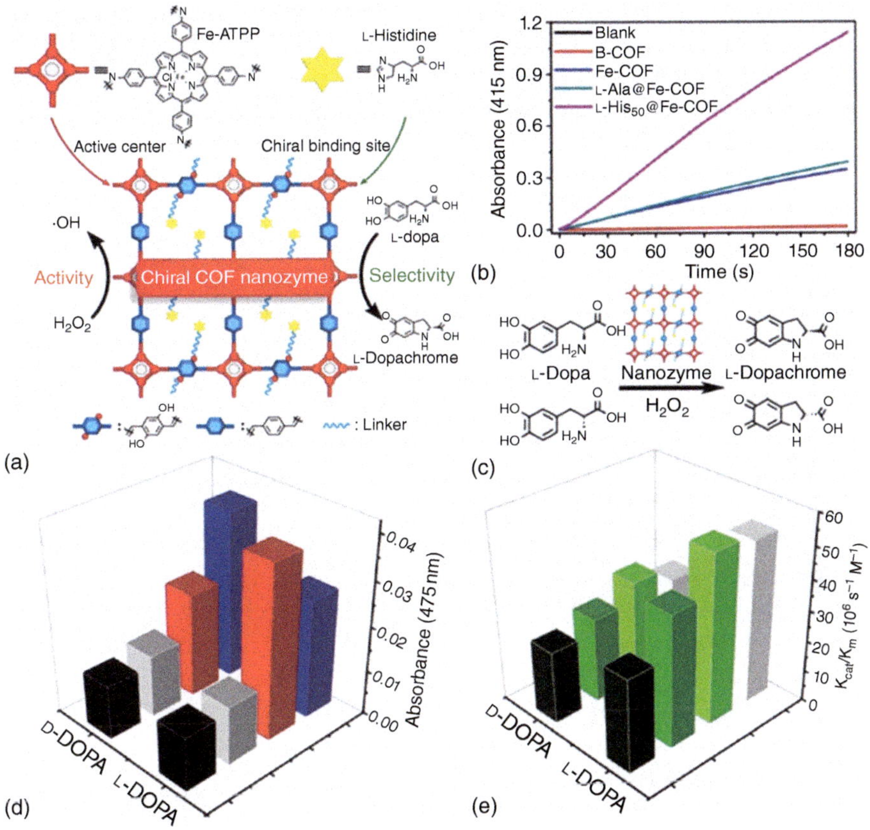


**Figure 6.3** (a) Schematic of construction of a chiral COF nanozyme. (b) The time-dependent absorbance changes of ABTS at 415 nm in the presence of different COF nanozymes. (c) Nanozyme-catalyzed $H_2O_2$-mediated oxidation of DOPA enantiomers to corresponding dopachrome. (d) The absorbance change in DOPA enantiomers with Fe-COF (black), L-Ala@Fe-COF (gray), L-His75@ Fe-COF (red), and D-His@Fe-COF (blue) at 300-second intervals. (e) $K_{cat}/K_m$ for L-Hisx@Fe-COF with different L-His content. The columns from dark to light are for L-His25@Fe-COF, L-His50@Fe-COF, L-His75@Fe-COF, and L-His100@Fe-COF, respectively. Source: Zhou et al. [42] / Royal Society of Chemistry.

the histidine coupled to the activation energy of the catalytic reactions and the substrate binding. The L-Ala-modified COF and -unmodified COF displayed lower catalytic activity compared to L-His covalent COF (Figure 6.3d). The L-His covalent COF displayed $H_2O_2$-mediated oxidation catalysis of DOPA enantiomers to their corresponding dopachrome. The $K_{cat}/K_m$ for L-His@Fe-COF displayed higher catalytic activity with different L-His content (Figure 6.3e).

Initial progress has been made in chiral nanomaterials for biocatalysis. However, it is still a great challenge to fabricate nanocatalysts with high enantioselectivity, which is a highly demanding requirement in chemical and biological applications. Regarding the natural enzyme structure of HRP, stereoselective reactions were achieved through the catalytic site (heme) using proximally arrayed L-amino acid residues. Chiral amino acids were usually employed as chiral binding selectors, which have been generally used for developing effective chiral nanocatalysts. However, the amino acids directly modified on nanoparticle surface may not be the best route for constructing high-enantioselective chiral nanocatalysts. Qu's group further fabricated a stereoselective nanocatalyst ($Fe_3O_4$@Poly(AA)) using ferromagnetic nanoparticle ($Fe_3O_4$) core as the catalytic active sites and amino acid–encoded chiral polymer shell as the chiral binding coupled selector [43]. The Poly(AA), polyacrylate(amino acid), was designed for selective access of chiral tyrosinol for the catalysis of active $Fe_3O_4$ cores, which enabled enantioselective catalysis.

The chiral nanocatalyst $Fe_3O_4$@Poly(AA) was fabricated through four steps: $Fe_3O_4$ NPs synthesis, silica shell coating over $Fe_3O_4$ ($Fe_3O_4$@$SiO_2$), chiral polymer shell modification ($Fe_3O_4$@$SiO_2$@Poly(AA)), and, finally, the silica shell was removed to obtain yolk–shell $Fe_3O_4$@Poly(AA). The structure showed that polymer shell existed as indicated by the rough surface, which was estimated to be 8.5 nm from dynamic light scattering (DLS) results. The silica shell of $Fe_3O_4$@$SiO_2$@Poly(L-Trp) can be removed by etching with concentrated NaOH solution, forming yolk–shell $Fe_3O_4$@Poly(L-Trp) nanoparticles. The catalytic activities of $Fe_3O_4$@Poly(AA) toward L-/D-tyrosinol were studied by time-dependent absorbance at 320 nm (Figure 6.4a–c). The initial reaction rates for oxidation of L- or D-tyrosinol enantiomers by $Fe_3O_4$@Poly(L-Trp), $Fe_3O_4$@Poly(D-Trp), and bare $Fe_3O_4$ revealed that $Fe_3O_4$@Poly(D-Trp) had better catalytic activity for D-tyrosinol. The ratio of initial rates of oxidation of D- and L-enantiomers by $Fe_3O_4$@Poly(D-Trp) was 2.38, while the ratio was 0.37 for $Fe_3O_4$@Poly(L-Trp), which could correlate with the first principles of stereochemistry.

$Fe_3O_4$@Poly(D-Trp) displayed the highest enantioselectivity, while L-Trp modification enabled high enantioselectivity against D-Trp. Based on the catalytic activity to D-/L-tyrosinol, the kinetic parameters, dialysis parameters, and molecular interaction simulations as well as activation energy results indicated the D-/L-Trp-appended polymer shell provided selectivity. The catalysis results showed that the as-made artificial peroxidases can perform high-enantioselective asymmetric catalysis to obtain D- or L-enantiomers, where the selectivity factor is as high as 5.38, even higher than that of HRP. The average adsorption free energy (AFE) and desorption free energy (DFE) of poly(D-/L-Trp) were calculated based on three binding models. The AFE of D-tyrosinol on the surface of poly(D-Trp) was much smaller than that of L-tyrosinol by 2.9 kcal $mol^{-1}$, which suggested that poly(D-Trp) was easier to adsorb D-tyrosinol compared to L-tyrosinol. It was also found that van der Waals interactions contributed majorly to the difference between AFE of D-tyrosinol and L-tyrosinol. From the hydrogen bonds, the electrostatic interactions were calculated, which were favorable for the adsorption of D-tyrosinol compared to L-tyrosinol on the outer surface of poly(L-Trp). The artificial peroxidases were further studied for their catalytic activity in biological medium of live yeast cells. FITC-tyrosinol$_L$/RhB-tyrosinol$_D$ as substrate tested for yeast cells were treated with different yeast using $Fe_3O_4$@poly(L-Trp) and $H_2O_2$. The green fluorescence of yeast cells was due to the tyrosine residues on the cell surface, while different labeling results on *Staphylococcus aureus*, *Escherichia coli*, and *Bacillus subtilis* were because

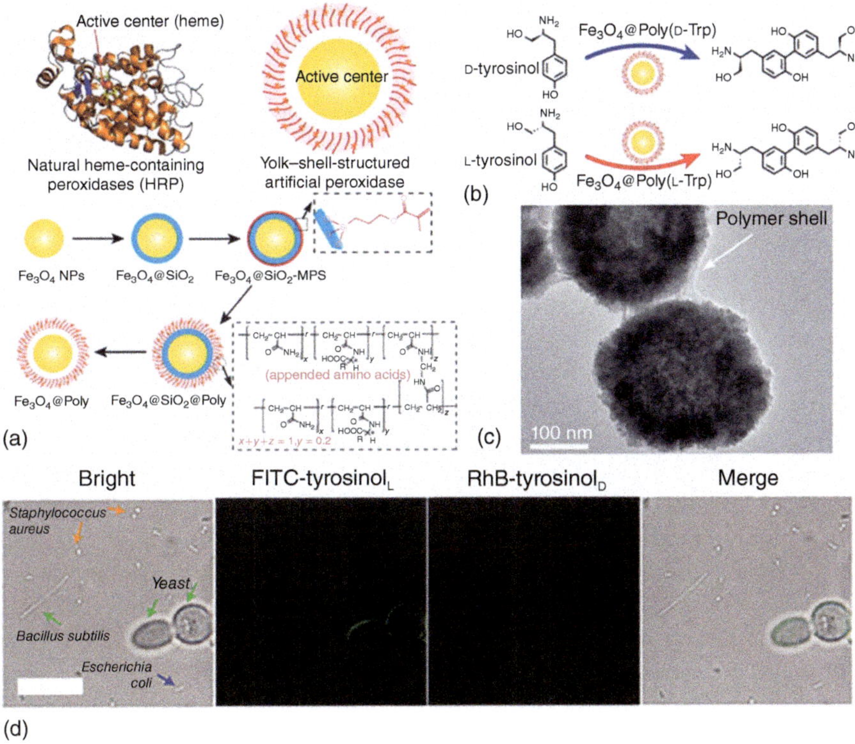


**Figure 6.4** (a) Representative structures of natural heme-containing peroxidase (left) and yolk–shell-structured artificial peroxidase (right), and the preparation procedures of the yolk–shell-structured artificial peroxidase. (b) The enantioselective oxidation of chiral tyrosinol catalyzed by $Fe_3O_4$@Poly(L-/D-Trp). (c) TEM images of yolk–shell $Fe_3O_4$@Poly(L-Trp). (d) $Fe_3O_4$@Poly(L-Trp) NPs catalyze FITC-L-tyrosinol to label yeast cell with specificity. Fluorescence images of different microbial cells (yeast, *Staphylococcus aureus*, *Escherichia coli*, and *Bacillus subtilis*) after treatment with $Fe_3O_4$@Poly (L-Trp), $H_2O_2$ (100 mM), FITC-tyrosinol$_L$ and RhB-tyrosinol$_D$. (Scale bars: 10 mm). Source: Zhou et al. [43] / with permission of Royal Society of Chemistry.

of different chemical components of the cell walls (Figure 6.4d). The cell wall of Gram-positive bacteria (*S. aureus*) is composed of teichoic/lipoteichoic acids, while the cell wall of Gram-negative bacteria (*E. coli* and *B. subtilis*) is made up of lipopolysaccharides. Fluorescent labeled FITC-tyrosinol$_L$ and RhB-tyrosinol$_D$ with green or red fluorescence were selectively catalyzed by the developed nanocatalyst in live yeast cells among yeast, *S. aureus*, *E. coli* and *B. subtilis* bacterial cells.

## 6.2 Biocatalysis of Chiral Drugs and Related Chemical Intermediates Based on Chiral Nanomaterials

The asymmetric catalysis is also important for (superpara) magnetic nanoparticles, which could display high enantioselectivity and reactivity with suitable recyclability. The magnetic nanoparticles in asymmetric catalysis are deemed as new promising nanomaterials in the asymmetric catalysis research field. A chiral MOF with synergic Lewis basicity with asymmetric properties was developed by chiral-encoded multi-functionalization strategies. A secondary L-proline@$Fe_3O_4$ nanoparticle was loaded in an MOF and formed the chiral hybrid nanomaterial [44]. The (L-proline@$Fe_3O_4$)@TMU-5 (LPF@5) composite was fabricated by modifying chiral L-proline on the $Fe_3O_4$ magnetic surface and encapsulating the (L-proline@$Fe_3O_4$) nanoparticles (LPF NPs) by TMU-5 in ultrasonication-assisted synthesis (Figure 6.5). The immobilization of L-proline on the $Fe_3O_4$ was characterized by FTIR through the formation of two peaks of asymmetric $\nu$as(COO−) and symmetric $\nu$s(COO−) stretching of the carboxyl groups at around 1400–1600 $cm^{-1}$. It was suggested that the adsorption of chiral L-proline ligands on the magnetic $Fe_3O_4$ NPs was based on the bridging bidentate bindings. The TEM results revealed that the nanoparticles had various particle sizes in the range of 50–100 nm.

The right selection of functionalities in the construction of hybrid nanomaterials is essential observation of the synergic effects between functional groups. Thanks to the Lewis basic amine group in the L-proline, it can work alongside the Lewis basicity of the azine groups of TMU-5 to allow synergic effects, which is important in asymmetric C—C bond formation in nitroaldol condensation. The initial amount of LPF NPs, conversion, selectivity to β-nitroalcohol, and the enantioselectivity toward S-β-nitroalcohol were optimized under gentle conditions. When the initial amount of LPF NPs was 0.15 g, a decrease in conversion was induced. The excessive use of LPF NPs led to extreme occupation of the surface and decreased the catalytic sites of TMU-5. The optimization results demonstrated that the LPF@5 hybrid complex with an initial amount of 0.1 g LPF NPs (0.1-LPF@5) was the optimum catalyst. Then this nanocatalyst was applied in asymmetric nitroaldol condensation. It was shown that LPF@5 had a higher selectivity toward β-nitroalcohol over the pristine TMU-5 and better enantioselectivity than the LPF NPs in asymmetric nitroaldol condensation. The high maximum enantioselectivity toward S-β-nitroalcohol was 98%, which clearly showed that the stabilized L-proline ligands on the $Fe_3O_4$ magnetic surface was an effective chiral nanocatalyst. Overall, the facile synthesized LPF@5 displayed remarkable results in asymmetric nitroaldol condensation. Both L-proline and azine sites are found important for asymmetric Henry condensation, in which it was thought that the catalysis occurred on the interface of LPF NPs and TMU-5 with both azine and L-proline simultaneously. It was shown that the nanocatalyst developed had kept its structure, magnetic character, and catalytic behavior at least after five cycles. The remarkable characteristics of asymmetric nitroaldol condensation are important for future fabrication of chiral amino acid ligands on the surface of $Fe_3O_4$ nanoparticles. Our study about this strategy and its application in the separation and sensing fields is ongoing.

Two-dimensional (2D) transition metal dichalcogenides (TMDs) have attracted great interest due to their extraordinary optical, physical, and chemical properties. Layered TMDs (e.g. $MoS_2$, $MoSe_2$, $WS_2$, and $WSe_2$), ranging from semiconductors to semimetals to true metals, display strong in-plane bonding interactions and enable exfoliation into monolayers or few 2D layers [45]. Generally, each TMD

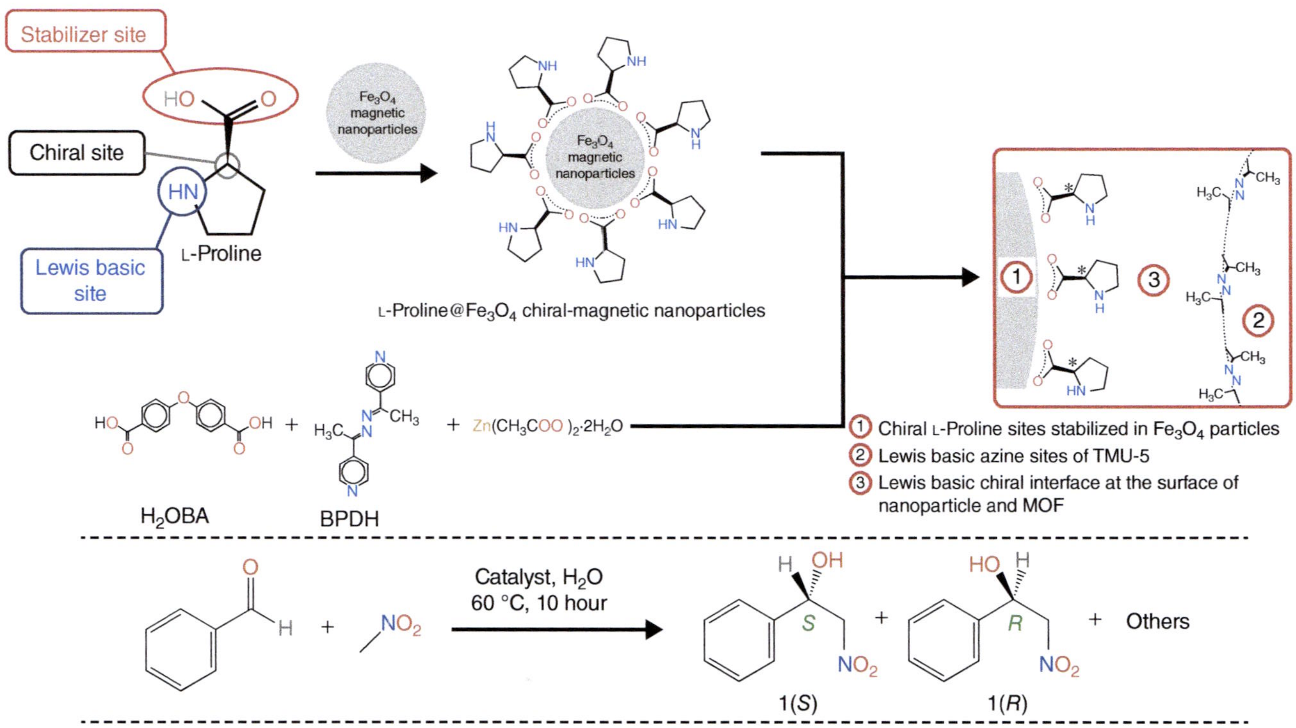


**Figure 6.5** L-Proline@$Fe_3O_4$ nanoparticle and (L-proline@$Fe_3O_4$)@TMU-5 nanocomposite. Application of the TMU-5-based nanocomposite in the asymmetric Henry condensation. Source: Razavi et al. [44] / Royal Society of Chemistry.

monolayer is composed of three atomic layers. Different from their native 2D forms, zero-dimensional (0D) TMD quantum dots (QDs) with sizes smaller than 10 nm displayed distinguished new properties owing to prominent edge and quantum confinement effects [30].

Weili Wei's group for the first time reported that the chiral TMD QDs of $MoS_2$ and $WS_2$ crystals functionalized with chiral ligands cysteine and penicillamine displayed enantioselective catalysis based on chiral substrates D- and L-tyrosinol [46]. The $MoS_2$ and $WS_2$ QDs were synthesized by the sonication treatment with NaOH in *N*-methyl-2-pyrrolidone solution, where chiral TMD QDs were fabricated using D-cys (or L-cys, D- or L-Pen). Further, D-cys-$MoS_2$ QDs/$Cu^{2+}$ were prepared by sonication with the D-cys-$MoS_2$ QDs and $CuCl_2$. The as-prepared $MoS_2$ QDs had uniform size distribution and were monodispersed with the average diameter of $2.03 \pm 0.48$ nm. The $MoS_2$ QDs displayed characteristic crystal structure with a lattice parameter of 0.25 nm, correlated well with the (103) plane of the bulk $MoS_2$ crystal. HRTEM results of the $WS_2$ QDs demonstrated that the lattice fringe spacing was 0.27 nm, which agreed well with the lattice spacing in (100) direction of $WS_2$. Different from $MoS_2$, the (002) diffraction peak at $2\theta = 14.4$ disappeared for $MoS_2$ QDs due to the thin-layer structure of the QDs. It was found that an amount of Mo was oxidized from $Mo^{4+}$ to $Mo^{6+}$ which was revealed by XPS spectra during the fabrication of $MoS_2$ QDs, and the molar ratio of $Mo^{4+}$ to $Mo^{6+}$ was about 8.63 : 1. The molar ratio of $W^{4+}$ to $W^{6+}$ was about 2.63 : 1 for $WS_2$ QDs. The chiroptical activity of D- and L-cys-$MoS_2$ QDs was interesting and displayed CD spectra with characteristic Cotton effect. The D- and L-cys-$MoS_2$ QDs exhibited CD bands at 212 and 265 nm with opposite sign in the classical bisignate line shape, which is due to preferential excitation of the split bands. The chiral origins for inorganic TMD QDs could be due to exciton -transition- related electronic interaction, which resulted in dipolar interaction between the excitations of QDs and chiral molecules. Overall, the CD mechanism was attributed to the electronic interaction effects between the adsorbed chiral molecules and the electron–hole states of QDs.

Further, the chiral selectivity was investigated for peroxidase-like activity. Chiral D- and L-cys-$MoS_2$ QDs were evaluated for peroxidase activity and enantioselectivity. With $Cu^{2+}$, the D- and L-cys-$MoS_2$ QDs both displayed high peroxidase-like activities against the achiral substrate TMB within 10 minutes. The achiral $MoS_2$ QDs/$Cu^{2+}$ complex did not reveal any oxidization catalytic activity against TMB. The catalytic activity for D- or L-cys-$MoS_2$ QDs/$Cu^{2+}$ complex displayed similar catalytic activity to natural enzymes on pH, temperature, and $H_2O_2$ concentration. It was found that both D- and L-cys-$MoS_2$ QDs/$Cu^{2+}$ complex followed the typical Michaelis–Menten reaction behavior against TMB and $H_2O_2$. The L-cys-$MoS_2$ QDs/$Cu^{2+}$ against TMB was almost unchanged after three times of recycle and slightly decreased after five times of recycle. The enantioselective catalytic activity of the D- and L-cys-$MoS_2$ QDs/$Cu^{2+}$ complex against D- and L-Tyr was investigated. It was found D-cys-$MoS_2$ QDs/$Cu^{2+}$ displayed higher catalytic activity than L-cys-$MoS_2$ QDs/$Cu^{2+}$ (Figure 6.6) for oxidation of D-Tyr. The L-cys-$MoS_2$ QDs/$Cu^{2+}$ had higher catalytic activity than D-cys-$MoS_2$ QDs/$Cu^{2+}$ in the oxidation of L-Tyr. For L-cys-$MoS_2$ QDs/$Cu^{2+}$, the $K_m$ values for L- and D-Tyr were $9.96 \times 10^{-4}$ and

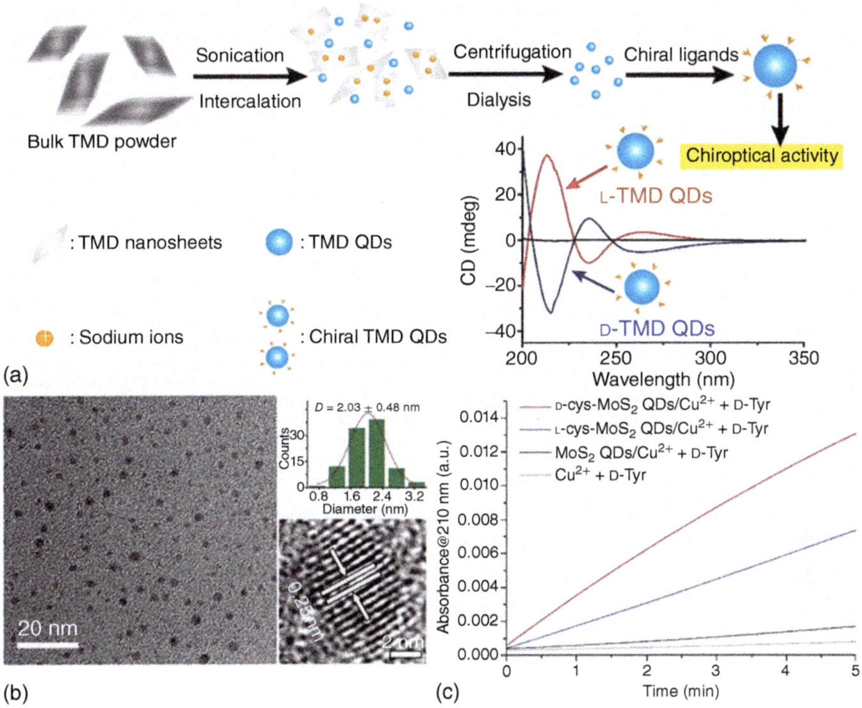


**Figure 6.6** (a) Schematic illustration for the sonication combined with ion intercalation method in the alkaline environment for preparation of TMD QDs, and the covalent modification with chiral ligands for preparing chiroactive TMD QDs. Typical CD spectra of the chiral TMD QDs are shown in the scheme. (b) TEM image of $MoS_2$ QDs; size distribution of $MoS_2$ QDs; and HRTEM image of $MoS_2$ QDs. (c) The time-dependent absorbance changes of D-Tyr in the presence of L-cys-$MoS_2$ QDs/$Cu^{2+}$, D-cys-$MoS_2$ QDs/$Cu^{2+}$, $MoS_2$ QDs/$Cu^{2+}$, and $Cu^{2+}$ system with $H_2O_2$ at 210 nm. Source: Zhang et al. [46] / with permission of American Chemical Society.

$6.61 \times 10^{-3}$ M, respectively, indicating its higher binding affinity to L-Tyr than D-Tyr. The $K_m$ values of D-cys-$MoS_2$ QDs/$Cu^{2+}$ against L- and D-Tyr were $5.46 \times 10^{-3}$ and $8.06 \times 10^{-4}$ M, respectively. The D- and L-cys-$MoS_2$ QDs/$Cu^{2+}$ conjugates showed opposite enantioselectivity, which were in good agreement with the first principles of stereochemistry. The chiral QDs possess enantioselective peroxidase-like activity for the oxidation of chiral Tyr substrates due to binding affinity differences between chiral QDs and Tyr substrates. The results provided great potential for developing novel enantioselective catalysts and gave rise to better design of chiral nanozymes.

Graphite oxide (GrO), graphene oxide (GnO), and their related sulfated derivatives are of great interest for various applications such as catalysis, optical and electronic devices. Proline-encoded GrO or GnO catalyst supports enabled the fabrication of Pro-GrO and Pro-GnO chiral catalysts. The enantioselectivities of the developed catalysts were comparable to proline in the homogeneous catalytic reaction under the exact optimal conditions (ee 71% R). The Pro-GrO and Pro-GnO chiral catalysts suggested that the prepared catalysts possessed superior catalytic activity for both conversion and enantioselectivity. Compared with the homogeneous reaction, the presence of the support materials led to improved catalytic activity in the reaction of aldol addition. The solid catalyst material could be reused for highly enantioselective catalysis, which was due to adsorption of proline for high catalytic activity. It was believed that the solid support played a role in the reaction for providing the effective solid material surface. However, deterioration of the surface by desorption of proline led to activity loss in the subsequent catalysis. The sulfated material groups formed on the edges allowed stronger bonding of proline that could provide stable catalytic sites for asymmetric aldol reaction. Herein, the developed Pro-GnO chiral catalysts provided high catalytic activity for asymmetric aldol reaction between 2-nitrobenzaldehyde and acetone with a selectivity of 90% and an enantioselectivity of 77%. Due to stable solid support and covalent-bonded proline, the most favorable experimental results obtained with the Pro-$GnO_3$s catalyst can be used for many cycles. Further research provided great potential for developing nanocatalysts that may be important for constructing several commercially available graphene-based catalysts.

Linear amino acids were explored as effective catalysts in asymmetric aldol reaction between ketones and aldehydes in the presence of water. M. Abid Zia's group used linear amino acids covalently bonded on graphene as catalysts for asymmetric aldol reaction between cyclohexanone and aldehydes [47]. The developed graphene-based nanocatalyst exhibited high yield as well as excellent ee value. The graphene sheet was first oxidized by Hummers and Offeman's method to develop (–COOH) and then treated with $SOCl_2$ to convert –COOH into –COCl groups. Then the graphene sheet with –COCl groups was treated with BOC-amino acid by refluxing with THF/pyridine. BOC-amino acid ensured that a covalent bond developed through the –COOH group of the amino acid. Then the amino acid was deprotected by hydrolysis with trifluoroacetic acid. In this covalent approach, the amino acids were immobilized on graphene and then used as heterogeneous asymmetric catalysts. The covalently immobilized amino acids on graphene sheet displayed typical FTIR peaks.

L-alanine, L-serine, L-valine, and L-arginine modified on graphene sheet were referred to as catalysts A, B, C, and D, respectively, which are capable of catalyzing asymmetric aldol reaction in aqueous medium. The aldol reaction of cyclohexanone and benzaldehyde was chosen as a model reaction, in which catalyst "A" was used as a chiral heterogeneous catalyst. The catalytic results indicated that the desired aldol adduct was formed with an 85% yield, 94% ee, and 6.7 : 1 dr. Catalysts "B" and "C" were used for the same reaction, and the results were: catalytic efficiency as 72% yield, 93% ee, and 6 : 1 dr for catalyst "B" and 84% yield, 91% ee, and 16.5 : 1 dr for catalyst "C". However, catalyst "D" did not display good catalytic results under the same conditions: a low yield of 58% and an ee value of 10%. The aldol reaction of cyclohexanone and *m*-nitrobenzaldehyde was also evaluated by these catalysts that afforded both high % yield and % ee value as shown in Figure 6.7. Under the same reaction conditions, catalyst "A" generated excellent results, 94% yield and 99% ee value. The nitro group attached to the benzaldehyde probably helped to induce remarkable results. These sumptuous results prompted to probe the efficiency of other catalysts for the same reaction. Catalysts "B", "C", and "D" displayed quite good results: 85% yield and 96% ee, 97% yield and >99% ee, and 96% yield and 94% ee, respectively. The graphene sheet immobilized on linear amino acids can effectively catalyze asymmetric aldol reaction of aldehydes and ketones through the enamine route. First, the donor ketone reacted with the immobilized amino acid and induced enamine linkage. Then, the acceptor aldehyde bound with the chiral enamine and formed an intermediate upon hydrolysis, forming the enantiomerically enriched aldol product. The water probably enhanced the reaction rate due to the hydrolysis of the intermediate state of enamine. These catalytic results demonstrated that acyclic amino acids bound to graphene can act as robust chiral organocatalysts and catalyze direct asymmetric aldol reaction with good reactivity and stereoselectivity. Throughout the study, water was the only solvent used. The catalysts showed good stability in water and were recycled by simple centrifuging or filtration that can be used for five runs without showing loss of catalytic efficiency.

Synthetic organic chemistry mostly relies on organic solvents as reaction media, while the asymmetric catalysis is mainly performed in water system [49]. A highly reactive and stereoselective catalyst is urgently required, which is especially important for developing new catalytic nanomaterials and also improving existing synthetic methods [29, 50]. The homogenized combination of nickel-based Lewis acid–surfactant-combined catalysts based on single-walled carbon nanotubes was designed to exhibit substantial activity in water. The well-dispersed Lewis acid–surfactant combined catalysts (LASC)–SWNT nanocomposites containing transition metals were fabricated with high yield, where the asymmetric addition reaction between benzaldoxime (1a) and non3-en-2-one was evaluated for its catalytic activities. The electronic properties of SWNTs enhanced the dispersion of the nanocatalyst. The uniform dispersion of SWNTs encoded with LASCs was fabricated through the ultrasonication method. In addition to the improved reactivity, stereoselectivity and long-term stability of the catalyst were enhanced in asymmetric addition reactions of aldoximes to chiral nitrones in high yields. When combined with chiral 2,2′-bipyridine ligand, the homogeneous dispersion of LASC–SWNT nanocomposites encoded with a vanadium ion was found ineffective for this reaction. Typical water-compatible Lewis acids such as iron(II), copper(II), and zinc(II) ions displayed low activity, whereas the homogeneous dispersed $Ni(OSO_3C_{12}H_{25})_2$–SWNT nanocomposite ($OSO_3C_{12}H_{25}$ = DS) was found to promote the conjugate addition reaction to afford the optically active nitrone in 79% yield with 97% ee. The effect of combining with other materials was also surveyed. Much smaller augmentation of yield and/or selectivity was found when double-walled carbon nanotubes (DWNTs) or multiwalled carbon nanotubes (MWNTs) were integrated with $Ni(DS)_2$. This result was due to the difference in electric properties among the dispersed carbon nanotubes. The ineffectiveness of fullerene and graphene nanoplatelets was due to their weak interaction with surfactants as their spherical and two-dimensional shapes easily lead to coagulation. This reaction did not happen in other organic solvents during such catalysis.

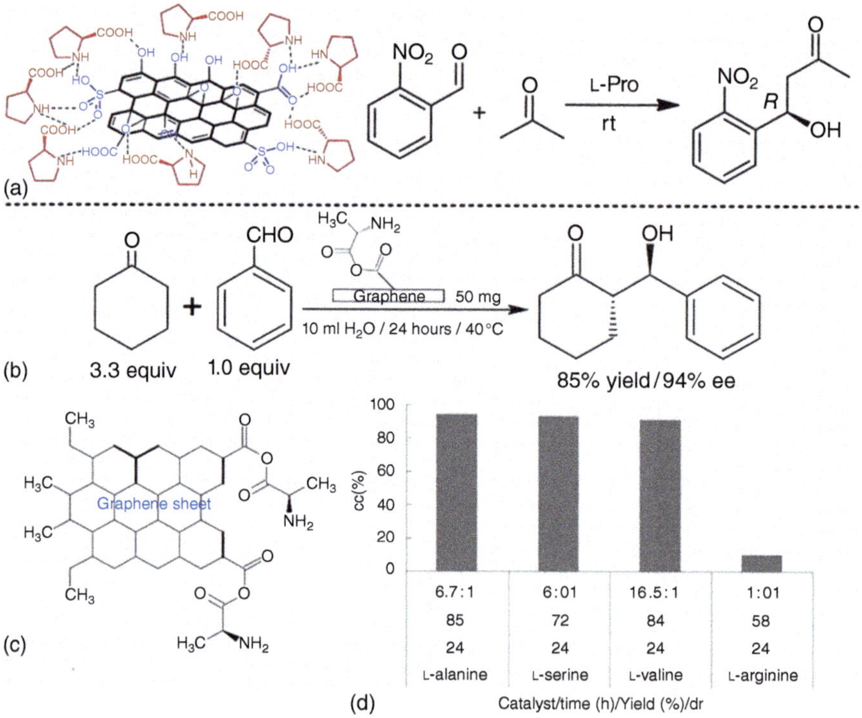


**Figure 6.7** (a) Binding possibilities of proline on GrO or GnO. Scheme of the studied asymmetric direct aldol addition reaction. Source: Szőri et al. [48] / Springer Nature. (b) Model reaction of cyclohexanone and benzaldehyde to 2-(hydroxy(phenyl) methyl) cyclohexanone, catalyzed by L-alanine covalently immobilized on graphene. (c) Schematic of covalently immobilized amino acids on graphene sheet. (d) Catalytic efficiency of L-alanine, L-serine, L-valine, and L-arginine covalently supported on graphene sheet for asymmetric aldol reaction between cyclohexanone and benzaldehyde. Source: Sadiq et al. [47] / Scientific Research Publishing Inc. / CC BY-4.0.

Consequently, substantially improved catalytic activity cannot be achieved without the uniform dispersion of SWNTs wrapped with $Ni(DS)_2$ in water system. The STEM images of the $Ni(DS)_2$–SWNT composite indicated that SWNTs were well dispersed. The EDS element results demonstrated that the $Ni(DS)_2$ molecules were anchored to the single line of SWNTs. The $Ni(DS)_2$ molecules were found to be stably located at the surface, and were very possibly embedded on the SWNT ropes. On the contrary, MWNTs failed to afford well-disentangled composites, and nickel was not anchored on the fullerene surface. The results demonstrated that $Ni(DS)_2$ molecules showed good catalytic activity only when strongly bound onto the SWNTs. The computer simulation results were used to understand the geometry of surfactant aggregates self-assembled on SWNTs. As mentioned, after saturation of the SWNT surface with $Ni(DS)_2$ molecules, the remaining $Ni(DS)_2$ molecules could participate in the formation of micelles. The tightly bound state of the composites helped to rationalize the persistence of the dispersion, which was stable for 30 days. The potential practical and straightforward real applications of the developed catalysts in water enabled an expedient, environmentally benign, and highly efficient pathway to access catalytic active nanocatalysts.

## 6.3 Biocatalysis of Peptides Based on Chiral Nanomaterials

The real applications of chiral NPs are deemed predominantly important. Ceramic NPs as major nanomaterials exist in nature and display robustness and other innate properties that offer great potential in catalysis, biomedical sciences, and optics. Nicholas A. Kotov's group reported the chiral ceramic NPs synthesized from tungsten oxide hydrate, $WO_{3-x}\cdot H_2O$ NPs [51]. The $WO_{3-x}\cdot H_2O$ NPs were synthesized in amino acid solutions with characteristic size, shape, composition, and UV–vis absorption spectrum. Proline (Pro) and aspartic acid (Asp)-mediated synthesized NPs had an average size of c. 1.6 nm with crystal lattice periodicity of 0.35 nm. The lattice spacing was calculated with periodicities of 0.35, 0.22, and 0.17 nm from electron diffraction results, which corresponded to the (111), (022), and (222) crystal faces of $WO_{3-x}\cdot H_2O$ NPs.

The NP solutions displayed intense vis–NIR band from 450 to 1100 nm with a maximum peak at 950 ± 50 nm and a shoulder at c. 580 nm. The amino acids were anchored to the surface through C—O—W bond formed from dissociated carboxyl groups and through amino groups that coordinated to the NP surface. Though having similarity in the geometry of the NPs, the dominant CD bands for NPs functionalized with Pro and Asp were quite different, which located at 400–700 nm and 500–1100 nm for Pro- and Asp-functionalized NPs, respectively. The differences in the spectral positions of the chiroptical band were associated with the molecular binding of amino acids to the NP surface. After bound to the NPs, the increased frequency for $\nu$(C=O) and decreased for $\nu$(C—O) corresponded to the formation of coordination bonds between the oxygen atom of the carboxyl group on the NP surface.

FTIR results revealed that the C=O and C—O stretching vibration bands located at 1725 and 1207 $cm^{-1}$ for Asp-NPs and at 1735 and 1220 $cm^{-1}$ for Pro-NPs. It was concluded that the carboxyl C—O unit coordinated with the tungsten ion on the NP surface as a monodentate ligand. Specifically, Asp has one additional C—O—W bond compared to Pro, induced stronger distortion of the inorganic crystal lattice and stronger CD bands correlated with the chiral inorganic core. The chiral distortions of atomic packing of $WO_{3-x}\cdot H_2O$ nanocrystals displayed characteristic chiral electronic transitions. The chiral $WO_{3-x}\cdot H_2O$ nanocrystals were identified by atomistic molecular dynamics simulations. Based on Forcite Plus module in Materials Studio, the theoretical model of $WO_{3-x}\cdot H_2O$ NPs with the lateral size of $1.9\times2.0\times2.2\,nm^3$ with a quasi-cubic shape was constructed (Figure 6.8a–c). The atomic geometry of the NPs was allowed to relax to the equilibrium states with L- and D-Asp and the energy of the structure remained constant within the simulation error. The obtained MD models of L- and D-Asp NPs displayed the distorted packing of tungsten oxide lattice after

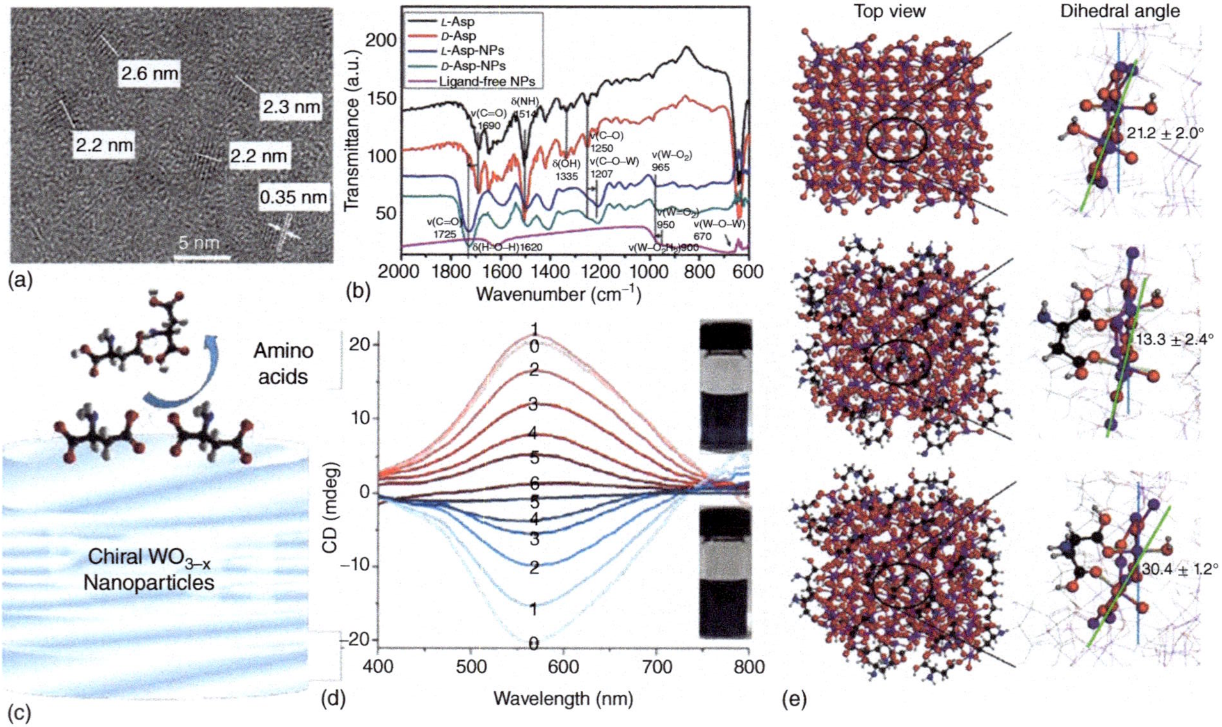


**Figure 6.8** (a) Representative high-resolution TEM image of L-Asp-NPs of $WO_{3-x}·H_2O$, $x = 0.03$. (b) FTIR spectra (2000–600 cm$^{-1}$) of L- and D-Asp, L-and D-Asp-NPs, and ligand-free $WO_{3-x}·H_2O$ NPs. (b) FTIR spectra (2000–600 cm$^{-1}$) of L- and D-Pro, L- and D-Pro-NPs, and ligand-free $WO_{3-x}·H_2O$ NPs. (c) The schematic for chiral $WO_{3-x}$ surface with amino acid. (d) The CD spectra for L- and D-Pro-NPs (red and blue, respectively) after addition of $H_2O_2$ in concentrations of 0,0 mmol L$^{-1}$;1, 0.06 mmol L$^{-1}$; 2,0.12 mmol L$^{-1}$;3, 0.18 mmol L$^{-1}$; 4, 0.24 mmol L$^{-1}$;5, 0.30 mmol L$^{-1}$; and 6, 0.36 mmol L$^{-1}$, respectively. (e) MD simulations of the atomic structure in the minimum energy state for (upper) ligand-free, (medium) L-Asp, and (bottom) D-Asp NPs. Source: Jiang et al. [51] / with permission of American Chemical Society.

L/D-Asp adsorbed on the $WO_3$ surface (Figure 6.8). The distortions were sufficient to affect the chiroptical activity associated with both plasmonic and polaronic states involved with the amino acid. The amino acids on the mineral surface were associated with the catalytic abilities of $WO_{3-x}{\cdot}H_2O$ nanocrystals, which led to the formation of peptide bonds, thereby resulting in Asp-Asp and Asp-Pro dipeptides. The chiroptical activity and chemical reactivity of tungsten oxide provide an advanced combination of properties relevant to chiral optics, chemical reactivities.

Supraparticles (SP) from NPs assembly is important, but it is unknown whether SPs can mimic their biochemical functions. It was reported that L-Tyr exhibited good enantioselective adsorption toward L-Pen, and D-Tyr displayed good enantioselective adsorption against D-Pen. Therefore, L-Pen-functionalized chiral metamaterials can be used for the biocatalysts of L-Tyr. Overall, chiral ZnS NPs self-assembled into 70–100 nm SPs and displayed sub-nanoscale porosity associated with interstitial spaces between constituent NPs. Similar to photosynthetic bacterial organelles, these SPs can serve as photocatalysts, enantioselectively converting L- or D-tyrosine (Tyr) into dityrosine (diTyr) (Figure 6.9a,b) [52]. The assembled chiral SPs displayed an average diameter of $d_{TEM} = 3 \pm 0.7$ nm, stabilized by L-or D-penicillamine (Pen). Pen-ZnS NPs were synthesized and assembled into SPs by incubation at 95 °C for three to six hours. The assembled SPs indicated a spherical morphology with

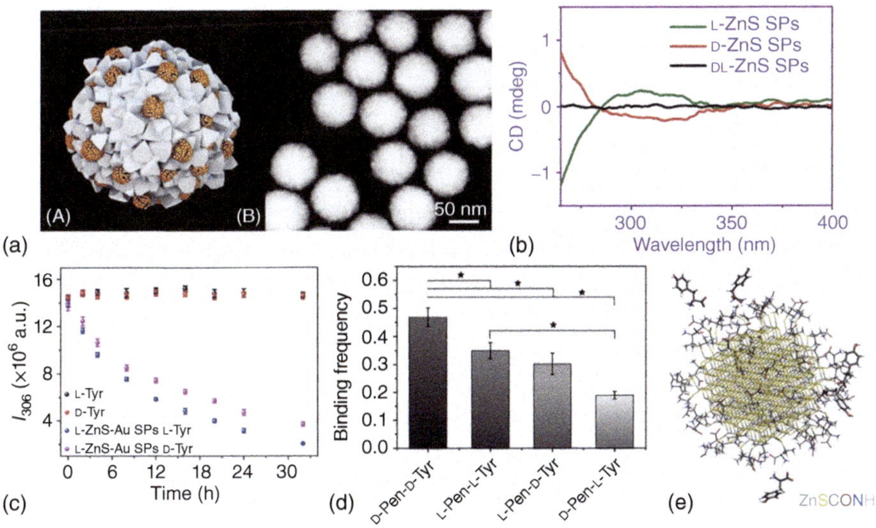


**Figure 6.9** (a) (A) Morphology of multicomponent L-ZnS–Au SPs. (B) HAADF-STEM images of multicomponent L-ZnS–Au SPs. (b) CD spectra of ZnS SPs from 265 to 400 nm. (c) The dependence of PL intensity at 306 nm on photocatalytic reaction time with L-ZnS–Au SPs for different Tyr enantiomers. (d) The binding frequency decreases in the order from D-Pen-D-Tyr, L-Pen-L-Tyr, L-Pen-D-Tyr, to D-Pen-L-Tyr. Error bars indicate the standard error from 5 ns block averages during 50 ns MD simulations. (e) Snapshot of D-Tyr binding on the D-Pen-ZnS NP surface. D-Tyr molecules are represented in "Licorice" style with thicker bonds. The 3-nm D-Pen-ZnS NP is represented in "Licorice" style with thinner bonds. Source: Li et al. [52] / with permission of Spinger Nature.

diameters of $d_{TEM} = 100 \pm 3$ nm and $d_{DLS} = 105 \pm 10$ nm, determined by TEM and DLS, respectively. They also displayed high monodispersity with a standard deviation of <6% indicating the quasi-equilibrium state of the SPs. The typical peaks at 3000–3300 $cm^{-1}$ were due to the stretching vibration of N–H of Pen attached to the surface of ZnS NPs. The peaks at 1450–1700 and 2200–2450 $cm^{-1}$ were due to C–H bending and S–H stretching vibrations of Pen. Compared with the CD spectra of Pen alone, a new CD peak appeared from 240 to 300 nm, which was due to electronic hybridization between the molecular orbitals of NPs and of chiral ligands. The attachment of Pen ligands induced local distortions of the ZnS surface and contributed to the CD spectra. Unlike the individual NPs, the chiroptical activity of the prepared SPs included an additional band in the 280–340 nm region.

As-prepared chiral SPs showed enantioselective photocatalytic ability toward Tyr at pH 8.0 under 315–400 nm light. L-Pen-modified ZnS NPs preferentially catalyzed L-Tyr, and D-Pen-modified ZnS NPs preferentially catalyzed D-Tyr, which was attributed to the enantioselective adsorption for L-Tyr/L-Pen and D-Tyr/D-Pen. Chiral ZnS-Au SPs demonstrated better photocatalytic behaviors than chiral ZnS SPs. Compared to ZnS SPs, the electronic mobility for ZnS-Au SPs was improved under light irradiation. The photogenerated electrons preferred the Au surface, and the electrons were mainly located on ZnS NPs, resulting in the separation of electrons and holes and accelerating the photocatalytic process. Due to the close interaction between Au and ZnS components, the photogenerated carriers could be transferred between different components, further preventing the recombination of electrons and holes and accelerating the photocatalytic reactions. This incorporation of Au NPs into ZnS SPs effectively absorbed light energy and promoted electron transport and electron–hole separation for photocatalytic conversion applications. Multi-component chiral SPs possessed better light absorption ability for the occurrence of plasmon–exciton coupling, and the existing different components enabled the transfer of hot electrons for applications in chemical reactions. More chiral SPs can be further fabricated by adjusting the composed metals, semiconductors ,and other NPs for the study of enantioselective catalysis [52].

Experimental data and molecular dynamics simulations indicate that the chiral bias of photocatalytic reactions is associated with the chiral environment of interstitial spaces and preferential partitioning of enantiomers into SPs, which can be further enhanced by coassembling ZnS with Au NPs. The conversion of the amino acid Tyr, which can be monitored using its strong photoluminescence (PL) peak at 306 nm, was used as a chiral catalytic substrate. When ZnS SPs were added to the Tyr solution, a strong increase in PL intensity was observed at 414 nm, which is characteristic of the accumulation of diTyr. DiTyr formation was accompanied by a decrease in PL intensity at 306 nm as expected considering the consumption of Tyr. While diTyr was the main product of the photocatalytic conversion of Tyr, the formation of DOPA may also be expected, which explained the redshift of the emission peak at 306 nm as the photolysis proceeded. Analysis of temporal profiles of diTyr emission, $I_{414}(t)$, allowed us to evaluate the chiral preferences of the SP catalysts. ZnS SPs carrying L-Pen preferentially catalyzed L-Tyr, while SPs made from D-Pen preferentially catalyzed D-Tyr (Figure 6.9c–e). The degree of enantioselectivity can be associated with preferential partitioning of enantiomers of Tyr into the SPs, which was confirmed by an enantioselective adsorption of Tyr enantiomers after exposure of rac-Tyr solution to L-or D-ZnS SPs. The individual ZnS NPs with L-or D-ligands did not show obvious catalytic preference for enantiomeric Tyr, and the catalytic activity was also lower than that of the SPs. Both ZnS NPs and ZnS SPs retained excellent stability in the course of photocatalysis. Importantly, the photochemical stability of chiral ligands on SPs was markedly better than that on NPs. In fact, no decomposition of Pen was observed in SPs after intense illumination for 36 hours (Figure 6.9d,f). When Tyr was illuminated in the presence of ZnS-Au SPs, the fluorescence intensity of the 306 nm band decreased dramatically during the illumination, which indicated a large enhancement of the efficiency of photocatalytic conversion of Tyr after incorporation of Au NPs. Note, however, that the rise of the 414 nm peak associated with diTyr was smaller in ZnS-Au than in ZnS SPs, pointing to the simultaneous enhancement of diTyr conversion into non-luminescent products. Besides replicating a specific function of biological nanoassemblies, these findings establish a path to enantioselective oxidative coupling of phenols for biomedical and other needs.

## 6.4 Biocatalysis of Biological Macromolecules Based on Chiral Nanomaterials

Accurate substrate specificity is one of the most important characteristics of enzymes and also the main gap between nanozymes and natural enzymes. Many enzymes are stereospecific; that is, when the substrates possess stereoisomers, the enzyme can only act on one of them. In addition, chiral discrimination on substrates' enantiomers is an important aspect of stereospecificity of enzymes. Enantioselective catalysis plays an important role in biomedical applications. Some drugs exhibit therapeutic effects only in one configuration, while other enantiomers may not work or even may be toxic. For instance, the dextral-enantiomer of thalidomide, a drug used to alleviate pregnancy response, has sedative effect, while the levo-enantiomer may cause teratogenicity. Many chemical synthetic molecules or natural enzymes have been used in chiral catalysis.

Considering the unique advantages of nanozymes, the development of efficient enantioselective nanozymes for asymmetric catalysis of racemic products is very important. Additionally, life activities are closely related to chirality. Most of the biomolecules in life exhibit homochirality. For example, most of the amino acids comprising proteins are left-handed and the riboses comprising DNA are right-handed. Therefore, enantioselective nanozymes may exhibit better performance in in vivo applications. Moreover, although it is still an unsolved mystery, most theories hold that the generation of homochirality is an important process in the origin of life, and it may also be an important factor in the initial generation of substrate specificity of natural enzymes. Thus, in conclusion, enabling nanozymes to be enantioselective may not only broaden their industrial and biomedical applications but also lay a good foundation for further improvement of the enzyme-like activity and selectivity of nanozymes. Indeed, we note that substantial progress has been made in recent attempts to improve enantioselectivity of nanozymes. Nanozymes themselves are mostly fabricated from achiral molecules, so it is difficult for them to show enantioselectivity. Recent studies on chiral nanozymes mostly focus on the coupling of chiral ligands on the surface of nanozymes, such as amino acids, nucleic acids, and some chiral chemical compounds. To capture this development, we herein introduce some typical chiral nanozymes proved to be effective on enantioselective and asymmetric catalysis. Because the studies in this field are still in their infancy, most of them are proofs of concept. So, the main purpose of this review is not to systematically summarize, but to highlight current strategies and advances to attract more attention and attempts from researchers on chiral nanozymes.

In particular, the chiral molecule–modified semiconductors achieved efficient utilization of RCP or LCP light. CdSe@CdS semiconductor nanorods (SNRs) with multiple post growth of platinum (Pt) and gold (Au) showed high photocatalytic activity and high efficiency of separation of $e^-$ and $h^+$ [53]. It has recently been reported that reactive oxygen species (ROS) are critical for both selective oxidation and DNA lesions. Chiral D- and L-cys-modified CdTe NPs were fabricated and displayed mirror CD spectra in the ranges of 350 and 410 nm. Interestingly, strong and different CD peaks occurred at 365, 415, 470, and 488 nm for the mixtures of L-cys CdTe NPs and DNA under RCP illumination for two hours, but opposite and mirror CD features appeared at the same CD peak positions for the mixtures of D-cys CdTe NPs and DNA under LCP illumination for two hours. Chiral cys-modified CdTe NPs exhibited nuclease-like behavior for the cleavage of DNA fragments. It was found that chiral cys-modified CdTe NPs could recognize the GAT′ATC of dsDNA with the base pairs exceeding 90 bp. Under laser illumination with either LCP or RCP light at 405 nm, DNA was selectively cut by chiral cys-modified CdTe NPs between the T and A bases of the restriction site GAT′ATC (Figure 6.10a). ROS (•OH) was generated by chiral cys-modified CdTe NPs under irradiation of CPL, and it was responsible for the cleavage activity to break the phosphodiester bond between T and A bases. L-cys CdTe NPs showed better absorption efficiency under RCP, and R-cys CdTe NPs showed better absorption efficiency under LCP for the production of a large number of ROS. Therefore, L-cys CdTe NPs had a higher cleavage rate under RCP than under LCP, and D-cys CdTe NPs rates were opposite. Under the same illumination time, different polarization directions obviously affected the ROS production by chiral NPs. This difference was attributed to the different absorption efficiency of the chiral NPs under CPL. The generation of high-yield ROS and the selectivity between the configuration of chiral CdTe NPs and the structures of DNA sequence both induced the cleavage activity [54]. Water-soluble L-cys CdTe NPs were synthesized and showed the average size of $4.21 \pm 0.4$ nm. L-cys CdTe NPs exhibited CD responses in the range of 200–450 nm; in particular, the CD peaks in the range of 200–310 nm were due to the chiral molecules of L-cys. Chiral CdTe QDs could be transferred to NR morphology, which increased the yield of ROS for antibacterial applications. Chiral

L-cys-modified CdTe NPs were mixed with salmon sperm DNA and then were illuminated by 405 nm RCP for two hours, resulting in significant morphology changes and high yields of ROS production for combating Gram-negative bacteria (Figure 6.10b). L-cys-modified CdTe NPs were fabricated and exhibited CD features in the ranges of 200–450 nm. In the presence of salmon sperm DNA, L-cys-modified CdTe NPs were transferred to NRs under irradiation of 405 nm RCP for two hours. For L-cys-modified CdTe NRs, the fluorescence peak changed from 610 to 617 nm, and five different CD peaks occurred at 365, 415, 448, 471, and 488 nm (Figure 6.10c). It was found that a large amount of ·OH was produced by L-cys-modified CdTe NRs under illumination of RCP for two hours, indicating that illumination triggered the production of ROS. Polycationic nonapeptides, KRKKRKKRK, were modified on the surface of L-cys-modified CdTe NPs for targeting Gram-negative bacteria (such as *Escherichia coli* and *Pseudomonas aeruginosa*), and the unique affinity was attributed to the replacement of the divalent cations of the lipopolysaccharide layer by the polycationic nanopeptide. 92.7% *E. coli* was dead in the presence of L-cys-modified CdTe NPs under RCP illumination for 30 minutes (Figure 6.10d); therefore, L-cys-modified CdTe NPs possessed better antibacterial behaviors. Furthermore, the unique fluorescence signals enabled PCNP-L-cys CdTe to be used as a fluorescence imaging agent [55]. Chiral semiconductors provided a new pathway for clinical treatment of bacterial infections.

Chiral semiconductors after RCP or LCP irradiation not only possessed DNA cleavage activity and antibacterial behaviors but also showed protein cleavage performance. Another research focused on protein cleavage performance assisted by protease-like chiral semiconductors. During the reaction, hydroxyl radicals are still as the reactive species that cause the cleavage of proteins. Nanocrystalline copper sulfide ($Cu_{2-x}S$) is a p-type semiconductor and a material of particular interest due to its promising applications in biomedicine and optoelectronic conversion. Chiral $Cu_{2-x}S$ QDs were fabricated by using D-/L-penicillamine (D-/L-Pen) as ligands (Figure 6.11). D- and L-Pen-modified $Cu_{2-x}S$ QDs displayed mirror CD responses at 608 and 863 nm with an anisotropy factor of 0.01. The CD peak intensity increased with the increasing concentration of Pen ligands. As-prepared chiral $Cu_{2-x}S$ QDs were applied for the cleavage of BSA under light excitation. It was shown that about 79%, 53%, and 20% BSA was cleaved by L-Pen-modified $Cu_{2-x}S$ QDs under 660 nm LCP, LP, and RCP light. The highest photocatalytic efficiency was achieved for L-Pen-modified $Cu_{2-x}S$ QDs under LCP light. This was because L-Pen-modified $Cu_{2-x}S$ QDs showed better adsorption toward LCP light for the generation of a large number of hot electrons, and OH· as the reactive species promoted the catalytic cleavage of BSA. The ROS content increased in response to irradiation over time. In comparison to D-Pen-modified NPs, L-Pen-modified $Cu_{2-x}S$ QDs possessed better affinity toward BSA, resulting in more ROS production under LCP light for BSA hydrolysis (Figure 6.11e,f). Furthermore, compared to sulfhydryl chiral ligands (cysteine, cystine, glutathione) and sulfhydryl achiral ligands (thioglycolic acid, cysteamine, 3-mercaptopropionic acid), chiral Pen ligand–modified $Cu_{2--x}S$ QDs exhibited better photocatalytic behaviors. These QDs also had a photocatalytic

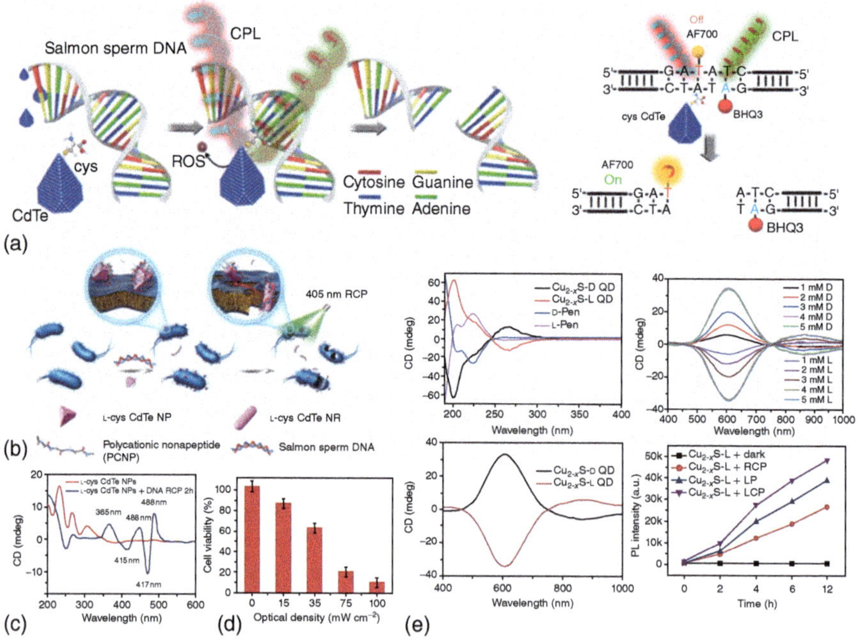

**Figure 6.10** (a) Schematic illustration of selective DNA cleavage by chiral cysteine-modified CdTe NPs. (b) Schematic diagram of chiral L-cys-modified CdTe NPs against Gram-negative bacteria under RCP irradiation. (c) CD spectra of L-cys CdTe NPs. (d) Cell viability analysis of *Escherichia coli* under different light intensity. (e) CD spectra of $Cu_{2-x}S$ QDs modified with different types of Pen and different concentrations of Pen. Source: (a) Reproduced with permission Sun et al. [54]. Copyright 2018, Springer Nature; (b–d) Reproduced with permission Wang et al. [55]. Copyright 2018, Wiley-VCH.

effect on other proteins such as human serum albumin (HSA), which showed the universality of this methodology [56].

## 6.5 Biocatalysis Based on Enzyme-Encoded Chiral Complex Nanomaterial System

The strategy of enzyme immobilization is sometimes expected to improve enzyme activity, selectivity, reusability, and resistance to inhibitors to a certain degree. Zhiming Rao's group reported a general and facile encapsulation of a chiral-forming oxidoreductase and a cofactor-recycling oxidoreductase as complex for improved catalysis. The selected enzymes were encoded in the mesoporous ZnO channels of a hybrid ZnO/carbon composite, in which immobilization of two oxidoreductases into mesoporous ZnO/carbon composite was achieved through a "kiwifruit-assembly" pattern. The ZnO is a typical inorganic semiconductor that can improve electron transfer between enzymes and the electrode in phenol or uric-acid-related solutions. The ZnO was potentially helpful in the oxidoreductase encapsulation for

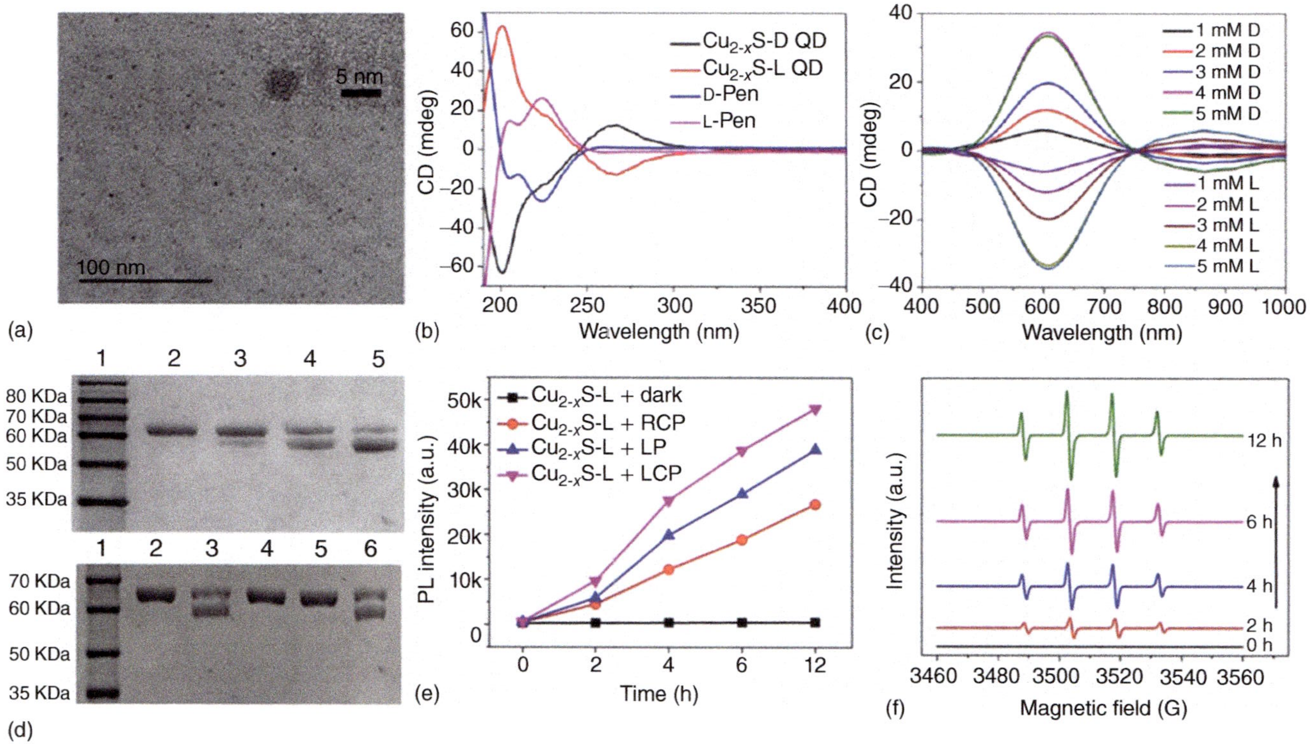


**Figure 6.11** (a) TEM image of L-QDs. (b) CD spectra of Pens and the QDs prepared using 5 mM Pens. (c) CD spectra of QDs prepared in the presence of different concentrations of chiral Pens. (d) The upper panel: the SDS-PAGE diagram shows the light dependence on the cleavage of BSA by L-QDs with exposure to different light conditions for 12 hours: lane 1, molecular marker; lane 2, BSA; lane 3, BSA + RCP; lane 4, BSA + LP; lane 5, BSA + LCP. The bottom panel: the SDS-PAGE shows the mechanistic results for the cleavage of BSA by L-QDs under LCP irradiation for 6 hours: lane 1, molecular marker; lane 2, BSA control; lane 3, BSA + QD; lane 4, BSA + QD + KI; lane 5, BSA + QD + ethanol; lane 6, BSA + QD + $NaN_3$. (e) Statistical chart of ROS production in QDs under different light conditions. (f) Detection of hydroxyl radicals by electron paramagnetic resonance (EPR) spectroscopy. Source: Hao et al. [56] / with permission of Wiley-VCH.

improving electron transfer between the enzyme active sites, the redox mediator of interest, and, therefore, the electrode surface. At the same time, the carbon nanodots or nanotubes could provide the same function as ZnO materials. In the meantime, it was important for avoiding enzyme adsorption on the surface or inner pores of unprotected ZnO nanomaterials, which may induce easy enzyme leaching from supports with environmental changes. It was also important for protecting ZnO nanostructures in nonacidic medium. Based on the encapsulating strategy, two oxidoreductases, Candida parapsilosis (*S*)-carbonyl reductase II (SCRII) and *Bacillus* sp. glucose dehydrogenase (GDH), were chosen to immobilize into the mesoporous ZnO/carbon composite (Figure 6.12). The pore size was around 7–12 nm and gave free sites to load SCRII–GDH, which had a diameter of around 4.1 nm. The characteristic diffuse selected area electron diffraction pattern showed successful formation of hexagonal nanocrystalline ZnO. The FTIR was used to characterize the synthesized nanocatalyst complex. The amide I band was from C=O stretching located in the protein backbone, and the amide II band was devoted to the N–H in-plane bending and C–N stretching vibrations of the peptide. The band at 1052 $cm^{-1}$ existed in the SCRII–GDHfree and SCRII–GDHnano spectra, however, not shown in the ZnO/carbon nanomaterial, which was due to the C=OH stretching vibration. The bands at 1658 and 1541 $cm^{-1}$ were found shifted toward each other in SCRII–GDHnano. The observed shift of the C=O stretching band suggested the interaction of enzymes with the ZnO matrix. The immobilized (*S*)-carbonyl reductase II-glucose dehydrogenase on nanoparticles (SCRII–GDHnano) exhibited 40–50% higher activity than the free enzyme, and stability of enzymes against heat, pH, and solvents was also significantly improved. The immobilized catalyst induced asymmetric catalysis of $75 \times 10^{-3}$ M substrate and exhibited a high yield of 100% with 99.9% enantioselectivity in one hour. At the same time, SCRII–GDHnano gave rise to over 72% yield and 99.9% enantioselectivity even when it was reused for 10 times. Furthermore, for a highly concentrated substrate of $400 \times 10^{-3}$ M, it displayed nearly 60% yield and 99.9% enantioselectivity in four hours. The coencapsulated oxidoreductases showed excellent catalytic activities, long-term stability, and good reusability and cofactor recycling, even at high substrate concentrations. It was demonstrated that SCRII–GDHnano had 4.5–8.0-fold higher yield in 2.0–8.0-fold shorter reaction times than the original free enzyme. Based on the radical proton-coupled electron transfer, the work opened an avenue in a general, facile, and unique approach for the immobilization of functional oxidoreductases with high catalytic efficiency, long-term and good recycling stabilities.

Self-assembly of proteins and inorganic nanoparticle complex assemblies induced a large family of uniformly sized hybrid colloids. Based on theoretical tools, the combined assembly of cadmium telluride nanoparticles with cytochrome C protein was reported, and spontaneous formation of spherical supraparticles with a narrow size distribution was observed [58]. Positively charged proteins assembled with positively charged NPs are of great importance. Conventionally, electrostatic attraction can drive the oppositely charged building blocks and induce self-assembly. The counterbalancing electrostatic repulsion and attractive interactions including

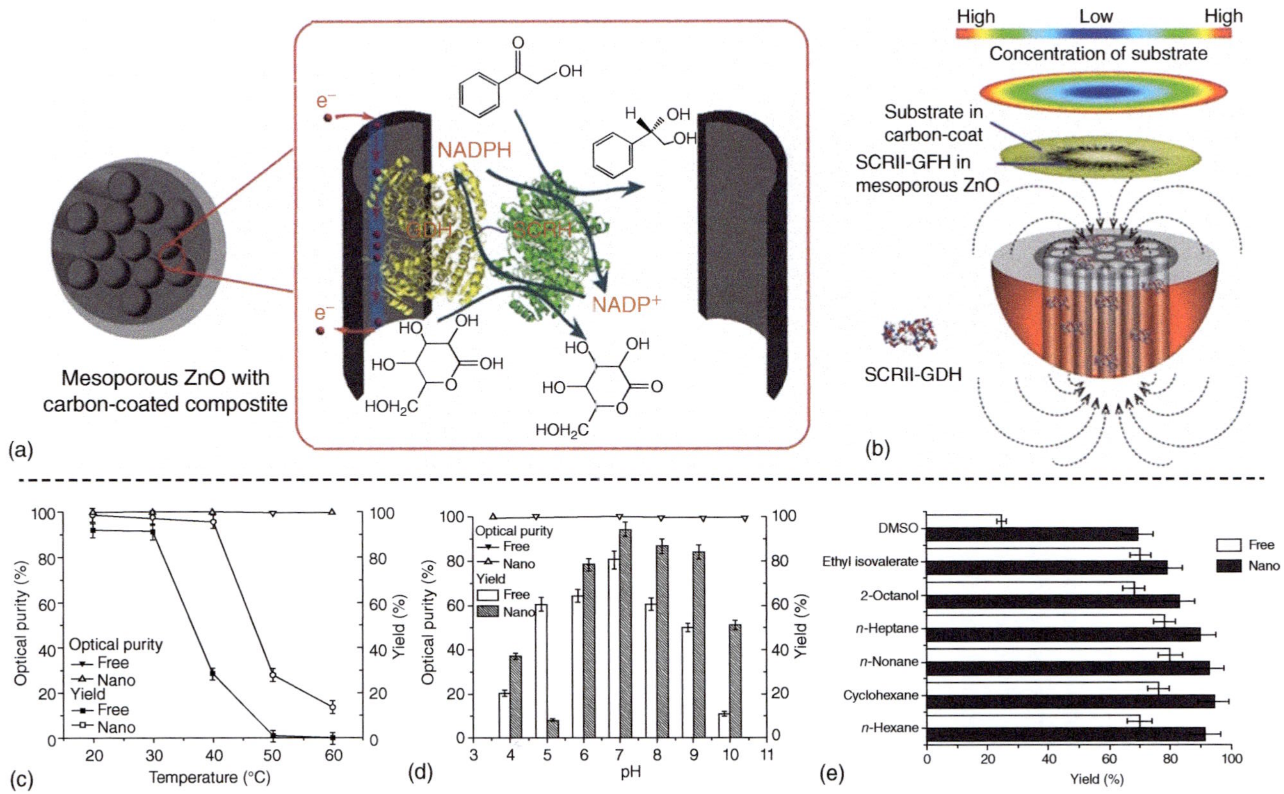


**Figure 6.12** (a) Principle of coencapsulation of SCRII–GDH on ZnO/carbon nanocomposites for asymmetric catalysis of (*S*)-PED. (b) "Kiwifruit-assembly" pattern of SCRII–GDHnano on mesoporous ZnO/C composites for high concentrated substrate. (c) Thermostability, (d) pH stability, and (e) solvent resistance of SCRII–GDHnano for asymmetric catalysis of (*S*)-PED. Nano, SCRII–GDHnano; Free, SCRII–GDH free. Source: Zhang et al. [57] / John Wiley & Sons.

dipolar, hydrogen bonding, hydrophobic, and van der Waals (vdW) forces are crucial for terminal assemblies. Besides surfactants, such self-limited assemblies are due to the counterbalance of attractive–repulsive forces. 6 mM NP dispersion at pH 5 was mixed with 6 mM CytC at pH 7. The stable NP-CytC dispersion displayed an electrokinetic potential (z) of +30 mV and pH 5.3 with 48–72 hours. The TEM results indicated the formation of uniform spherical SPs with a TEM diameter of $d_{TEM} = 94 \pm 5.6$ nm. The size distribution was narrow with a standard deviation less than 7%. DLS of 1 : 1 NP-CytC dispersion indicated the SPs with $d_{DLS} = 99 \pm 19.0$ nm. The self-assembled spheres were assembled with NPs and CytC in a 1 : 1 starting ratio. When NPs are the dominant components, the assemblies tend to form 2D structures. When proteins are dominant, large irregular aggregates are formed. The intermediate stages of the self-assembly process indicated the path by which the SPs form. The assembled particles with $d_{DLS}$ = $126 \pm 9.7$, $66 \pm 15.0$, $72 \pm 9.4$, and $100.6 \pm 17.3$ nm were observed at 1, 24, 48, and 72 hours. This was accompanied by a gradual disappearance of signals with $d_{DLS}$ = $5 \pm 0.4$ and $20 \pm 1.4$ nm, due to original NPs and early NP-CytC clusters. The CD data indicated that the CytC molecules incorporated in SPs kept their folded state. Original CytC molecules had negative (−) CD peaks at 208 and 220 nm, corresponding to β-sheets and α-helices. In the SPs, a positive CD peak induced by α-helices at 225 nm dominated, and a new negative peak appeared at 340 nm (Figure 6.13). The overall attractive potential between the similarly charged CdTe NPs and CytC can be confirmed by calculations based on classical and extended Derjaguin, Landau, Verwey, and Overbeek (DLVO) theory. The interactions such as dipole–dipole, charge–dipole, and hydrophobic interactions with the pair potential becomes attractive for all separation distances. The non-monotonic nature of these interactions and their nonadditivity at short separation distances are likely to further shift the balance toward the attractive interactions. Such self-limiting assemblies were due to the competition between electrostatic repulsion and noncovalent attractive interactions. Similar to micelles, supraparticles incorporated with other biological components can mimic nitrate reductase activity.

Nicotinamide adenine dinucleotide phosphate (NADPH) is a sacrificial electron donor, which is usually added to the SP dispersion. The incorporation of NADPH in the SPs was observed from XEDS data. The direct electron transfer from CdTe to NRed was slow and thought to be mediated by CytC. When the SPs was illuminated at 470 nm, the rate of $NO_2^-$ production by SP-NRed in the presence of NADPH increased fourfold. In the absence of light, the presence of the SPs has virtually no effect on the activity of the enzyme. The control experiments confirmed the significance of the NPs and SPs for the photoenzymatic $NO_3^-$ reduction, which included the enzyme reaction (i) either with CdTe NPs or (ii) with CytC upon illumination. None of the control experiments showed enzyme activity as high as for the illuminated SP-NRed. Tight packing of nanoscale components induced effective charge and exciton transport in supraparticles and bionic combination of properties as demonstrated by enzymatic nitrate reduction initiated by light absorption in the nanoparticles.

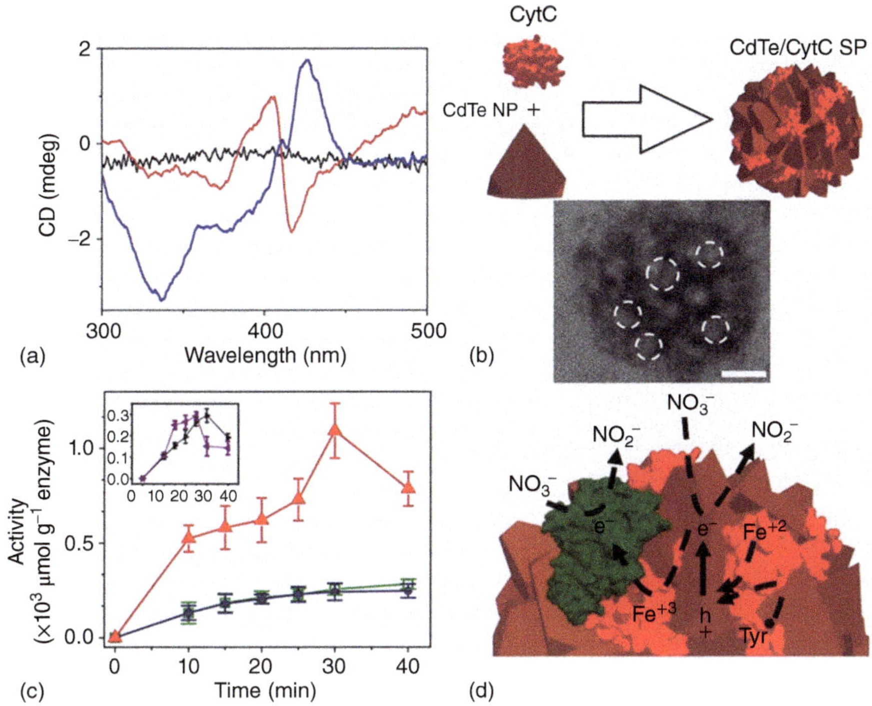


**Figure 6.13** (a) CD spectra for CytC (red, 6 mM), CdTe NPs (black, 6 mM), 1 : 1 mixture of CdTe/CytC after 72 hours (blue). (b) Schematics of the formation of CdTe-CytC SPs. (c) Formation of nitrite for SP-nicotinamide adenine dinucleotide phosphate (NADPH)-nitrate reductase (NRed) (red) excited at 470 nm and for NADPH-NRed (green) and SP-NADPH-NRed in dark (blue). Inset: formation of nitrite for NADPH-NRed being excited at 470 nm in the presence of only one of SP components: either CdTe NPs (black) or CytC (purple) when no hybrid SPs were formed. Errors bars are based on triplicate experimental measurements. (d) Schematic of the reactions on the photoexcitation of SP-NADPH-NRed. Nitrate reductase is represented as a surface model from PyMol software, PDB entry is 2BIH. Source: Park et al. [58] / with permission of Spinger Nature.

## 6.6 Chiral Nanomaterials Based Biocatalysis In Situ for Emerging Biological Applications

Chiral nanomaterials have become preferable bio-nanocatalysts due to their special structure and chemical composition advantages. It is worth noting that the biocatalysts based on chiral nanomaterials, as emerging materials, not only have the general properties of many other nanocatalysts but also have good selectivity and biocompatibility. The biological applications of these emerging materials have received great attention. Current research has proven their potential applications in biological catalysis, such as protein or DNA cleavage, peptide synthesis, tumor therapy, and in many other biomedical fields [59].

Targeted bioorthogonal catalysis has great potential for biological localized drug preparation. Biological active chiral drug in situ synthesis by biorthogonal reactions in living systems is of vital importance in the clinic. A powerful alternative approach for enhancing therapeutic efficacy and alleviating adverse reactions is greatly important to transform a prodrug as bioactive species at desired biological sites. Transfer hydrogenation that can use hydrogen donors such as sodium formate and reduced nicotinamide adenine dinucleotide has attracted great attention on living cells.

Qu's group reported chiral Pd catalysts immobilized on mesoporous silica nanoparticles (MSN-Pd) for the asymmetric transfer hydrogenation (ATH) reaction in situ chiral drug synthesis in inflamed area based on the reductant sodium formate. The TEM and nitrogen adsorption–desorption analysis indicated that the size and pore width of the MSN were c. 98 and 10.6 nm, respectively. The diameter of Pd nanoparticles was 2.9 nm. The chiral enantiomers including cinchonidine (CD), cinchonine (CN), quinidine (QD), and quinine (QN) were used as chiral ligands to introduce MSN-Pd catalyst with enantioselectivity. Mediated by chemotaxis of neutrophil membrane, the inflammation site-selective chiral drug synthesis in living cells was achieved. Based on the enantiomerically modified Pd-catalyzed ATH reaction with the inflammation-targeting neutrophil membrane, the site-selective synthesis of chiral ibuprofen (IBU) in situ was achieved to suppress inflammation in vivo. First, conversion of the achiral compound atropic acid (AA) to chiral hydratropic acid (HAA) was introduced by chiral MSN-Pd nanocatalyst as the model reaction (Figure 6.14a).

The $^{1}H$ NMR results showed that the MSN-Pd/CD accelerated the reduction with the hydrogen source of HCOONa (Figures 6.14b). Integrals of the peaks at $d = 6.23–5.96$ ppm ($CH_2{=}C$) for reagent AA and $d = 1.36$ ppm ($CH_3$) for product HAA showed high substrate conversion. The $^{1}H$ NMR integrals were used for the calculation of initial reaction rate ($r_0$) for the first 20 minutes, and the enantioselectivity was analyzed by HPLC (Figure 6.14c–e). The MSN-Pd/CD and MSN-Pd/CN displayed superior catalytic activity against MSN-Pd/QD and MSN-Pd/QN, while the MSN-Pd/CD- and MSN-Pd/QN-catalyzed hydrogenation provided S-HAA in excess.

It was only a slight decrease in hydrogenation activity after five cycles of MSN-Pd/CD, and there was no obvious crystallographic change after five cycles. Flow-cytometric analysis showed that the purity of the activated neutrophils was higher than 90%. The synthesized chiral drug ibuprofen in situ was able to alleviate inflammation in the mouse paw. The neutrophil membrane-functionalized chiral Pd catalysts displayed inflammation-targeting ability and enantioselectivity in the anti-inflammatory reaction. Based on MSN-Pd/CD@Neu- and MSN-prodrugs@Neu-treated paws, inflammatory cell infiltration obviously decreased and fibroblasts were preserved in the tissue architecture. LPS injection caused acute inflamed paw were dramatically reduced by the treatment of MSN-Pd/CD@Neu and MSN-prodrugs@Neu, ROS and PGE2 levels. The LPS-triggered pyrexia was also substantially relieved by MSN-Pd/CD@Neu and MSN-prodrugs@Neu. The overall results indicated that the availability of in situ chiral IBU synthesis by targeted ATH reaction for anti-inflammation in vivo. The developed strategy provided a novel route for bioorthogonal catalysis in targeted prodrug activation.

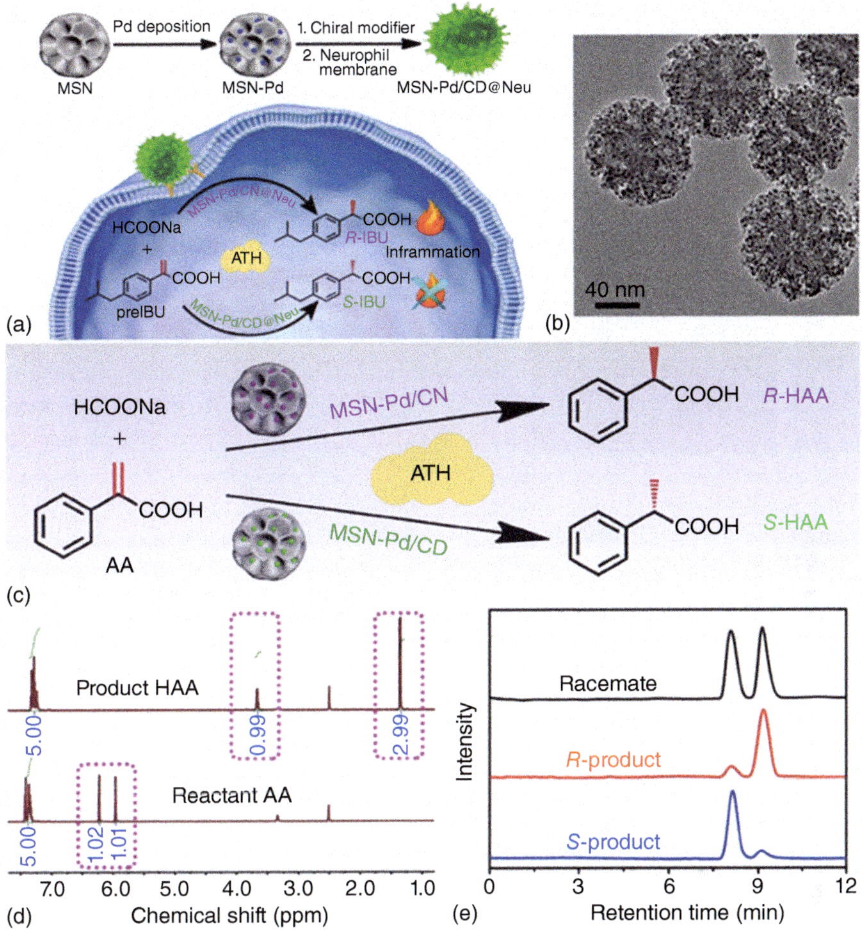


**Figure 6.14** (a) Schematic illustration of the preparation of chiral catalyst MSN-Pd/CD@Neu. Bottom: neutrophil-membrane-directed ATH in living organisms for in situ synthesis of chiral IBU to alleviate inflammation. (b) TEM images of MSN-Pd/CD@Neu. (c–e) ATH reaction promoted by chiral modified MSN-Pd catalysts. (c) Synthesis of chiral HAA with HCOONa as the hydrogen donor. (d) 1HNMR spectrum (600 MHz, DMSO-d6) of reactant AA and product HAA via MSN-Pd/CD catalysis. (e) HPLC chromatograms of methyl ester of standard racemate HAA, the *R*-product, and the *S*-product, of which the last two were catalyzed by MSN-Pd/CN and MSN-Pd/CD, respectively. Source: Du et al. [60] / with permission of Wiley-VCH.

In particular, chiral semiconductors exhibit high performance in active oxygen generation for biocatalysis applications. ROS can regulate the physiological functions of organisms, and it is also the basis for the characteristics of cancer [61, 62]. Therefore, cancer can be treated by boosting ROS production [63, 64]. Penicillamine as a surface ligand was used to prepare chiral semiconductors. A similar method had been used to synthesize chiral semiconductors functionalized by D, L-cysteine (cys), for example chiral $Cu_xCo_yS$ NPs [65], chiral $Fe_xCu_ySe$ NPs [66], and chiral ZnS NPs [52]. Cell senescence is a stable state for cell cycle arrest, and the monitoring

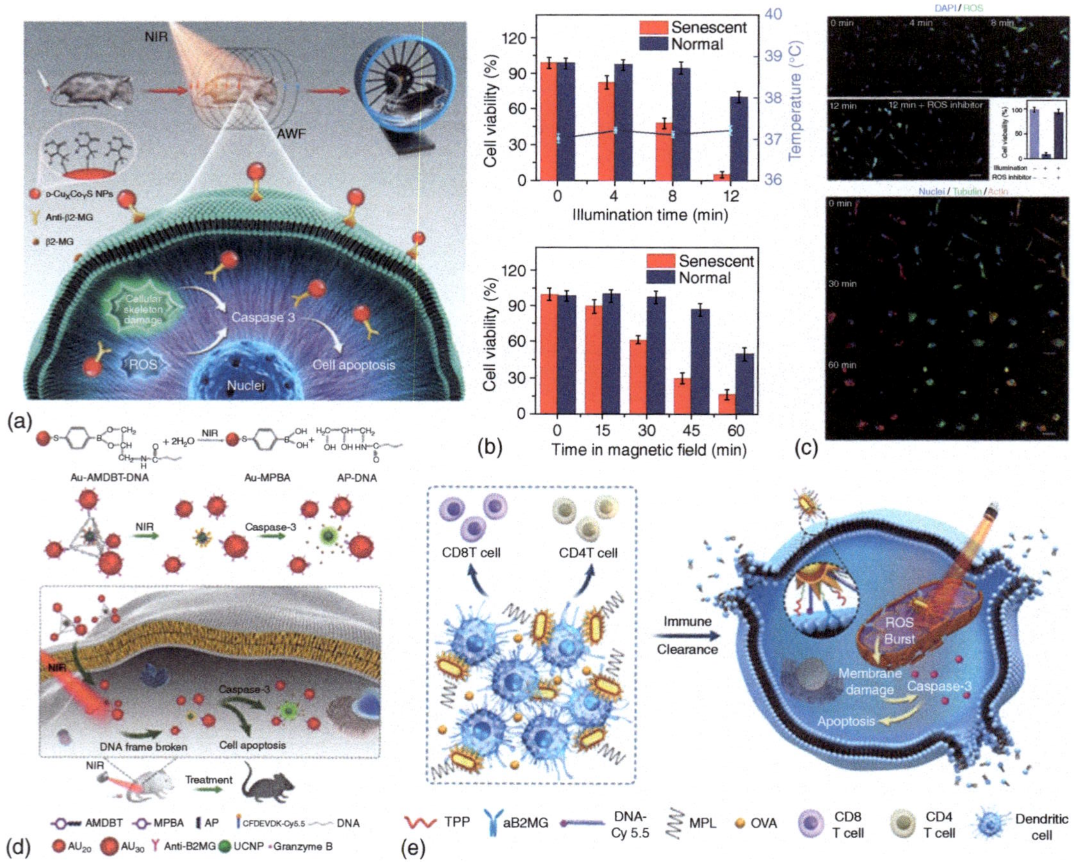


**Figure 6.15** (a) Chiral $Cu_xCo_yS$ NPs produce ROS under the synergistic action of AMF and NIR to eliminate senescent cells. (b) Senescent cell viability under different NIR irradiation time and AMF treatment time. (c) Changes in ROS production and cytoskeleton damage over time. (d) Schematic illustration of UAuTE structures for the recognition and elimination of senescent cells. (e) Au@AgAu Cs NRs modified with β2-MG produce ROS to eliminate senescent cells. Source: (a–c) Reproduced with permission Li et al. [65] / with permission of Wiley-VCH.; (d) Reproduced with permission Qu et al. [67]. Copyright 2020, Wiley-VCH; (e) Reproduced with permission from Lu et al. [68] / Copyright 2020, Wiley-VCH.

of cell senescence can provide some information about the age-related diseases. Chiral metamaterials were developed to selectively activate the apoptotic process in senescent cells under NIR photon illumination. In the process of cell senescence, beta-2-microglobulin (β2-MG) would be expressed on the surface of senescent cells; therefore, β2-MG antibody–modified chiral metamaterials were prepared for specific recognition of senescent cells. For example, chiral $Cu_xCo_yS$ NPs with the sizes of $2 \pm 0.5$ nm were prepared by using D- and L-penicillamine as ligands and exhibited intensive CD responses in the region from 400 to 700 nm. In particular, D-$Cu_xCo_yS$ NPs showed better uptake ability of senescent cells compared to L-$Cu_xCo_yS$ NPs and D/L-$Cu_xCo_yS$ NPs, owing to the higher affinity between D-$Cu_xCo_yS$ NPs and the lipids or proteins on the cell membrane. D-$Cu_xCo_yS$ NPs displayed better biological performances in vitro and in vivo and therefore were functionalized by the antibody of β2-MG for the elimination of senescent cells. β2-MG antibody–modified D-$Cu_xCo_yS$ NPs could recognize β2-MG expressed on the surface of senescent cell membranes. Under excitation of 980 nm laser, ROS were generated owing to the superior photon catalytic ability of D-$Cu_xCo_yS$ NPs and damaged the intracellular microenvironment, resulting in the apoptosis of senescent cells without damaging the activities of normal cells (Figure 6.15a). The photon illumination dominated the production of ROS, and magnetic treatment was mainly responsible for the cellular skeleton damage. The ROS levels in the senescent cells after combination treatment increased much higher than those in cells after individual treatment, and the same pattern was seen for cellular skeleton change. Eight minutes of photon illumination and 30 minutes of magnetic treatment were chosen to study the ability of D-$Cu_xCo_yS$ NPs to regulate senescent cells with a negligible decrease in the viability of normal cells (Figure 6.15b,c). It was concluded that the synergistic effect could reduce the necessary number of internalized NPs for senescent cells than that with individual treatment. D-$Cu_xCo_yS$ NPs could successfully eliminate senescent cells and counteract senescence-induced dysfunction of organs in vivo [65]. Besides, chiral plasmonic NP-UCNP assemblies were fabricated and modified by the antibody of β2-MG for the recognition and elimination of senescent cells. The generation of ROS could induce the clearance of senescent cells, and delivering Granzyme B to senescent cells with high activity was also an effective strategy for the clearance of senescence. Kuang et al. used Au NP-UCNP assemblies to deliver Granzyme B for the clearance of senescent cells. 30 nm Au NPs and 20 nm Au NPs were modified by β2-MG antibody and 4-(4-(aminomethyl)-1,3,2-dioxaborolan-2-yl)benzenethiol (AMDBT), and DNA was coupled at the end of AMDBT and further modified on the surface of Au NPs. Au NPs and UCNPs were assembled into chiral tetrahedron by using DNA skeletons, which exhibited CD signals at 527 nm. Under 980 nm illumination, the boronic ester linkage was broken, resulting in the disassembly of chiral tetrahedron and generating decreased CD signals. At the same time, granzyme B and CFDEVDK-Cy5.5 coupled UCNPs were released, and the recovered 660 nm UCPL peak under 980 nm light would transfer to Cy5.5, inducing the appearance of 736 nm PL signal of Cy5.5 through the FRET process. The changed CD responses and PL signals both indicated the disassembly of chiral tetrahedron. Granzyme B exposed on UCNPs induced the apoptosis of senescent cells via activation of caspases (Figure 6.15d). NIR-light-engineered chiral metamaterials show potential prospects for applications in senescent cell clearance [67]. Au@AgAu CS NRs were synthesized and coated with polydopamine (PDA), and Au spikes were formed through the damage of PDA shell. The average length and diameter of Au@AgAu CS NRs were $118 \pm 3$ nm and $55 \pm 2$ nm, respectively, and the aspect ratio was about 2.1 for Au@AgAu CS NRs. Amino-poly(ethylene glycol)-thiol was used to modify on the surface of Au@AgAu CS NRs through Au—S bonds, and then the antibody of β2-MG and triphenylphosphonium (TPP) were further modified on the surface of Au@AgAu CS NRs through amide bonding. As-modified Au@AgAu CS NRs could specifically combine on the surface of senescent cells depending on the unique binding between the antibody of β2-MG and β2-MG overexpressed on senescent cells, and the presence of TPP enabled Au@AgAu CS NRs for targeting the mitochondria in senescent cells. Upon 808 nm light irradiation for five minutes, ROS were generated owing to the intensive plasmonic coupling between CS NPs and the formed spiked morphology with strong hot spots. The production of ROS for Au@AgAu CS NRs was 2.6 times higher than that of Au NRs, which further caused the mitochondrial damage and apoptosis of senescent cells with relatively low NIR light power (Figure 6.15e) [68].

ROS produced by chiral metamaterials not only can be applied for the elimination of senescent cells but also can be used to prevent β-amyloidopathy in vivo. Chiral semiconductors $Fe_xCu_ySe$ NPs were prepared by using D- and L-penicillamine (Pen) as ligands and exhibited mirror CD responses at three peaks of 435, 515, and 780 nm. In comparison to D-Pen-modified $Fe_xCu_ySe$ NPs, L-Pen-functionalized $Fe_xCu_ySe$ NPs showed little effects on the aggregation of Aβ42 under 808 nm NIR illumination. D-Pen-functionalized $Fe_xCu_ySe$ NPs exhibited superior binding ability toward Aβ42 than L-Pen-modified $Fe_xCu_ySe$ NPs, which was attributed to the selective affinity between D-Pen and pentapeptide from Aβ42. The pentapeptide may be a significant binding site in the interaction of the Aβ fibrils with the D-NPs. At the same time, chiral $Fe_xCu_ySe$ NPs displayed superior affinity to chiral $Cu_{2-x}Se$ NPs as well, and the achiral $Fe_xCu_ySe$ NPs and $Cu_{2-x}Se$ NPs showed weak affinity for the Aβ42 protein. D-Pen-modified $Fe_xCu_ySe$ NPs triggered the disaggregation of Aβ42 fibrils in dense structures to become looser monomers. It was because D-Pen-modified $Fe_xCu_ySe$ NPs generated $^1O_2$ and ·OH under excitation of 808 nm laser, inducing the dissociation of Aβ fibrils (Figure 6.16a). The amount of ROS in the protein solution increased in response to light over time. A gradual accumulation of $^1O_2$ and •OH increased with the time of irradiation (Figure 6.16b). Higher ROS concentrations occurred for chiral $Fe_xCu_ySe$ NPs than for chiral $Cu_{2-x}Se$ NPs under 808 nm irradiation for 10 minutes. Chiral $Fe_xCu_ySe$ NPs produced higher amounts of ROS than chiral $Cu_{2-x}Se$ NPs, decreasing the stability of β-sheet conformation of Aβ42 fibrils and preventing the fibrillation of monomers [66].

The specific affinity of chiral molecules toward dsDNA and cell membrane enabled chiral nanostructures with good selectivity for biocatalysis applications through adjusting the generation or consumption of ROS. For example, chiral molecules were used as carbon source for the preparation of chiral CDs. L- and D-cys were synthesized to 5–10 nm L-CDs and D-CDs. L-CDs and D-CDs showed

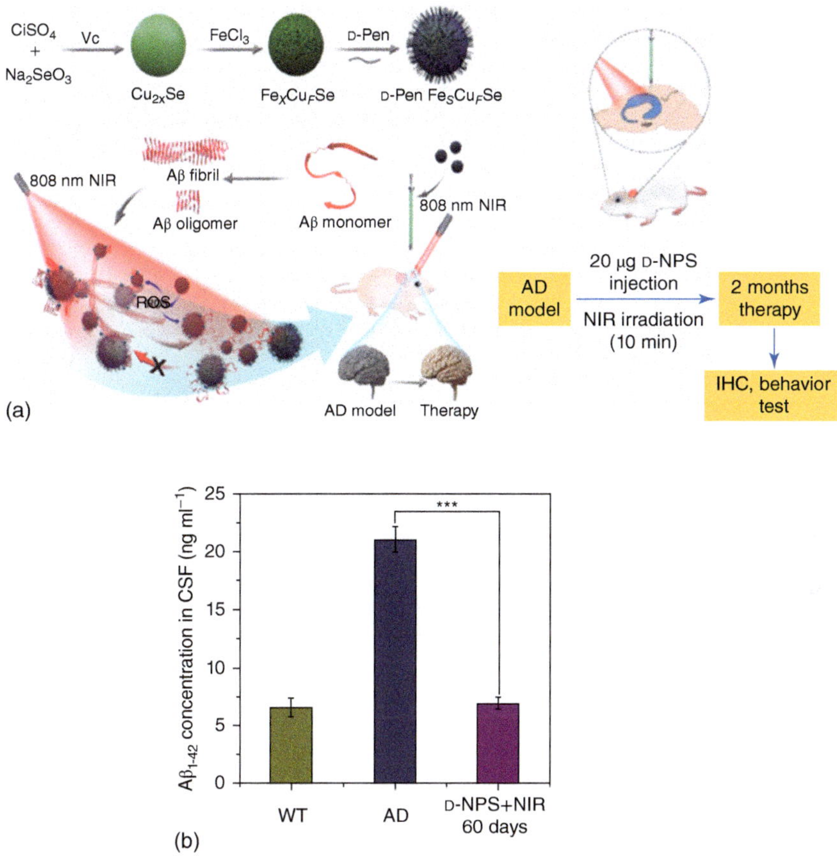


**Figure 6.16** (a) Schematic illustration of D-Pen $Fe_xCu_ySe$ NPs generating ROS under excitation of 808 nm laser to dissociate Aβ fibrils. (b) D-Pen $Fe_xCu_ySe$ NPs for the treatment of AD mice and the therapeutic effects after 60 days. Source: (a, b) Reproduced with permission Zhang et al. [66]. Copyright 2020, Wiley-VCH.

mirror CD peaks. It was explored that D-CDs largely mediated the topological rearrangement, in comparison to L-CDs. Because D-CDs exhibited higher affinity toward dsDNA than L-CDs, D-CDs generated a large number of ·OH to cleave the phosphate backbone from dsDNA. D-CDs demonstrated topoisomerase I mimicking activity (Figure 6.17a) [69].

The chiral ligands modified on NPs and NP assemblies not only provided nanostructures with transferred chiral properties but also provided nanostructures with good configuration identification in biocatalysis due to the unique enantioselective adsorption of chiral molecules. Furthermore, the strong absorption of chiral metamaterials toward light also enabled the production of a large number of $e^-/h^+$ and even more ROS for the enhancement of biocatalysis applications.

Furthermore, D-GSH as the unnatural amino acid, was not used in cellular metabolism, and D-GSH-modified chiral structures would accumulate in the cytoplasm, inducing the oxidative stress in living cells. Chiral metamaterials

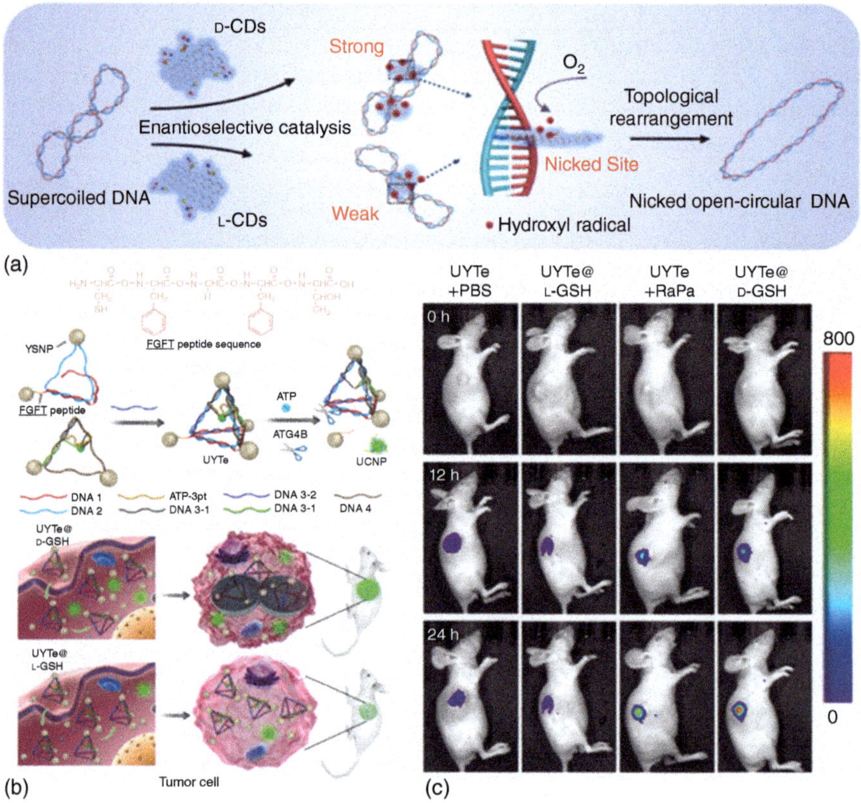


**Figure 6.17** (a) Schematic illustration of D-CDs stimulating enzyme activity to mediate topological rearrangement. (b) Schematic illustration of chiral UYTe structures used to monitor autophagy and ATP concentration. (c) UCL imaging of autophagy induced by chiral UYTe structures in vivo. Source: (a) Reproduced with permission Li et al. [69]. Copyright 2020, Wiley-VCH; (b) Sun et al. [70] / Springer Nature / CC BY-4.0. (c) Sun et al. [70] / with permission of Springer Nature.

were used for the direct observation of autophagy in vivo. ATG4B was regarded as the autophagic biomarker and could hydrolyze FGFE peptide (sequence: cys-Phe-Gly-Phe-Thr). Therefore, FGFE as linkers were modified on the surface of one of YSNPs. YSNPs and UCNPs were assembled into UCNP-centered YSNP tetrahedron (UYTe) structures by using DNA tetrahedron as backbone. The UYTe structures showed strong CD signal at 511 nm and quenched UCPL signals. The presence of ATG4B would hydrolyze FGFE peptide, inducing the disassembly of UYTe structures. One of the YSNPs was released from UYTe structures. The CD signals decreased and the UCPL peak intensity was almost unchanged. The autophagy process also accompanied with the production of ATP in living cells. The presence of ATP further combined with the aptamers on UCNPs, inducing the release of UCNPs from UYTe structures. The CD peak was not changed, whereas the UCPL signals were recovered. These different changes in CD responses and UCPL peak intensity can be used for monitoring autophagy and the ATP concentration (Figure 6.17b). D- and L-GSH were further modified on the surface of UYTe structures. In particular, D-GSH UYTe structures demonstrated better autophagy effects than L-GSH UYTe structures and UYTe structures, and at the same time, ATP was largely produced as the degree of autophagy increased. This was because that D-GSH UYTe structures were accumulated in living cells and further enhanced the oxidative stress as well as accelerated the autophagy process (Figure 6.17c) [70].

The generation of ROS by chiral nanostructures could be used for the treatment of venous thromboembolism (VTE). VTE is a disease with high morbidity and mortality. In the traditional medical treatment, antiplatelet therapy and anticoagulant therapy are generally adopted, and the antithrombotic drugs used may have side effects. To reduce the side effects of drugs and improve the therapeutic effects, Xu et al. developed a D-/L-cysteine-modified chiral $Co_3O_4$ SPs, which could produce ROS under the electromagnetic field (MF), for the treatment of thromboembolism (Figure 6.18a). Chiral $Co_3O_4$ SPs were prepared by using D- and L-cysteine as ligands, and they had an average diameter of about 165 nm. The CD spectra of D-$Co_3O_4$ SPs and L-$Co_3O_4$ SPs showed obvious symmetry, and the maximum g-factor of 0.02 reached at 556 nm (Figure 6.18b). More importantly, D/L-$Co_3O_4$ SPs showed a strong CD signal in the ultraviolet, visible, and near-infrared regions (190–800 nm) (Figure 6.18c), which was originated from $Co_3O_4$ NPs with lattice distortions. D/L-$Co_3O_4$ SPs also demonstrated paramagnetic behaviors. When the platelet membrane (PM) extracted from mouse platelets was coated on the surface of $Co_3O_4$ SPs, not only the diameter of the SPs increased by about 80 nm, but also the CD signal in the ultraviolet region was

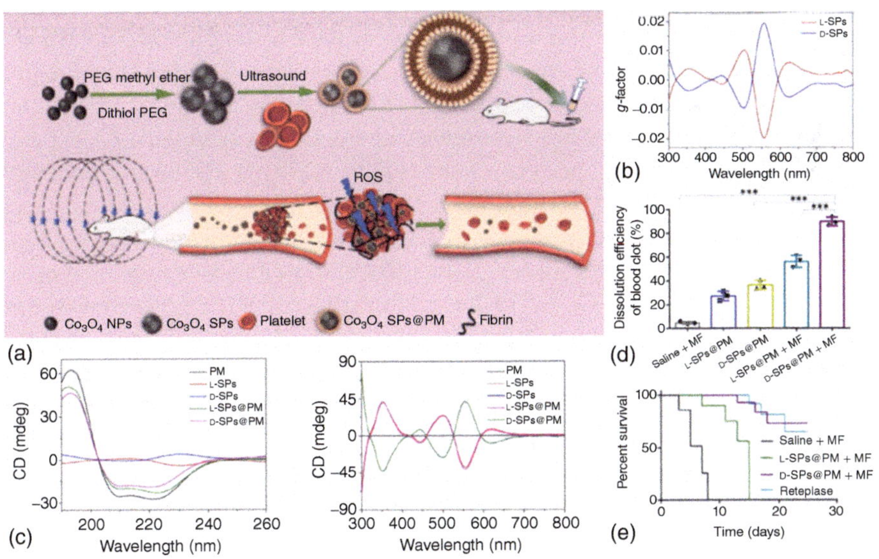


**Figure 6.18** (a) Preparation of $Co_3O_4$ SPs@PM and its application in thrombolytic therapy. (b) *g*-factor of L-/D-SPs. (c) CD spectra of PM, L-/D-SPs, and L-/D-SPs@PM. (d) Thrombolysis efficiency under different treatment conditions. (e) Survival rate of mice with venous thrombosis under different treatment conditions. Source: (a–e) Reproduced with permission Wang et al. [71] Copyright 2021, Wiley-VCH.

enhanced. In the in vitro experimental study, the influence of different MF frequencies and intensities on the fibrinolytic efficiency were compared, and the optimal MF frequencies and intensities were determined to be 100 G and 50 Hz, respectively. Then, the thrombolytic efficiency of D-$Co_3O_4$ SPs and L-$Co_3O_4$ SPs was compared, and it was found that the thrombolytic efficiency of D-$Co_3O_4$ SPs was significantly higher than that of L-$Co_3O_4$ SPs. In addition, D-SPs@PM+MF also showed the best thrombolytic efficiency under the synergistic effect of MF and D-cysteine modification (Figure 6.18d). Similarly, in vivo experiments in mice, using $FeCl_3$ solution to create the thrombus phenomenon, found that D-SPs@PM had a good effect on thrombolytic therapy under the action of MF. This was because ROS was produced when D-/L-SPs interacted with blood clots, and D-SPs produced more ROS than L-SPs. Under the action of MF, the level of ROS produced by chiral SPs was also significantly increased, and ROS promoted thrombolysis. In addition, the chiral-induced spin selectivity (CISS) effect could effectively transfer electrons under MF, which also provided an explanation for the high thrombolytic efficiency of D-SPs. Thromboembolic therapy with D-SPs increased the survival rate to more than 70% in mice (Figure 6.18e) [71].

Chiral nanostructures were also applied for the prevention of hepatitis B virus (HBV) infection due to the generation of ROS. HBV is a hepatotropic DNA virus that causes human hepatitis diseases such as acute and chronic viral hepatitis and has been a severe threat to public health. In particular, HBV core antigen (HBcAg) induces the assembly of HBV and further triggers the onset of a new infection. Therefore, controlling HBcAg is important to prevent the HBV infection. Recently, Xu et al. used chiral $Cu_2S$ NPs to selectively cleave HBcAg and achieved the effective blocking of HBV assembly and the prevention of HBV infection both in vitro and in vivo (Figure 6.19a). Chiral $Cu_2S$ NPs with average sizes of $4 \pm 0.5$ nm were prepared in the presence of L-or D-peptide penicillamine–phenylalanine–tryptophan (Pen-Phe-Trp) ligands. L/D-$Cu_2S$ NPs showed characteristic symmetrical CD peaks with 46 mdeg chiroptical property at 530 nm (Figure 6.19b). Under 808 nm light with an intensity of 200 mW $cm^{-2}$, chiral $Cu_2S$ NPs achieved the cut of HBcAg, which was confirmed by the results of sodium dodecyl sulfate polyacrylamide gel electrophoresis (SDS-PAGE) (Figure 6.19c). In particular, L-$Cu_2S$ NPs possessed better cleavage activity than the D-$Cu_2S$ NPs. It was found that chiral $Cu_2S$ NPs could specifically bound with the functional domain from phenylalanine$_{23}$ ($F_{23}$) to leucine$_{30}$ ($L_{30}$) of HBcAg primary sequence owing to the high specificity of the tripeptide (Pen-Phe-Trp)-tailored chiral $Cu_2S$NPs for HBcAg, and the cutting site was between amino acid residues $F_{24}$ and proline$_{25}$ ($P_{25}$). ROS were produced in the presence of chiral $Cu_2S$ NPs, and L-$Cu_2S$ NPs generated 1.5-fold higher ROS concentration than D-$Cu_2S$ NPs under 808 nm light (Figure 6.19d). The intracellular HBcAg concentration was reduced by 95%, and in HBV transgenic mice, the levels of HBV surface antigen (HBsAg) and HBV DNA were decreased by 93% and 86%, respectively (Figure 6.19e). Chiral nanostructures showed potential prospects for HBV control and viral infection [72].

It has been reported that chiral nanostructures also possess antioxidant behaviors. The excellent affinity between L-GSH and cell membrane induced specific

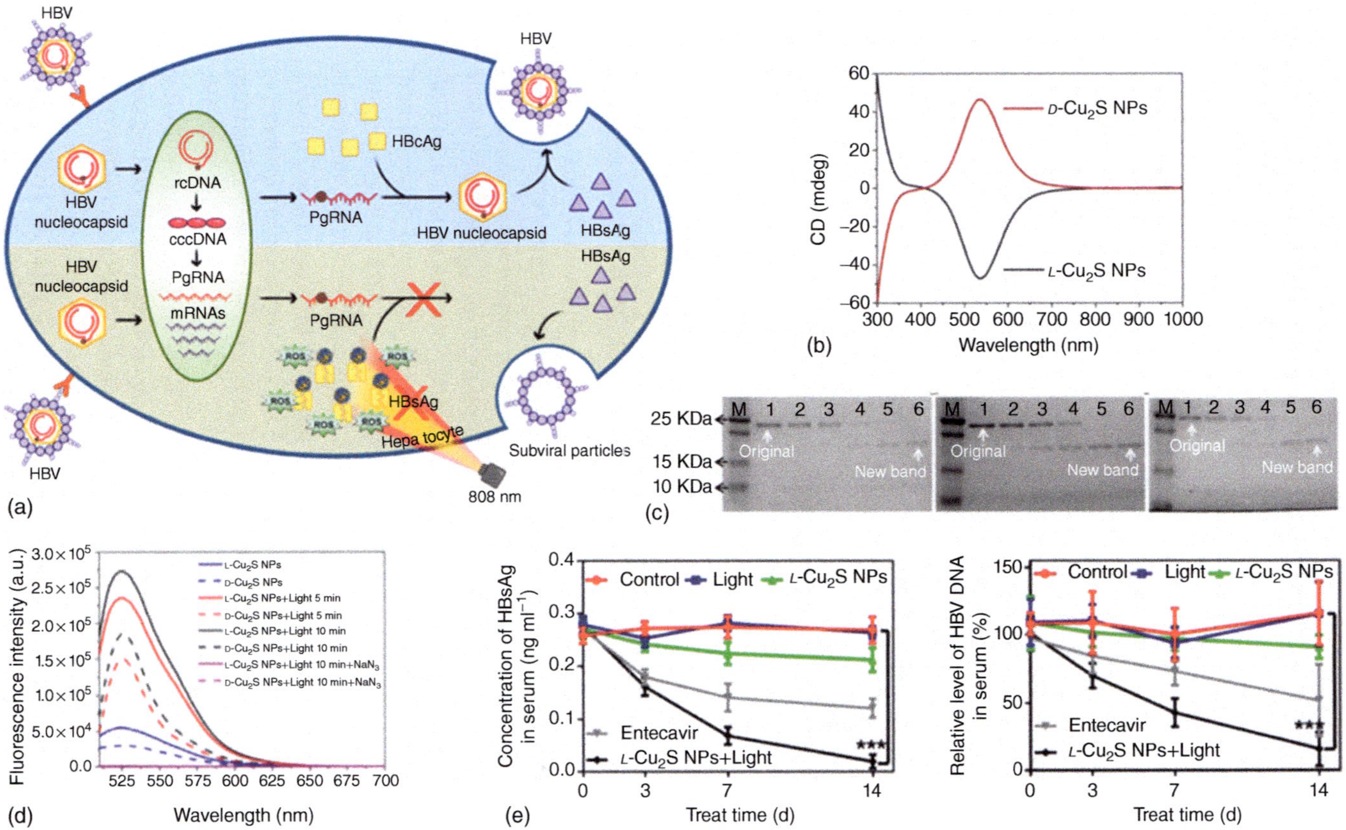


**Figure 6.19** (a) Mechanism of chiral $Cu_2S$ NPs blocking HBV infection under 808 nm laser irradiation. (b) CD spectra of L-/D-$Cu_2S$ NPs. (c) SDS-PAGE gel electrophoresis of HBcAg under different conditions. (d) ROS produced by L-/D-$Cu_2S$ NPs under different conditions. (e) The concentration of HBcAg and the relative level of HBV DNA in the serum of mice treated with different conditions. Source: (a–e) Reproduced with permission Guo et al. [72] / with permission of Wiley-VCH.

recognition of chiral L-GSH nanostructures and cells. Chiral Se NPs were prepared by using D-, L-, and DL-formed GSH as ligands and showed mirrored chiral responses in the regions of 200–400 nm. Due to the higher cell adhesion ability of L-GSH than D-GSH, L-GSH Se NPs were mainly located in the liver and kidneys, whereas D-GSH Se NPs and D/L-GSH Se NPs suffered rapid renal clearances. This enabled L-GSH Se NPs to be an alternative candidate for exploring the antioxidant effects. It was found that L-GSH Se NPs possessed better antioxidant behaviors against PA-caused oxidative damage. The stronger adherence of L-GSH Se NPs on the membrane of cells than D-GSH Se NPs and D/L-GSH Se NPs induced higher cellular uptake efficacy of L-GSH-Se NPs (Figure 6.20a) [73]. Furthermore, chiral molecules with different structures could adjust the morphology of chiral NPs for antioxidation applications. Copper-based nanostructures, such as Cu, $Cu_2O$ NPs, and CuO NPs, are cheap and easy to be synthesized and have been used for biosensing and photocatalysis [76]. Chiral molecules were applied for the fabrication of porous $Cu_xO$ nanoclusters (NCs) with good biocompatibility. These chiral molecule–mediated porous $Cu_xO$ NCs demonstrated excellent antioxidant activities, which mimicked the activity of peroxidase (POD), SOD, catalase (CAT), and glutathione peroxidase (GPx). This would eliminate excess ROS to protect tissues from oxidative stress for applications in the treatment of Parkinson's disease. By using different chiral amino acids (such as Tyr, Asp, Glu, and Phe), the morphology of $Cu_xO$ NCs was changed, which would largely affect enzyme-mimicking behaviors. In comparison to ellipsoid $Cu_xO$ NPs by using Tyr/Asp as ligands and $Cu_xO$ NRs by using Glu as ligands, porous $Cu_xO$ NCs were formed by using Phe as ligands, due to the different affinities between copper ions and different amino acids. As-prepared porous $Cu_xO$ NCs not only were stable and possessed low cytotoxicity but also exhibited multienzyme-like activities, owing to the high surface area, porous configurations as well as the Phe ligands. $Cu_xO$ NCs displayed superoxide dismutase–mimicking activity and served as scavengers of $O_2 \bullet^-$; they showed catalase-like behavior for the decomposition of $H_2O_2$ to $H_2O$ and $O_2$, as well as demonstrated glutathione peroxidase–like activity for the decomposition of $H_2O_2$ and the oxidation of GSH to GSSG. Three different cell models were established to verify the universality of $Cu_xO$ NCs as ROS scavengers. After treatment with $Cu_xO$ NCs, the cell viability was significantly improved and ROS-mediated cell apoptosis also decreased. Therefore, $Cu_xO$ NCs can be used as ROS scavengers in multiple cell models. $Cu_xO$ NCs could eliminate ROS and possessed antioxidation behavior for treating Parkinson's disease (Figure 6.20b) [74]. It was also reported that PtFe@$Fe_3O_4$ NPs possessed dual enzyme–like catalytic activities for ROS generation in tumor catalytic therapy under NIR illumination. PtFe@$Fe_3O_4$ NPs displayed POD-like activity for the catalytic transformation of $H_2O_2$ to superoxide anions and hydroxyl radicals and exhibited CAT-like activity for the catalytic transformation of $H_2O_2$ to $O_2$ (Figure 6.20c). A large amount of ROS induced 80.2% cell death under 808 nm laser, due to the increased lipid peroxidation of membranes, and reduced the activity of proteins as well as damaged DNA and chromosomes (Figure 6.20d) [75]. Au nanoclusters also possessed catalase-like activity, which could catalytic $H_2O_2$ to $O_2$. The self-supply of $O_2$ achieved anticancer therapy. Natural enzymes

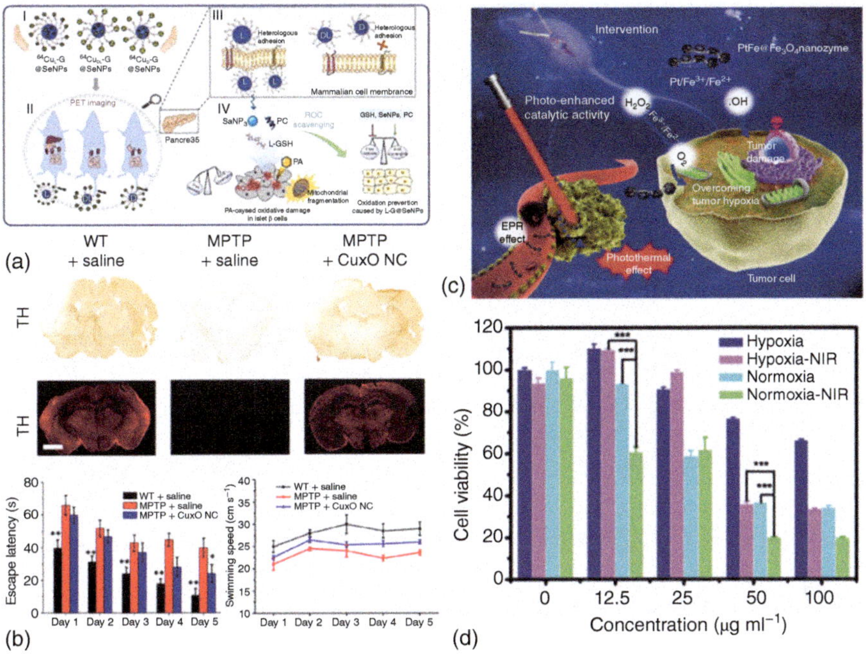

Figure 6.20 (a) Schematic illustration of antioxidant activity of chiral G@Se NPs. (b) TH expression in the brain of PD mice under different conditions and effect of $Cu_xO$ NCs on PD mice. (c) Schematic illustration of tumor catalytic therapy of PtFe@$Fe_3O_4$ NPs under near-infrared irradiation. (d) Cell viability of cells incubated with different concentrations of PtFe@$Fe_3O_4$ NPs in the presence or absence of oxygen and NIR. Source: (a) Reproduced with permission Huang et al. [73] Copyright 2020, Wiley-VCH; (b) Huo et al. [74] / with permission of American Chemical Society; (c, d) Reproduced with permission Li et al. [75] Copyright 2019, Wiley-VCH.

suffer from sensitive environmental conditions, especially for the acidic tumor microenvironment, and may become inactive under pathological conditions. Artificial chiral nanoenzymes featured with multifunctional enzyme-like activities show potential applications in various biomedical processes. In addition, it was reported that a mesoporous copper aluminum mixed metal oxide (MMO) [77], a defect-rich gadolinium (Gd)-doped zinc oxide (D-$ZnO_x$:Gd) semiconductor [78], and Au-$Bi_2S_3$ nanoparticles [79] produce ROS.

Advanced chiral nanoparticle construction techniques have enabled fabrication of complex catalytic systems, which provide great potential to induce robust asymmetric catalysts [9]. Owing to diverse choice of chiral modifiers, homogeneous metal complex catalysis and bio-related reactions catalyzed by chiral metal nanoparticles were dramatically explored [80–82]. Several reports proved that not only small molecules but also chiral polymers would be good candidates as chiral selector for nanocatalysts [83, 84]. Some catalytic active examples indicated good enantioselectivities with broad substrate generality and recyclability, which should be useful for heterogeneous catalysts [85–87]. Moreover, chiral biomolecule-modified nanoparticles usually display unique catalytic behaviors that are different from the corresponding homogeneous complex systems or original chiral ligands [88–91]. It was notable that active species in nanoparticle catalysts were not leached from molecular complexes although true active species were unclear in most cases [92–97]. It has been an emerging area of research, and we expect that further research will achieve more efficient and unique chiral nanoparticles for asymmetric biocatalysis.

## References


1 Zaramella, D., Scrimin, P., and Prins, L.J. (2012). *Journal of the American Chemical Society* 134: 8396.
2 Szollosi, G., Gombkoto, D., Mogyoros, A.Z., and Fulop, F. (2018). *Advanced Synthesis & Catalysis* 360: 1992.
3 Masuda, R., Yasukawa, T., Yamashita, Y., and Kobayashi, S. (2021). *Angewandte Chemie International Edition* 60: 12786.
4 Ranganath, K.V.S., Kloesges, J., Schafer, A.H., and Glorius, F. (2010). *Angewandte Chemie International Edition* 49: 7786.
5 Wang, J.Z., Zhao, L.W., Shi, H.M., and He, J. (2011). *Angewandte Chemie International Edition* 50: 9171.
6 Yasukawa, T., Saito, Y., Miyamura, H., and Kobayashi, S. (2016). *Angewandte Chemie International Edition* 55: 8058.
7 Liu, R., Jin, R., An, J. et al. (2014). *Chemistry – An Asian Journal* 9: 1388.
8 Yang, Q., Han, D., Yang, H., and Li, C. (2008). *Chemistry – An Asian Journal* 3: 1214.
9 Yasukawa, T. and Kobayashi, S. (2020). *Nanoparticles in Catalysis* (ed. S. Kobayashi), 279. Cham: Springer International Publishing.
10 Chen, X., Xu, L., Wang, A. et al. (2019). *Journal of Chemical Technology & Biotechnology* 94: 236.
11 Zhang, R.F., Zhou, Y.L., Yan, X.Y., and Fan, K.L. (2019). *Microchimica Acta* 186: 11.
12 Lin, Y., Huang, Y., Ren, J., and Qu, X. (2014). *NPG Asia Materials* 6: e114.
13 Meng, J.J., Chang, F.W., Su, Y.C. et al. (2019). *ACS Catalysis* 9: 8693.
14 Yasukawa, T., Miyamura, H., and Kobayashi, S. (2016). *ACS Catalysis* 6: 7979.
15 Bhowmick, S., Zhang, L., Ouyang, G.H., and Liu, M.H. (2018). *ACS Omega* 3: 8329.
16 Tang, Z.Y., Wang, W.Y., Pi, Y.B. et al. (2019). *ACS Sustainable Chemistry & Engineering* 7: 17967.
17 Choudary, B.M., Jyothi, K., Roy, M. et al. (2004). *Advanced Synthesis & Catalysis* 346: 1471.
18 Fan, Q.-H. and Ding, K. (2011). *Asymmetric Catalysis from a Chinese Perspective* (ed. S. Ma), 207. Berlin, Heidelberg: Springer Berlin Heidelberg.
19 Ding, C., Wei, W.L., Sun, H.J. et al. (2014). *Carbon* 79: 615.
20 Farokhi, A., Berijani, K., and Hosseini-Monfared, H. (2018). *Catalysis Letters* 148: 2608.

**21** Fedorova, O.V., Titova, Y.A., Vigorov, A.Y. et al. (2016). *Catalysis Letters* 146: 493.
**22** Berijani, K., Morsali, A., and Hupp, J.T. (2019). *Catalysis Science & Technology* 9: 3388.
**23** Candu, N., Rizescu, C., Podolean, I. et al. (2015). *Catalysis Science & Technology* 5: 729.
**24** Jing, L.Y., Zhang, X.M., Guan, R.Q., and Yang, H.Q. (2018). *Catalysis Science & Technology* 8: 2304.
**25** Xia, Y., Zhou, Y., and Tang, Z. (2011). *Nanoscale* 3: 1374.
**26** Wang, F., Zhang, Y., Du, Z. et al. (2018). *Nature Communications* 9: 1209.
**27** Savitha, G., Saha, R., and Sekar, G. (2016). *Tetrahedron Letters* 57: 5168.
**28** Zheng, J., Wu, Y., Deng, K. et al. (2016). *ACS Nano* 10: 8564.
**29** Zhang, F., Sun, Y., Tian, D. et al. (2016). *Chemical Communications (Camb)* 52: 12685.
**30** Yuan, C., Wu, X., Gao, R. et al. (2019). *Journal of the American Chemical Society* 141: 20187.
**31** Gross, E., Liu, J.H., Alayoglu, S. et al. (2013). *Journal of the American Chemical Society* 135: 3881.
**32** Voitko, K.V. (2019). *Journal of Nanoscience and Nanotechnology* 19: 5074.
**33** Wang, C., Zheng, M., and Lin, W. (2011). *Journal of Physical Chemistry Letters* 2: 1701.
**34** Pieters, G. and Prins, L.J. (2012). *New Journal of Chemistry* 36: 1931–1936.
**35** Dubbeldam, D., Calero, S., and Vlugt, T.J.H. (2014). *Molecular Simulation* 40: 585.
**36** Li, X., Wu, J., He, C. et al. (2019). *Small* 15: 1804770.
**37** Qiu, X., Zhang, Y., Zhu, Y. et al. (2021). *Advanced Materials* 33: 2001731.
**38** Wu, C.-D. and Zhao, M. (2017). *Advanced Materials* 29: 1605446.
**39** Sun, Y., Zhao, C., Gao, N. et al. (2017). *Chemistry* 23: 18146.
**40** Zhou, Y., Sun, H., Xu, H. et al. (2018). *Angewandte Chemie International Edition* 57: 16791.
**41** Zhang, Q.-Y. and Zhao, L. (2018). *Tetrahedron Letters* 59: 310.
**42** Zhou, Y., Wei, Y., Ren, J., and Qu, X. (2020). *Materials Horizons* 7: 3291.
**43** Zhou, Y., Wei, W., Cui, F. et al. (2020). *Chemical Science* 11: 11344.
**44** Razavi, S.A.A., Berijani, K., and Morsali, A. (2020). *Catalysis Science & Technology* 10: 8240.
**45** Toth, P.S., Velický, M., Bissett, M.A. et al. (2016). *Advanced Materials* 28: 8256.
**46** Zhang, H., He, H., Jiang, X.M. et al. (2018). *ACS Applied Materials & Interfaces* 10: 30680.
**47** Sadiq, M., Aman, R., Saeed, K. et al. (2015). *Modern Research in Catalysis* 04: 43.
**48** Szőri, K., Réti, B., Szöllősi, G. et al. (2016). *Topics in Catalysis* 59: 1227.
**49** Yue, Y.N., Zeng, S., Wang, H. et al. (2018). *ACS Applied Materials & Interfaces* 10: 23055.
**50** Lin, Y., Ren, J., and Qu, X. (2014). *Accounts of Chemical Research* 47: 1097.
**51** Jiang, S., Chekini, M., Qu, Z.B. et al. (2017). *Journal of the American Chemical Society* 139: 13701.
**52** Li, S., Liu, J., Ramesar, N.S. et al. (2019). *Nature Communications* 10: 4826.

**53** Ma, W., Mao, J., Hao, C. et al. (2019). *Applied Catalysis B: Environmental* 245: 691.
**54** Sun, M., Xu, L., Qu, A. et al. (2018). *Nature Chemistry* 10: 821.
**55** Wang, W., Hao, C., Sun, M. et al. (2018). *Advanced Functional Materials* 28.
**56** Hao, C., Gao, R., Li, Y. et al. (2019). *Angewandte Chemie International Edition* 58: 7371.
**57** Zhang, R., Jiang, J., Zhou, J. et al. (2018). *Advanced Materials* 30: 1705443.
**58** Park, J.I., Nguyen, T.D., de Queirós Silveira, G. et al. (2014). *Nature Communications* 5: 3593.
**59** Hao, C., Xu, L., Kuang, H., and Xu, C. (2020). *Advanced Materials* 32: e1802075.
**60** Du, Z., Liu, C., Song, H. et al. (2020). *Chem* 6: 2060.
**61** Yang, B., Chen, Y., and Shi, J. (2019). *Chemical Reviews* 119: 4881.
**62** Kwon, S., Ko, H., You, D.G. et al. (2019). *Accounts of Chemical Research* 52: 1771.
**63** Yang, Z., Dai, Y., Yin, C. et al. (2018). *Advanced Materials* 30: e1707509.
**64** Sies, H. and Jones, D.P. (2020). *Nature Reviews. Molecular Cell Biology* 21: 363.
**65** Li, S., Sun, M., Hao, C. et al. (2020). *Angewandte Chemie International Edition* 59: 13915.
**66** Zhang, H., Hao, C., Qu, A. et al. (2020). *Angewandte Chemie International Edition* 59: 7131.
**67** Qu, A., Wu, X., Li, S. et al. (2020). *Advanced Materials* 32: e2000184.
**68** Lu, M., Qu, A., Li, S. et al. (2020). *Angewandte Chemie International Edition* 59: 8698.
**69** Li, F., Li, S., Guo, X. et al. (2020). *Angewandte Chemie International Edition* 59: 11087.
**70** Sun, M., Hao, T., Li, X. et al. (2018). *Nature Communications* 9: 4494.
**71** Wang, X., Sun, M., Qu, A. et al. (2021). *Angewandte Chemie International Edition* 60: 18240–18246.
**72** Guo, X., Sun, M., Gao, R. et al. (2021). *Angewandte Chemie International Edition* 60: 13073.
**73** Huang, Y., Fu, Y., Li, M. et al. (2020). *Angewandte Chemie International Edition* 59: 4406.
**74** Hao, C., Qu, A., Xu, L. et al. (2019). *Journal of the American Chemical Society* 141: 1091.
**75** Li, S., Shang, L., Xu, B. et al. (2019). *Angewandte Chemie International Edition* 58: 12624.
**76** Xin, Y., Yu, K., Zhang, L. et al. (2021). *Advanced Materials* n/a: 2008145.
**77** Dissanayake, D., Achola, L.A., Kerns, P. et al. (2019). *Applied Catalysis B: Environmental* 249: 32.
**78** Liu, Y., Wang, Y., Zhen, W. et al. (2020). *Biomaterials* 251: 120075.
**79** Wang, X., Zhang, C., Du, J. et al. (2019). *ACS Nano* 13: 5947.
**80** Zhang, H.Y., Yang, W.T., and Deng, J.P. (2015). *Polymer* 80: 115.
**81** Song, J.X., Zhang, H.Y., and Deng, J.P. (2015). *Reactive & Functional Polymers* 93: 10.

**82** Ranaivoarimanana, N.J., Habaki, X., Uto, T. et al. (2020). *RSC Advances* 10: 37064.

**83** Moutet, J.-C., Duboc-Toia, C., Ménage, S., and Tingry, S. (1998). *Advanced Materials* 10: 665.

**84** Wei, H. and Wang, E. (2013). *Chem Soc Rev* 42: 6060.

**85** Barath, E. (2020). *Synthesis-Stuttgart* 52: 504.

**86** Deiana, L., Ghisu, L., Cordova, O. et al. (2014). *Synthesis-Stuttgart* 46: 1303.

**87** Zhang, D.Y., Zhang, H.Y., Song, C. et al. (2012). *Synthetic Metals* 162: 1858.

**88** Sharifzadeh, Z., Berijani, K., and Morsali, A. (2021). *Ultrasonics Sonochemistry* 73: 105449.

**89** Butcha, S., Assavapanumat, S., Ittisanronnachai, S. et al. (2021). *Nature Communications* 12: 1314.

**90** Han, B.W., Li, Y.J., Yu, Y., and Gong, L. (2019). *Nature Communications* 10: 3804.

**91** Vieille-Petit, L., Suss-Fink, G., Therrien, B. et al. (2005). *Organometallics* 24: 6104.

**92** Hu, A.G., Yee, G.T., and Lin, W.B. (2005). *Journal of the American Chemical Society* 127: 12486.

**93** Li, X. and Zhao, Y. (2019). *Journal of the American Chemical Society* 141: 13749.

**94** Samal, M., Chercheja, S., Rybacek, J. et al. (2015). *Journal of the American Chemical Society* 137: 8469.

**95** Sarkar, S.M., Yusoff, M.M., and Rahman, M.L. (2015). *Journal of the Chinese Chemical Society* 62: 177.

**96** Zhang, D.Y., Ren, C.L., Yang, W.T., and Deng, J.P. (2012). *Macromolecular Rapid Communications* 33: 652.

**97** Cheng, T.Y., Long, J., Liang, X.H. et al. (2014). *Materials Research Bulletin* 53: 1.

# Index


*a*

achiral thiol-coated CdTe 7
aflatoxin B1 (AFB1) 180
aggregation-induced emission (AIE) 216
Ag shell of Au@Ag NPs 45
aldol reaction 254, 255, 256
allosteric effect 117, 120
alpha-fetoprotein (AFP) 156, 159
amino acids 11, 28, 30, 31, 35, 38, 40, 41, 59, 62, 129, 139, 150, 155, 184, 186, 187, 189, 190, 237, 241–244, 246–248, 250, 254–257, 259–262, 275, 278, 280
4-(4-(aminomethyl)-1,3,2-dioxaborolan-2-yl)benzenethiol (AMDBT) 273
anti-TROP monoclonal antibodies (mAb) 160
asymmetric transfer hydrogenation (ATH) reaction 270
atomic packing 4, 6, 257
Au@Ag@Au NP-UCNP heterodimers 181, 182
AuCuAu NRs 203
Au/(DNA-Ag) core-shell NPs 44–46
Au@gap@AgAu NPs 178
Au-gap-Ag NPs 233, 236
Au NP-Ag NP dimers 157, 158
Au NP helices 103, 104
Au NP heterotrimers 160
Au NPs metallic core 42
Au NR dimer@UCNP structures 225, 226
Au NR helices 103, 104
Au NR SBS dimers 82, 83, 85
$Au_{38}(SCH_2CH_2Ph)_{24}$ NP 5
Au-thiolate ligand 5
azobenzene-hosted nanostructures 82

*b*

base complementary pairing 153
beta-2-microglobulin (beta 2 MG) 273
bimetallic NPs 178, 189
binaphthyl compound (BINAP) 10, 60, 61, 81
biocatalysis
  biological applications 269–282
  biological macromolecule 261–265
  drugs 249–257
  encoded chiral complex nanomaterials system 264, 266–269
  peptide 257–261
biological macromolecule 11, 70, 173, 261–265
biological transparency windows 80
biomarkers bioanalysis 156–164
biomolecules 11, 12, 28, 38, 40, 89, 149, 150, 162, 184, 193, 244, 261, 281
bisphenol A (BPA) 175–177
Brunauer-Emmett-Teller (BET) surface 116

*c*

cadmium telluride (CdTe) NPs 7, 11, 29, 33, 135, 139, 262, 263, 264, 266, 268, 269
CdSe nanorods (NRs) 32, 35
CdSe/ZnS nanocrystals 17
CdS QDs 181–183
cell cycle progresses 169
cell death (Ferroptosis) 200, 209
cell models 280

Ce NPs 242, 243
Ce NPs oxidase-like activity 242
ceramic nanohelices 69–70
ceramic NPs 257
chemodynamic therapy 206
chiral Au NP heterodimers 152, 153, 156, 157, 159, 175, 176
chiral carbon dots (CDs) 29, 182, 183, 190, 274
chirality
  characterization methods
    computational studies 14–17
    computer-modeling study 17–20
    optical spectra 12–14
  definition 1
  materials
    biological nano- and mesoscale compounds 11
    hierarchical chirality 11–12
    inorganic and organic molecular 10–11
  molecules 2
  nanoparticles 2
  origin of nanomaterials
    geometries 2–6
    structural origins 6–10
chiral mesoporous silica layer 41
chiral nanoassemblies
  core-satellite nanoassembly 113–129
  helices 100–111
  nanochains 111–113
  nanoparticle dimers 85–90
  nanoplatelets and chiral films 129–134
  supraparticles (SPs) 134–139
  tetramer assemblies 90–100
  twisted nanorod pairs 80–85
chiral nanocrystals
  chiral metal nanostructure
    metal nanohelices 50–54
    metal nanoparticles/nanorods 41–50
    metal nanorings and superstructures 54–59
  metallic compound nanoparticles
    ceramic nanohelices 69–70
    oxide nanoparticles 59–64
    sulfide/selenide nanoparticles 64–68
  semiconductor helicoids 38–39
  semiconductor nanoparticles 29–38
  synthetic methods 28
chiral nanomaterials
  biocatalysis 241–249
  multimodal therapy methods 220–234
  photochemical effects 233, 235–237
  photodynamic therapy (PDT) 209–220
  photothermal therapy 199–209
chiral nanomaterials based enantioselectivity 241–249
chiral nanostructures
  analysis strategy 149–150
  biomarkers bioanalysis 156–164
  DNA detection 150–156
  enzymes bioanalysis 164–170
  metal ions bioanalysis 170–175
  recognition 184–193
  small molecules detection 175–184
chiral plasmonic NPs 40, 233, 236, 237, 273
chiral recognition 27, 184–193, 244, 246
chlorin e6 (Ce6) 127, 128, 209, 210, 211, 225, 226
circular dichroism (CD) 6, 8, 9, 12–19, 27, 28, 30–38, 41–53, 55, 57, 60–62, 65, 66, 68, 80–104, 106–114, 116–119, 121, 122, 124–127, 129, 131, 133, 135, 137, 138, 150, 152, 153, 155, 156, 158, 160, 161–165, 167, 169, 170, 171, 173, 175, 177, 178, 180–182, 184, 186, 187, 189, 190–192, 203, 221, 233, 235, 236, 252, 253, 257, 259, 260, 262, 263, 264, 265, 268–271, 273–279
circularly polarized light (CPL) 13, 49, 107, 108, 109, 110, 111, 117, 127, 174, 175, 201, 203, 221, 235, 236, 237, 262
circularly polarized luminescence (CPLm) 27, 108, 109
circular polarized light (CPL) 13, 49, 107, 108, 109, 110, 111, 117, 127, 174, 175, 201, 203, 221, 235, 236, 237, 262
circulating tumor cells (CTCs) 156
classical hot-injection technique 35
cogel formation 108

complementary ssDNAs 126
copolymer polystyrene-block-poly (acrylic acid) (PS-b-PAA) 127
core-satellite nanoassembly 80, 113–129
core-satellite nanostructures 119, 122, 127, 164
Cotton Effect 43, 50, 53, 60, 65, 101, 131, 189, 252
Coulomb dipole-dipole interactions 30
covalent organic frameworks (COFs) material 246
crystalline porous materials 246
$Cu_{2-x}Se$-Au coupling 207
$Cu_{2-x}Se$-Au Janus NPs 207, 208
Cys-decorated CdTe NPs 33
cystine enantiomers 187, 188

## d

D- and L-penicillamine (Pen) 273, 274
D-$Cu_xCo_yS$ NPs 273
Density Functional Theory (DFT) 6, 15–17, 41
deoxyribonuclease I (DNase I) 42, 165, 168
deoxyribonucleotide sequences 164
D-gelators 108
D-GSH 32, 57, 96, 275, 277, 280
diastereomers 2, 28
diffuse-reflectance circular dichroism (DRCD) 53, 54
3,4-dihydroxyphenylalanine (DOPA) 242–247, 260
1,3-diphenylisobenzofuran (DPBF) 229
DL-dithiothreitol (DTT) 82
D/L-12-hydroxystearic acid (HSA) 52, 264
D/L-penicillamine (D/L-Pen) 4, 41, 122, 123, 125, 189, 191
D/L-penicillamine ligands 4
DNA-bound Au cores (Au NPs-DNA) 44
DNA detection 100, 150–156
DNA driven chiral nanomaterials 165
DNA endonuclease 164
DNA functionalization 79
DNA methyltransferase 164–168
DNA origami 81, 82, 101–103, 107
DNA-origami-organized NPs 101
DNA origami templates 81, 82, 103
DNAzyme 127, 173, 175
double hydrophilic block copolymers (DHBC) 69
double-layer DNA origami template 82
D-penicillamine 31, 131, 259
D-tartaric acid functionalized Au NPs (D-Au NPs) 187
D-tyrosinol 248
dual-mode optical biosensor 170, 181
dynamic light scattering (DLS) 260
dynamic NP dimer 88

## e

effective core potential (ECP) 15–17, 19
electrostatic repulsion 80, 89, 101, 129, 131, 266, 268
enantiomeric ligand functionalized silver NPs 41
enantiomers 1, 2, 5–7, 9, 12, 13, 28, 41, 47, 53, 56–58, 61, 80, 93, 100, 101, 113, 137, 184, 186–191, 242–248, 259–261, 270
enzyme bioanalysis 164–170
epidermal growth factor receptor 2 (HER2) 156, 162, 163
epithelial cell-adhesion molecule (EpCAM) 162, 163
expanded mesoporous silica (EMSN) 162, 244, 245

## f

Fenton reaction 213
ferroptosis 208, 209
Fourier transform (FT) IR spectroscopy 242, 244, 250, 254, 257, 258, 266

## g

gelators N-(9-fluorenylmethoxycarbonyl)-protected glutamic acid (Fmoc-Glu) 108
generalized gradient approximation (GGA) 19
gold nanorods (Au NRs) 11, 41, 50, 51, 80–83, 85, 89, 100, 103, 104, 107, 111–114, 119, 120–123, 126, 127, 129, 135, 150–152, 154, 155, 159, 160, 165, 167, 170–172, 174, 175, 186, 187, 200–204, 218, 220, 222, 223, 225, 226, 229, 232, 274
gold (III) tetrachloride acid 40

gold-upconversion nanoparticle (Au-UCNP) pyramids 92, 94
gold yolk-shell nanoparticles 89
Gram-positive bacteria (S. aureus) 249
graphene quantum dots (GQDs) 12, 29

### h

Hausdorff chirality measure (HCM) 7, 19
HBV core antigen (HBcAg) 278
Hepatitis B virus (HBV) infection 278, 279
heterochains 111, 112, 152
high-resolution transmission electron microscopy (HRTEM) 33, 34, 36, 37, 46, 51, 57, 59, 63, 66, 91, 116, 126, 252, 253
hot-spot CD amplification 86
hydrogels 107–109, 111
8-hydroxy-2′-deoxyguanosine (8-OHdG) 156–158
hypoxia environment 212

### i

inner filter effect (IFE) 181
intensive plasmonic coupling 189, 201, 225
intensive plasmonic properties of NPs 201
iron dependent lipid peroxidation 209
iron 5,10,15,20-tetrakis(40-tetraphenylamino) porphyrin unit (Fe-ATPP) 246
*N*-isobutyryl-cysteine 4

### k

Kohn-Sham orbitals 16, 17

### l

L-aspartic acid (L-Asp) 62, 190, 258
L-cysteine methylester 4
left circular polarized (LCP) light illumination 129
Lewis acid-surfactant combined catalysts (LASC)-SWNT 255
Lewis basicity 250
L-gelators 108
L-GS-bimane 42, 43
linear amino acids 254, 255
local density approximation (LDA) 19
localized surface plasmon resonance (LSPR) effect 65, 121, 199, 206, 207, 232
L-penicillamine 4, 31, 40, 41, 122, 131, 189, 263, 273, 274
LPF@5 hybrid complex 250
L-proline@$Fe_3O_4$ nanoparticle 250, 251
L-tartaric acid modified Au NPs (L-Au NPs) 187
L-tyrosinol 248, 249, 252
luminescence resonance energy transfer (LRET) 92, 164, 170

### m

macromolecules 11, 70, 173, 175, 261–264
magnetic nanoparticles 249, 250
2-mercaptobenzoimidazole-5-carboxylic acid (MBIA) 180
16-mercaptohexadecanoic acid (16-MHDA) 160
mesoporous silica nanoparticles (MSN) 270, 271
metal-semiconductor hybrid nanoassemblies 89
metal ions bioanalysis 170
metal organic frameworks (MOFs) 10, 115, 201, 216, 246
metal-semiconductor composite nanostructures 232
metal-to-ligand charge transfer (MLCT) transitions 12
Michaelis-Menten reaction 252
microcystins 175, 177
microRNA (miRNA) 82, 85, 92, 94, 118, 119, 150, 156, 162, 163, 164, 170
$MnO_2$@Au NPs structures 207, 208
$MoO_x$ NPs 225, 229, 230
multimodal therapy methods 220–233
mycotoxins 114, 178, 180, 181

### n

nanochains 80, 111–114

nanoclusters 6, 7, 8, 29, 40, 42, 44, 122, 280
nanographene oxides (NGO)-capsulate AIE photosensitizers 216
nanohelical chirality 111
nanozymes 33, 209, 246, 247, 253, 261, 262
natural enzymes 241, 244, 246, 247, 252, 261, 280
near-infrared (NIR) laser induced phototherapy 225
neural stem cells (NSCs) 117, 118, 236
NGP-TPEred 216, 217, 218
nicotinamide adenine dinucleotide phosphate (NADPH) 268–270
β-nitroalcohol 250
non-racemate 2, 3
1:1 NP-CytC dispersion 268
NPs plasmonic frequency 87

*o*
ochratoxin A (OTA) 113, 178–180
optical rotation dispersion (OR or ORD) spectra 12
ORD spectrum 12–13
organic and inorganic helicoids 101
organogelators 52

*p*
peroxidase-like activity 244, 252, 253, 280
photochemical effects 233–237
photodynamic therapy (PDT) 127, 181, 209–220, 209–223, 225–229, 231, 232
photosensitizers (PSs) synthesis 216
photothermal conversion efficiency 200, 201, 204, 206, 207, 209, 212, 221, 223, 227, 229, 231, 232, 233
photothermal therapy 199, 202, 203, 205, 208
plasmon hybridization terminology 81
plasmonic effects 33, 80
plasmonic enhancement effects 6
plasmonic metal NP-semiconductor 229, 231
plasmonic metal-semiconductor heterostructures 206
plasmonic molecules 131
plasmonic nanomaterials 220
plasmonic NPs 6, 8, 9, 40, 180, 190, 199, 221, 225, 229, 233, 237
plasmonic NR dimers 82, 83
plasmonic walking device 82
plasmon-induced resonant absorption 43
platelet membrane (PM) 138, 277
poly(cyclopentadithiophene-alt-benzothiadiazole) 212, 213
poly(styrene-*b*-acrylic acid) (PS-PAA) 82, 85
polycationic nonapeptide 263
polymerase chain reaction (PCR) approach 100
polymyxin B 181, 183
polyphosphoester-based nanocarrier (NP-PPE) 210
polyvinylpyrrolidone (PVP) 47, 61, 114
powder X-ray diffraction (XRD) 242
Pro-GnO chiral catalysts 253, 254
prostate specific antigen (PSA) 156, 158–160
protein biosensing 104
protoporphyrin IX (PpIX) 213
PS-free nanostructures 217

*q*
quantum dots (QDs) 12, 18, 25, 29, 31, 33, 35, 38, 65, 88, 89, 93, 107–109, 111, 118, 119, 135, 139, 181–184, 219, 252, 253, 262–265, 270

*r*
racemic mixtures 2, 35, 50
reactive oxygen species (ROS) 68, 117, 175, 200, 207, 209, 210, 212, 213, 217, 219–221, 223, 226, 227, 229–234, 262, 263, 265, 270–275, 277–281
RT-PCR method 150

*s*
(S)-β-nitroalcohol 250
(S)-carbonyl reductase II-glucose dehydrogenase on nanoparticles (SCRII-GDHnano) 266

SELEX (Systematic Evolution of Ligands by Exponential Enrichment) 213
self-limiting method 134, 135
semiconducting polymer NP core 212
semiconductor nanorods (SNR) 30, 35, 37, 38, 122, 262
side-by-side (SBS) dimers 80, 82, 83, 85
sodium borohydride ($NaBH_4$) 40, 189
sodium citrate 40, 80
sphalerite (ZB) 32
stereoselectivity 242–244, 246, 255
sulfamethazine (SMZ) 181, 182
sulfide/selenide nanoparticles 64–68
supraparticles (SPs) 11, 27, 134–139, 259, 266, 268
surface-enhanced Raman scattering-chiral anisotropy (SERS-ChA) effect 58
surface ligands 2–6, 8, 11, 12, 28, 29, 40, 41, 44, 61, 130, 271
synthetic organic chemistry 255

*t*

targeted bioorthogonal catalysis 269
tetramer assemblies 79, 80, 90–100
three-dimensional (3D) distortion 29
TMD quantum dots (QDs) 252
transition metal sulfides 227
transmission electron microscopy (TEM) 31–34, 37, 39, 45, 47–49, 51, 52, 55–58, 61–68, 70, 80, 81, 87, 90, 91, 93–97, 102, 103, 105, 109, 111–116, 118–121, 124–126, 128, 129, 131, 133, 136, 138, 200, 242–245, 249, 250, 253, 258, 260, 265, 268, 270, 271
triphenylphosphine (TPP)-grafted dextran (Dex-TPP) 213
triphenylphosphonium (TPP) 274
tropomyosin (TROP) 160, 161
twisted nanorod pairs 80
two dimensional (2D) transition metal dichalcogenides (TMDs) 250
type 2 chirality of NPs 4, 5

*u*

UCNP@ZIF-$NiS_x$ assemblies 116
upconversion nanoparticles (UCNPs) 89, 92, 114, 115

*v*

van der Waals (vdW) forces 101, 104, 131, 149, 248, 268
vascular endothelial growth factor (VEGF) 156
venous thromboembolism (VTE) 277
vibrational circular dichroism (VCD) 8, 12, 15, 18

*w*

$WO_{3-x}H_2O$ nanocrystals 257–259
wurtzite (WZ) 32, 35, 136

*x*

X-ray diffraction (XRD) 15, 57, 66, 67, 130, 132, 242, 244

*y*

yolk-shell NPs (YSNPs) 95, 229
yolk-shell NPs (YSNPs) dimers 95
yolk-shell NPs tetrahedron (UYTe) 95
yolk-shell NRs (YSNRs) 122, 173, 174

*z*

zeolitic imidazolate framework-8 (ZIF) 114–116, 209
zeptomolar DNA detection 153
zinc selenium (ZnSe) 69, 136
Zr-based porphyrinic MOF (ZrMOF) NPs 214–216